Edited by
Nobuyoshi Yasuda

The Art of Process Chemistry

Related Titles

Ravina, E.

The Evolution of Drug Discovery

From Traditional Medicines to Modern Drugs

2011

ISBN: 978-3-527-32669-3

Blaser, H.-U., Federsel, H.-J. (eds.)

Asymmetric Catalysis on Industrial Scale

Challenges, Approaches and Solutions

Second Edition

2010

ISBN: 978-3-527-32489-7

Dunn, P. J., Wells, A. S., Williams, M. T. (eds.)

Green Chemistry in the Pharmaceutical Industry

2010

ISBN: 978-3-527-32418-7

Zhang, D., Zhu, M., Humphreys, W. G. (eds.)

Drug Metabolism in Drug Design and Development

2008

ISBN: 978-0-471-73313-3

Chorghade, M. S. (ed.)

Drug Discovery and Development

2 Volume Set

2007

ISBN: 978-0-471-39846-2

Wesselingh, J. A., Kiil, S., Vigild, M. E.

Design and Development of Biological, Chemical, Food and Pharmaceutical Products

2007

ISBN: 978-0-470-06154-1

Pollak, P.

Fine Chemicals

The Industry and the Business

2007

ISBN: 978-0-470-05075-0

Edited by Nobuyoshi Yasuda

The Art of Process Chemistry

WILEY-
VCH

WILEY-VCH Verlag GmbH & Co. KGaA

The Editor

Dr. Nobuyoshi Yasuda
Process Research Department
Merck & Co. Inc.
126, E. Lincoln Ave.
Rahway, NJ 07065
USA

Library of Congress Card No.: applied for

British Library Cataloguing-in-Publication Data
A catalogue record for this book is available from the British Library.

Bibliographic information published by the Deutsche Nationalbibliothek
The Deutsche Nationalbibliothek lists this publication in the Deutsche Nationalbibliografie; detailed bibliographic data are available on the Internet at <http://dnb.d-nb.de>.

Cover Design Grafik-Design Schulz, Fußgönheim
Typesetting Toppan Best-set Premedia Limited, Hong Kong
Printing and Binding Strauss GmbH, Mörlenbach

Printed in the Federal Republic of Germany
Printed on acid-free paper

ISBN: 978-3-527-32470-5

Contents

The Art of Process Chemistry. Edited by Nobuyoshi Yasuda

ISBN: 978-3-527-32470-5

Preface

What is the definition of "Art"? According to Wikipedia, "Art is the process or product of deliberately arranging elements in a way that appeals to the senses or emotions."

Music is one of the great art forms and provides listeners powerful emotions by twisting all ranges of human feelings, from earthy to heavenly and from physics to metaphysics. However, this principal applies in many human activities. When the appeal of a subject to the senses or emotions increases beyond a certain threshold, people find beauty in it and it becomes "Art". For example, when Olympic athletes run in a 100 meter race, we feel the excitement of their performance and we sense the amazing movements of the human body, finding beauty in them. That is "Art".

Of course, this definition can be applied to science and technology as well. In another example, as the shape of automobiles becomes more streamlined to increase speed, it becomes more attractive and awakens our emotions as we find beauty in it. Many people find beauty even inside the car.

All of this is also true of organic synthesis. As syntheses become highly innovative, creative and effective, the syntheses gain appeal to the senses and emotions of chemists who find beauty in them. In that moment, organic synthesis becomes "Art".

It is logical to discover "Art" more frequently at the frontier of science, where most innovation and creativity takes place. For organic synthesis, pharmaceutical research is on one of the frontiers. In pharmaceutical research laboratories, synthetic organic chemistry plays a major role in two departments, namely Medicinal and Process Chemistry.

The objective for Medicinal Chemistry is the identification of the chemical structures for potential new medicines. Eventually, these new medicines will be launched into the market to address unmet medical needs and to improve the quality of life for all human beings. The marketing of new medicines is the lifeblood of the pharmaceutical industry. Due to the broad impact Medicinal Chemistry has in the drug discovery process, it is recognized as a top job for synthetic organic chemists.

To prepare the target compounds, Medicinal Chemists leverage their knowledge and skill in synthetic organic chemistry, but an understanding of pathology, phar-

The Art of Process Chemistry. Edited by Nobuyoshi Yasuda

ISBN: 978-3-527-32470-5

macology, and physiology are also important for making decisions on which compounds should be evaluated. Currently, Medicinal Chemists prepare small amounts of new chemicals for bio-assays and ADME (absorption, distribution, metabolism, and excretion) studies to identify the drug candidates through quantitative structure–activity relationships, and so on. With the advancement of computational biochemistry, we can imagine a time when Medicinal Chemists may only need to visualize chemical structures for *in silico* tests rather than prepare real substances for *in vivo* and/or *in vitro* studies. For the Medicinal Chemist, synthetic organic chemistry is only one of many competencies for their job.

The objective for Process Chemistry is to establish clean cost-effective manufacturing processes for new medicines identified by Medicinal Chemistry in a timely manner. At an absolute minimum, the reproducibility of the process and the quality of the final products has to meet established standards, such as the ICH guidelines.

To reach the ultimate goal, a process chemist seeks to *reduce manufacturing costs of medicines and ensure the speed of supply of drug candidates to facilitate the drug discovery and development processes.*

How does the Process Chemist reduce manufacturing cost? Manufacturing cost is made up of two components: operational cost and raw material cost. Operational cost consists of redemption of capital equipment, labor cost, overhead, vendor's profits, and so on. Reducing the number of chemical steps in a process is directly tied to lower operational costs. A more convergent synthetic route is generally more efficient than a linear route. Keeping this in mind, details such as reaction time and work-up time (the so-called overall cycle time) are additional factors which affect the operational cost. Another important contributor to operational cost is associated with waste disposal. All waste from manufacturing processes must be disposed of properly. In order to protect our environment, the enforcement of laws regarding waste disposal is becoming more stringent with time and waste disposal cost is expected to increase year by year. Therefore, the concept of "Green Chemistry" is critical to modern Process Chemistry. The most straightforward solution to reduce the waste disposal cost is reduction of the amount of waste from a manufacturing process. The relative amount of waste versus product generated is measured by either the e-factor or PMI (process mass intensity). These indicators are critical benchmarks for the Process Chemist. Use of hazardous reagents not only costs more for their proper disposal but also adds more burden to analysis of products to ensure the quality of products under ICH guidelines. Again, this all leads to increased operational cost.

Lowering the starting material costs can be achieved by improving overall yield. The higher the overall yield, the less starting materials are required and the lower the raw materials cost. Furthermore, Process Chemists must collaborate with a procurement department to lower the supply cost. If the raw materials could be prepared in a simple process from commodity chemicals, in the long term, the raw material cost would simply depend on material demands. If demand is created, the price of the raw material can fall dramatically. One good example of these phenomena is the price of *tert*-butyldimethylsilyl chloride. Today, it is a common reagent available at very affordable prices. This low price is due to the high demand for acetoxyazetidinone, the key starting material for several carbapenem antibiotics.

Moreover, the Process Chemist can also have a major impact on supply cost through the development of better synthetic methods. This research by Process Chemists can impact the cost of raw materials.

Evidently, to create the most cost efficient process, the process chemist must utilize the most advanced organic chemistry, if not devise new transformations, to address all these competing concerns.

How does the Process Chemist ensure speed of drug candidates to facilitate the drug discovery and development processes? In the big picture, this objective could also be closely related to cost. To support all preclinical and clinical studies, including Phase I to III studies, the Process Chemist must prepare drug candidates under GMP guidelines. Timing for delivery of a drug candidate is critical for the development timeline. If the drug candidate is supplied earlier, it can be marketed sooner, resulting in benefits to patients as well as the company. The patent life of a new drug starts when a patent from Medicinal Chemistry is filed. The sooner the delivery is made, the faster clinical studies can be completed and the longer the patent coverage of the medicine during the marketing phase. If the development of the candidate is terminated early for any reason, the pharmaceutical company can avoid spending additional, unnecessary developmental costs. Thus, the quicker the supply of the drug candidate is available, the more cost effective the project.

What does "quicker" mean in terms of drug supply? How can the Process Chemist provide a drug candidate more quickly? Is it good enough to scale up the original Medicinal Chemistry route, despite problems with length or cost, simply because it has been demonstrated on a small scale? The answer differs from case to case. The Process Chemist must have keen chemical insight into which route could be suitable for optimization and which could be a potential manufacturing route. Time and effort spent on optimization of unsuitable routes are practically meaningless – a waste of resources. To conserve resources, this judgment should be made in a very short period of time, balancing short term goals and longer ones. This critical judgment clearly depends on the quality of organic chemists.

As this discussion makes clear, the demands of the drug development process for the Process Chemist are quite different from those of the Medicinal Chemist. The role of Process Chemistry is to devise and fully understand the most cost efficient total syntheses of new medicines with the most advanced methodologies. By far, synthetic organic chemistry is the most important skill for a Process Chemist. Synthetic organic chemistry impacts all parts of the job and guides all decision making in Process Chemistry. In a way, there is little difference between a Process Chemist in industry and a Synthetic Organic Chemist in academia. On a scientific level, their goals are the same and, therefore, Process Chemists must be innovative Synthetic Organic Chemists, striving for new, more efficient chemistry.

In this book, there are nine chapters, each of which is devoted to the synthetic chemistry of one candidate project. Some of these molecules have already become marketed drugs. Each chapter consists of two parts which reflects the two fundamental roles of Process Chemistry; *the establishment of cost effective process and the discovery of new more effective chemistry*. In Section 1 of each chapter, titled "Project Development", the author(s) will discuss the first phase of Process Chemistry research. In each chapter, the Medicinal Chemistry route to the target compound is analyzed. To overcome the potential problems of this Medicinal Chemistry route, the original route can be optimized, new routes can be considered or some novel chemical transformations can be proposed. The shape of the process route may evolve depending on where the drug candidate is in the drug development process. Some chapters describe the manufacturing processes of marketed medicines. The process is reshaped to meet the ultimate goal of the drug development program. Through this optimization, innovations in the process will raise the synthesis to the level of "Art".

As stated previously, these activities are only part of the job of the process chemist. As described in Section 2 of each chapter, titled "Chemistry Development", the author(s) will focus on the advancement of synthetic organic chemistry discovered during the process development. In order to satisfy the Process Chemist's scientific curiosity and to advance synthetic organic chemistry, further optimization followed by investigation of the scope and limitations of these reactions is explored. In order to ensure the robustness of the reaction and to optimize it in a more scientific way, elucidation of the reaction mechanism is undertaken. Mechanistic studies are very beneficial in improving our synthetic organic chemistry skills and provide opportunities to raise these reactions to a further dimension, again that of "Art".

In recent years, the rate of change in the pharmaceutical industry has accelerated dramatically. Declining revenue growth due to patent expirations and the lower success rate for new medicines has forced the industry to make cost efficiency a top priority. Tighter research and development budgets may seem restrictive at first glance but have provided the opportunity to reshape research, making it more efficient. By further driving new research to higher levels of efficiency, the research becomes a form of "Art".

This book is quite unique in addressing the major objectives of Process Chemistry in every chapter in two aspects. Please enjoy the projects described herein which I believe have attained the status of "Art".

May 2010 *Nobuyoshi Yasuda*

List of Contributors

Cheng Chen
Guy R. Humphrey
Artis Klapers
Jeffrey T. Kuethe
Zhiguo Jake Song
Lushi Tan
Debra Wallace
Nobuyoshi Yasuda
Yong-Li Zhong
Merck Research Laboratories
Process Research
P.O. Box 2000
Rahway, NJ 07065
USA

Michael J. Williams
Merck Research Laboratories
Process Research
770 Sumneytown Pike
P.O. Box 4
West Point, PA 19468
USA

The Art of Process Chemistry. Edited by Nobuyoshi Yasuda

ISBN: 978-3-527-32470-5

1
Efavirenz®, a Non-Nucleoside Reverse Transcriptase Inhibitor (NNRTI), and a Previous Structurally Related Development Candidate

Nobuyoshi Yasuda and Lushi Tan

There are a few key enzymes for the proliferation of human immunodeficiency virus (HIV). Reverse transcriptase is one of them since HIV is a member of the DNA viruses. Efavirenz® (**1**) is an orally active non-nucleoside reverse transcriptase inhibitor (NNRTI) and was discovered at Merck Research Laboratories [1] for treatment of HIV infections. Efavirenz® was originally licensed to DuPont Merck Pharmaceuticals which was later acquired by Bristol-Myers Squibb.1) The typical adult dose is 600 mg once a day and **1** is one of three key ingredients of the once-a-day oral HIV drug, Atripla® (Figure 1.1).

Efavirenz® (**1**) is the second NNRTI development candidate at Merck. Prior to the development of **1**, we worked on the preparation of the first NNRTI development candidate **2** [2]. During synthetic studies on **2**, we discovered and optimized an unprecedented asymmetric addition of an acetylide to a carbon–nitrogen double bond. The novel asymmetric addition method for the preparation of **2** also provided the foundation for the process development of Efavirenz®. Therefore, in this chapter we will first discuss chemistries for the preparation of **2** in two parts; process development of large scale synthesis of **2** and new chemistries. Then, we will move into process development and its chemistries on Efavirenz®.

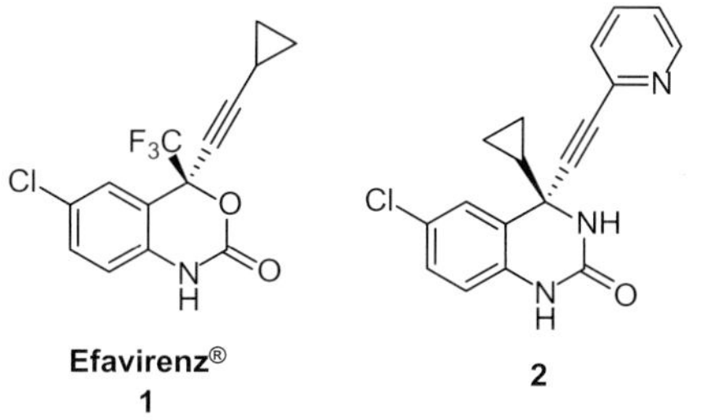

Figure 1.1 NNRTI candidates.

1) Currently, Bristol-Myers Squibb is marketing Efavirenz® under their brand name of Sustiva® and Merck under the brand name of Stocrin®.

The Art of Process Chemistry. Edited by Nobuyoshi Yasuda

ISBN: 978-3-527-32470-5

1.1
First Drug Candidate 2

1.1.1
Project Development

1.1.1.1 Medicinal Route

The first NNTRI drug candidate **2** was selected for development in 1992. Compound **2** exhibits very potent antivirus activity of IC_{50} = 12 nM (inhibition HIV-1 RT using rC-dG template/primer). The Medicinal Chemistry original preparation route is depicted in Scheme 1.1 [2].

Medicinal chemists at Merck prepared **2** in eight linear steps with an overall yield of 12%. Their starting material, 4-chloro-3-cyanoanline (**3**), was reacted with

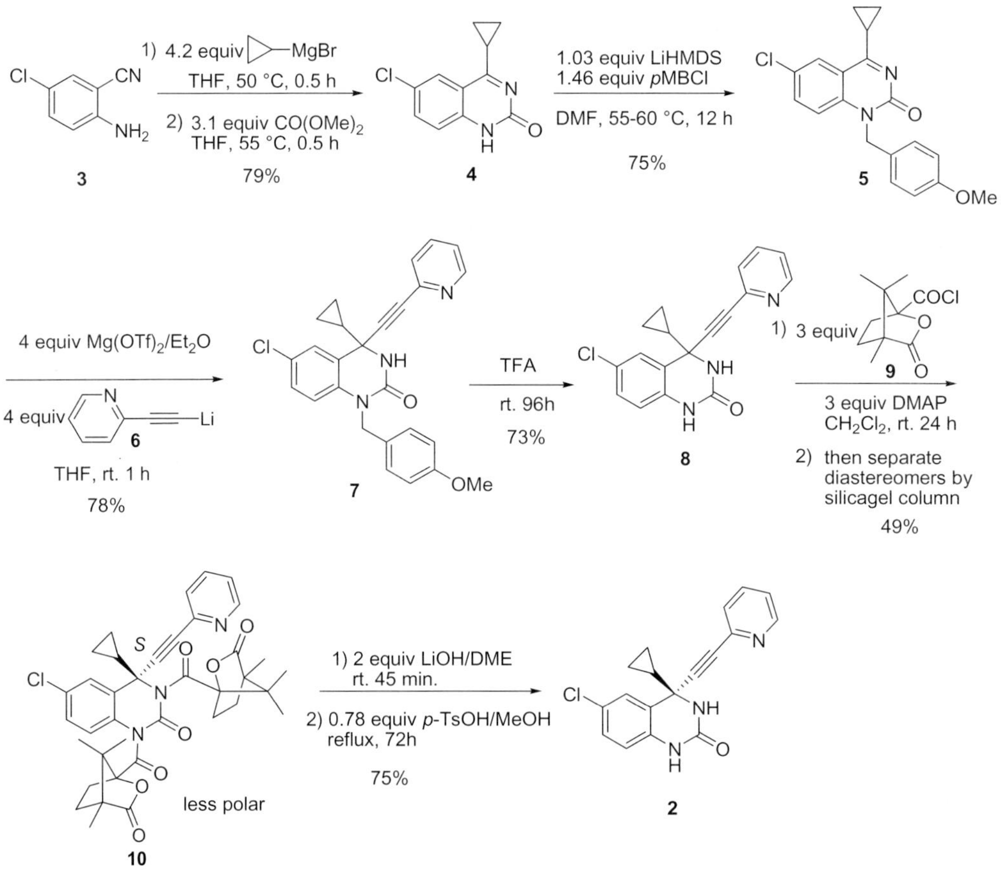

Scheme 1.1 Medicinal original route.

4.2 equiv of cyclopropyl Grignard without protection of the aniline. The resulting imidate was trapped *in situ* with dimethoxycarbonate in THF at 55–60 °C to provide quinazolin-2(1*H*)-one **4** in 79% yield. The free nitrogen of **4** was protected with a *p*-methoxybenzyl (*p*MB) group in 75% yield by treatment with $LiN(TMS)_2$ and *p*MB chloride in DMF at 55–60 °C for 12 h. 1,2-Addition to the carbon-nitrogen double bond in **5** required 4 equiv of lithium 2-pyridylacetylide (**6**) in the presence of 4 equiv of $Mg(OTf)_2$. A racemic mixture of adduct **7** was obtained in 78% yield. TFA treatment of **7** provided the target molecule **8** as a racemic mixture in 73% isolated yield. Reaction of **8** with 3 equiv of camphanyl chloride **9** and DMAP provided a diastereomeric mixture of bis-camphanyl imidate **10** and its diastereomer, which was separated by silica gel column chromatography. The less polar isomer **10** had the desired stereochemistry and afforded **2** after solvolysis. The absolute stereochemistry of **2** was determined as *S* from the single crystal X-ray structure of the enatiomer of **10** (the more polar isomer).

1.1.1.1.1 Problems of the Original Route

Several limitations of the original method were identified at the beginning of the project as follows;

1) When we started this project, the starting material **3** was not commercially available on a large scale (currently, large amounts of **3** are available for around $1000 per kg).
2) A large excess of cyclopropyl Grignard was required.
3) Chiral separation of the racemic product required silica gel separation of bis-camphanyl derivatives.
4) Furthermore, camphanyl chloride is quite expensive ($113.5 per 5 g from Aldrich) and resolving a racemic mixture at the final step of the preparation is not an efficient method for large scale synthesis.

1.1.1.2 Process Development

Even though there are a few drawbacks, as mentioned above, we felt that the Medicinal Chemistry route was straightforward and we should be able to use the original synthetic scheme for a first delivery with modifications as follows;

1) Our starting material had to be changed due to the limited availability of **3**. Our new starting material was readily available and was converted to **4**, where our new route intercepted the original synthetic Scheme 1.1.
2) Protection of the nitrogen in **4** faced the classical *N*- versus *O*-alkylation selectivity issue, which was solved by selection of the solvent system. The original protecting group, *p*MB, was replaced with 9-anthrylmethyl (ANM), which provided the best enantioselectivity with the newly discovered asymmetric addition to the ketimine.
3) Asymmetric acetylene addition should be pursued to avoid the tedious final enantiomer separation by silica gel column after derivatization with an excess of expensive camphanyl chloride.

4) The final deprotection step must be modified to accommodate the new protective group (ANM) and an isolation method for a suitable crystalline form of **2** had to be developed.

1.1.1.2.1 **Selection of the Starting Material**

The starting material for the Medicinal route, 4-chloro-2-cyanoaniline (**3**), was difficult to obtain on a large scale. We decided to use affordable and readily available 4-chloroaniline (**11**), as our starting material [3] and we envisioned introduction of a ketone function by using *ortho*-directed Friedel–Craft acylation of a free aniline, which was reported by Sugasawa *et al*, in 1978 [4], as shown in Scheme 1.2. After optimization of the Sugasawa reaction based on the elucidated reaction mechanism as described later, the desired *ortho*-acylated aniline **13** was isolated in 82% yield from 4-chlorobutyronitrile (**12**) with 2 equiv of **11**, 1.3 equiv of BCl_3 and 1.3 equiv of $GaCl_3$ at 100 °C for 20 h. The resulting chloro-ketone **13** was cyclized to the corresponding cyclopropyl ketone **14** in 95% yield by treatment with KO*t*-Bu. Reaction with **14** and 2.5 equiv of potassium cyanate in aqueous acetic acid nicely intercepted the same intermediate **4** in the original route, in 93% yield. It was important to use the corresponding HCl salt of **14**, instead of a free base, as the starting material, as shown in Scheme 1.2. When the free aniline was used for the cyclization reaction, ~10% of *N*-acetyl impurity **15** was generated under the same conditions.

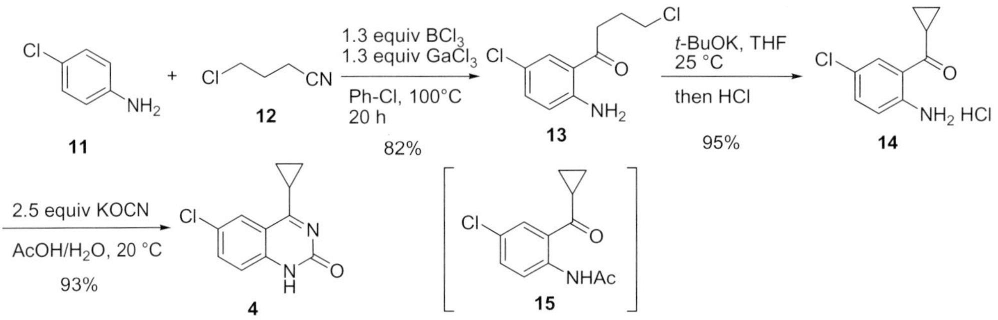

Scheme 1.2 Selection of starting material.

1.1.1.2.2 **Protection of Nitrogen in 4**

At the beginning of the project, we had studied the introduction of the *p*MB group to **4** as a nitrogen protecting group, as used in the Medicinal Chemistry route. There was a classical regioselectivity problem, *O*- versus *N*-alkylation. Under the Medicinal Chemistry conditions, the desired *N*-alkylated product **5** was mainly formed, but around 10–12% of the corresponding *O*-alkylated product **16** was also generated in DMF. The desired **5** was isolated in only 75% yield after triturating the crude product mixture with diethyl ether. Theoretically, *N*-alkylation is favored over *O*- when nonpolar solvents are used. The reaction in THF (instead of DMF) was extremely slow but formation of *O*-alkyl **16** was suppressed to about 2%, as expected. Ultimately, it was found that reaction in THF with 8 to 10 vol% DMF proceeded at a similar rate to straight DMF and the formation of **16** was suppressed to about 3%. A methanol swish of the crude product mixture was highly efficient, obtaining **5** with a high purity in an excellent yield. The isolated yield of **5** was increased from 75% to 90% by a combination of these modifications (Scheme 1.3).

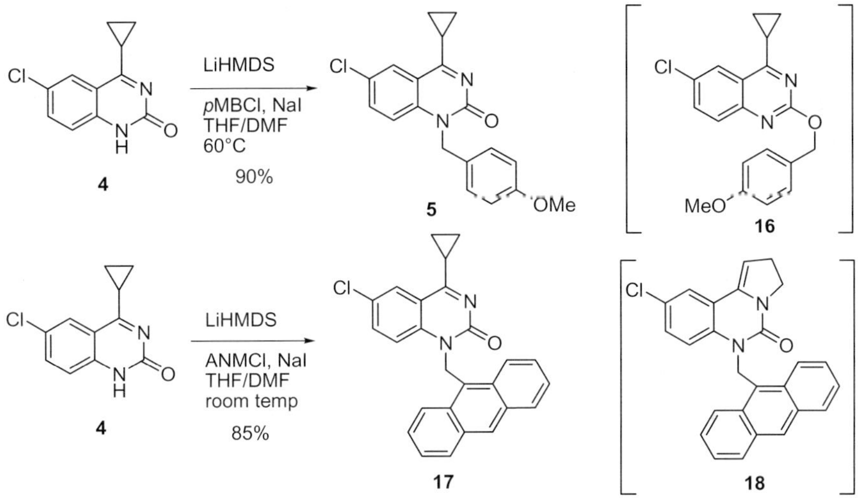

Scheme 1.3 Protection of nitrogen.

Later, we discovered that the nitrogen protecting group of **4** had a strong influence on the enantioselectivity of the newly discovered asymmetric addition of acetylides to the ketimines. After screening potential protective groups, the 9-anthranylmethyl (ANM) group was selected as the most suitable protective group and provided the best ee, as high as 97%, in the next asymmetric addition step. The reaction conditions for protection with the ANM group were modified slightly from those with *p*MB. The reaction temperature was lowered from 60 °C to room temperature to avoid generation of impurities. The desired ANM derivative **17** was obtained in 85% yield as a crystalline compound after swishing the crude product sequentially with chlorobutane and methanol. It was noted that compound **17** was not thermodynamically stable and rearranged into a by-product **18** upon heating in toluene.

1.1.1.2.3 Addition of Acetylene and Early Development of Final Product Isolation

Acetylide addition in the racemic version Originally, 4 equiv of lithium 2-pyridylacetylide (**6**) in THF/hexane was added to a mixture of **5** and 4 equiv of $Mg(OTf)_2$ in Et_2O at room temperature. Precoordination with $Mg(OTf)_2$ and **5** was reported to be essential to prevent reduction of the carbon–nitrogen double bond in **5** [2]. However, it turned out that precoordination was unnecessary for this reaction, as shown in Scheme 1.4, and racemic adduct **7** was obtained in 86% yield by treatment with 1.3 equiv of **6** at −15 °C in THF without $Mg(OTf)_2$.

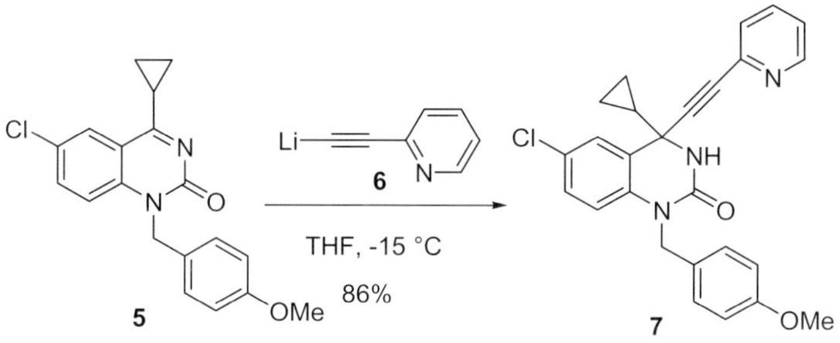

Scheme 1.4 Racemic addition of acetylene.

Classical chiral resolution with camphorsulfonic acid, followed by removal of *p*MB It is always a good idea to have some back-up synthetic scheme which is workable, especially with tight project timelines, if at all possible. Of course, asymmetric addition of acetylide is the ideal solution for the project, but at the beginning of the project we investigated a "quick fix", classical chiral resolution [5] (Scheme 1.5).

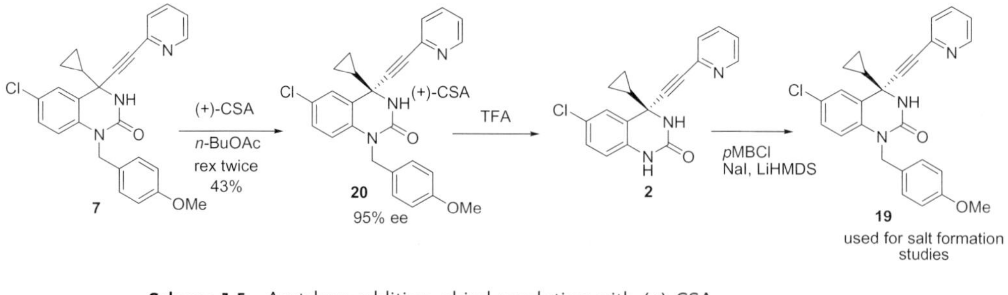

Scheme 1.5 Acetylene addition, chiral resolution with (+)-CSA.

Our approach for chiral resolution is quite systematic. Instead of randomly screening different chiral acids with racemic **7**, optically pure *N*-*p*MB **19** was prepared from **2**, provided to us from Medicinal Chemistry. With **19**, several salts with both enantiomers of chiral acids were prepared for evaluation of their crystallinity and solubility in various solvent systems. This is a more systematic way to discover an efficient classical resolution. First, a (+)-camphorsulfonic acid salt of **19** crystallized from EtOAc. One month later, a diastereomeric (-)-camphorsulfonic acid salt of **19** also crystallized. After several investigations on the two diastereomeric crystalline salts, it was determined that racemic **7** could be resolved nicely with (+)-camphorsulfonic acid from *n*-BuOAc *kinetically*. In practice, by heating racemic **7** with 1.3 equiv (+)-camphorsulfonic acid in *n*-BuOAc under reflux for 30 min then slowly cooling to room temperature, a crude diastereomeric mixture of the salt (59% ee) was obtained as a first crop. The first crop was recrystallized from *n*-BuOAc providing 95% ee salt **20** in 43% isolated yield. (The optical purity was further improved to ~100% ee by additional recrystallization from *n*-BuOAc and the overall crystallization yield was 41%). This chiral resolution method was more efficient and economical than the original bis-camphanyl amide method.

Deprotection of the *p*MB group from **20** proceeded smoothly in TFA to provide the drug candidate **2**. The isolation conditions of a suitable crystal form of **2** for development were optimized later since we had to change the protective group of the nitrogen of **4** for the subsequent asymmetric addition reaction.

Asymmetric addition of 2-pyridylacetylide to ketimines 5 and 17 Even though the chiral resolution was much more efficient than the chromatographic method, we felt this resolution method was still not efficient enough for larger scale preparation of **2**. However, this resolution method provided us some assurance for investigation of the unprecedented asymmetric addition of the acetylide, since upgrades of ee of adduct **7**, even low ee, had been achieved upon recrystallization with (+)-camphorsulfonic acid.

There are many reports on the asymmetric addition of nucleophiles to carbon–nitrogen double bonds [6]. However, the majority of these reports are based on substrate control and rely on chiral auxiliaries in imines. Moreover, almost all of these reports are just for aldo-imine cases [7].

Regarding the reagent control asymmetric addition to imines, there were three reports with aldo-imines. Based on our best knowledge, no asymmetric addition to ketimine was reported prior to our work (*vide infra*).

Taking Tomioka's pioneering work [8] as a precedent, we have screened β-amino alcohols as chiral modifiers [9] in the nucleophilic addition of lithium 2-pyridinylacetylide **6** to the *p*MB protected ketimine **5**. We were pleased to discover that when **5** was treated with a mixture prepared from 1.07 equiv each of quinine and 2-ethynylpyridine by addition of 2.13 equiv of *n*-BuLi in THF at −40 to −20 °C, the desired adduct **19** was obtained in 84% yield with maximum 64% ee. Soon after, we found selection of the nitrogen protective group had great influence on the outcome of the asymmetric addition and the ANM (9-anthranylmethyl) derivative **17** gave us the best result (97% ee in high yield). On a large scale, 2.63 mol of **17** was reacted with 1.4 equiv of 2-ethynylpyridine, 1.5 equiv of quinine, and 2.98 equiv of *n*-BuLi in THF at −25 °C for 14 h. The assay yield of the organic layer, after aqueous quench, was 87% with >97% ee. The product **21** was isolated as a (+)-camphorsulfonate salt in 84% yield with >99%ee (HPLC area% at 220 nm was 99 A%), as shown in Scheme 1.6 [10].

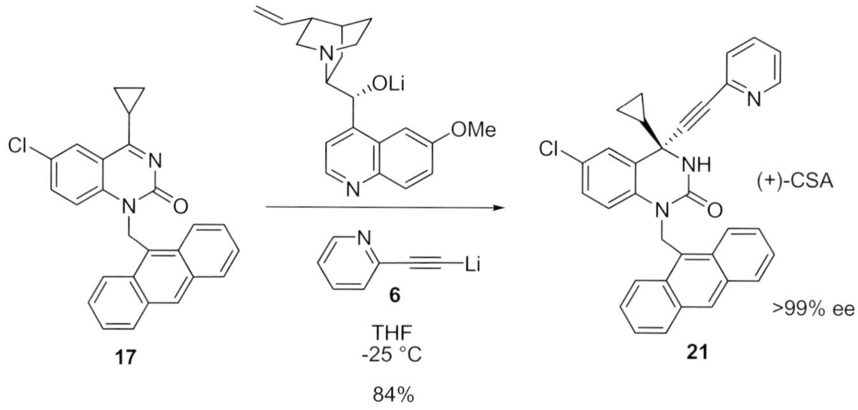

Scheme 1.6 Asymmetric addition of 2-pyridinylacetylide.

1.1.1.2.4 **Deprotection and Isolation of the Drug Candidate 2**

The ANM group in **21** could be removed under conditions similar to those for the removal of the *p*MB group, and the reaction was faster than that of *p*MB, since anthranylmethyl cation is more stable than *p*MB cation. However, the anthranylmethyl cation also reacted with the product **2** under the reaction conditions. Therefore, we had to add cation-trap reagents, such as anisole or thioanisole, to the reaction mixture. Both reagents were equally effective but anisole was selected due to easier handling and benign smell. The reaction proceeded smoothly with (+)-camphorsulfonate salt **21** in 1 volume of anisole and 1.5 volume of TFA at room temperature overnight and the assay yield of **2** was almost quantitative. However, the work-up was a little more complicated than we anticipated. It was found that the anthranylmethyl cation was successfully trapped with anisole to form a major by-product **22**. Moreover, a portion of compound **22** further reacted with anisole under the reaction conditions, to generate anthracene (**23**) and bis-anisyl-methane (**24**), as depicted in Scheme 1.7. Direct crystallization of **2** from the crude mixture failed because **2** tends to cocrystallize with **23**.

The work-up process was optimized for large scale preparation. The reaction mixture was concentrated *in vacuo* and the residue was dissolved in EtOAc, which was washed with aqueous NaOH (adjusted to pH 8.5). The solvent of organic

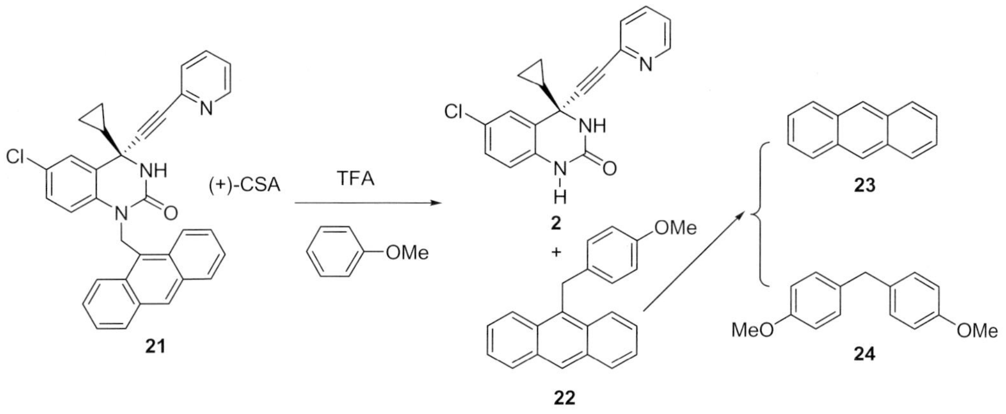

Scheme 1.7 Deprotection of ANM group from **21**.

extract was switched from EtOAc to MeOH. The residual water amount in the MeOH solution was adjusted to 2% by addition of water. The major impurity **22** was precipitated out from the solution and was removed by filtration. Anthracene (**23**) was removed by passing though SP206 (polystyrene resin; 30 volumes based on assay yield of **2**) with elution of 98% MeOH/H_2O (anthracene remained on the resin). The rich cut (typically 1.5 bed volumes) was concentrated and the solvent was switched to EtOAc. Compound **2** was crystallized as a EtOAc solvate, with ~13% loss to the mother liquor. Isolation of EtOAc solvate was performed to ensure removal of trace amounts of anisole from the product. EtOAc was removed by co-distillation of water from a suspension of EtOAc solvate of **2** in water and compound **2** was isolated as its monohydrate in 99.9 A% with 100% ee and overall isolated yield was 78%. It is noted that the X-ray diffraction pattern of EtOAc solvate and monohydrate are almost identical. Thus, EtOAc and water would share the same position in its crystal lattice. Isolation as EtOAc solvate might be eliminated with further development and the isolated yield is expected to be improved, if **2** were selected for late stage development.

1.1.1.2.5 **Overall Preparation Scheme**

Thus, our developed process route is depicted in Scheme 1.8 and process improvements are summarized as follows:

1) Development of drug candidate **2** was supported by providing sufficient amounts of the bulk in a short period of time.
2) Target compound **2** was prepared in six chemical steps in 41% overall yield.
3) Our starting material was changed from non-commercially available 2-cyano-4-chloroaniline (**3**) to readily available 4-chloroaniline (**11**).

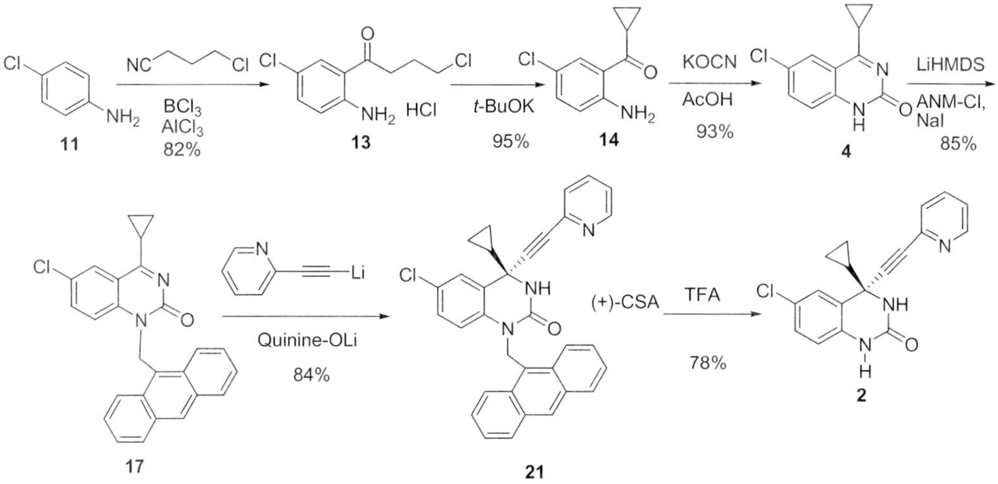

Scheme 1.8 Developed process for preparation of **2**.

4) The Sugasawa reaction (*ortho*-acylation of aniline) was optimized for this route using a combination of $BCl_3/GaCl_3$.
5) Installation of an *N*-protecting group was optimized to suppress formation of *O*-benzylation.
6) A classical chiral resolution method was established, prior to investigation of the asymmetric addition of lithium acetylide to the ketimine **5**.
7) The novel asymmetric nucleophilic substitution to the ketimine was discovered and optimized for this preparation.
8) The ANM group was selected as the nitrogen protecting group for the novel asymmetric nucleophilic substitution providing the optimum enantioselectivity.
9) The deprotection process was optimized and unexpectedly generated anthracene was removed by resin treatment.

1.1.2
Chemistry Development

The large scale preparation of the drug candidate **2** was accomplished via the Sugasawa reaction (an *ortho*-selective Friedel–Craft acylation on anilines) and the asymmetric addition to ketimines. *Understanding the reaction mechanism and reaction parameters is the only way to gain confidence that the reactions will perform as required upon scale up.* Below we discuss both subjects in detail.

1.1.2.1 **Sugasawa Reaction**

The first time we encountered the Sugasawa reaction was in the early 1990s, when we worked on anti-MRSA carbapenem projects. We were very interested in this unique reaction and started to investigate it in detail. Generally speaking, Friedel–Craft reaction on anilines is very difficult even though anilines are electron-rich aromatic rings. The reaction requires Lewis acids to activate electrophiles. However, Lewis acids are more prone to coordinate aniline nitrogen instead of electrophiles, and, as a result, the Lewis acid coordinated anilines become electron-deficient aromatic rings and shut down the desired reaction [11]. Thus, to progress the Friedel–Craft reaction with anilines, the nitrogen atom in anilines has to be protected. For example, Kobayashi, *et al.*, reported *para*-selective Friedel–Craft acylation with acetanilide in the presence of a catalytic amount of $Ga(OTf)_3$ [12].

In 1978, Sugasawa *et al.*, at Shionogi Pharmaceutical Co. reported *ortho*-selective Friedel–Craft acylation with free anilines with nitrile derivatives [4]. Sugasawa reported that the reaction requires two different Lewis acids (BCl_3 and $AlCl_3$) and does not proceed when *N,N*-dialkyl anilines are used. He proposed that boron bridging between nitriles and anilines led to exclusive *ortho*-acylation but a conclusive mechanism was not elucidated. The report did not offer any reason why two different Lewis acids were required and why the reaction did not progress with *N,N*-dialkyl anilines. Therefore, we initiated mechanistic studies.

1.1.2.1.1 NMR Studies on the Mechanism of the Sugasawa Reaction

Elucidation of the reaction mechanism of the Sugasawa reaction was initiated under the initiative of Dr. Alan Douglas who was the head of our NMR group [13]. The results are summarized in Scheme 1.9.

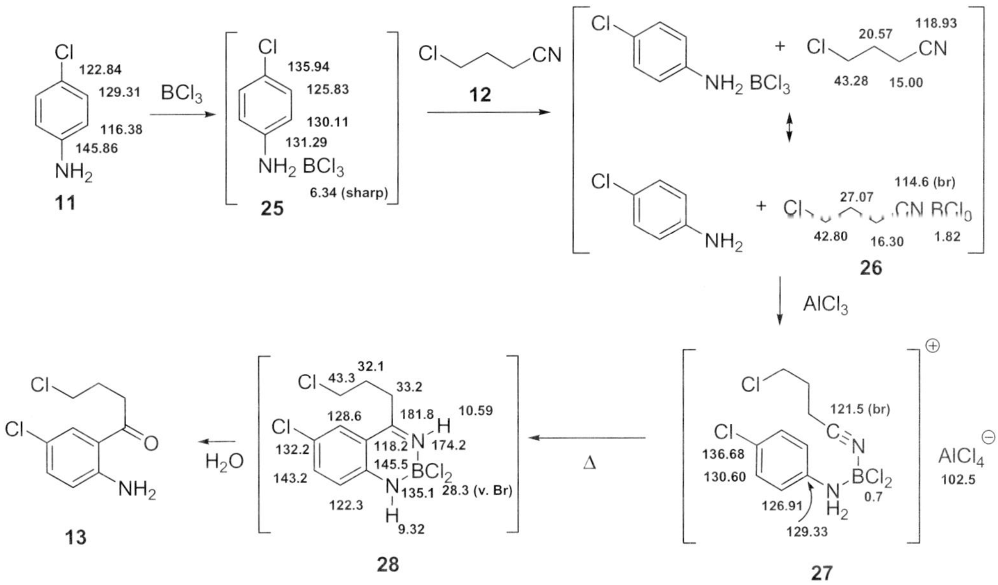

Scheme 1.9 NMR studies on the Sugasawa reaction.

By addition of BCl_3 to aniline **11** in an NMR tube, formation of a boron complex **25** was confirmed by high-field shifts of the α- and γ-carbons of the anilines and low-field shifts of the β- and δ-carbons in ^{13}C NMR, as indicated in Scheme 1.9. When nitrile **12** was added to the mixture, an equilibrium mixture of **25** and the boron complex **26** of the nitrile was observed. The structure of **26** was also confirmed by the similar ^{13}C NMR chemical shift changes. Next, $AlCl_3$ was added to this mixture. The most striking observation was the formation of the sharp NMR signal of Al. The NMR signal of Al atom is typically broad due to the tendency to form dimeric (or polymeric) complexes. The observed sharp signal indicated that the environment around the Al atom should be highly symmetrical, and the ^{27}Al chemical shift (102.5 ppm) was identical to that reported for $AlCl_4^-$. These data indicated that the Al atom existed as aluminum tetrachloride anion. Based on ^{13}C NMR and ^{11}B NMR, a structure of a so-called "supercomplex" **27** was elucidated. In **27**, both the aniline nitrogen and the nitrile nitrogen were simultaneously coordinated to the boron, which lost one of three chlorine atoms to $AlCl_3$. No cyclization of **27** was observed when the reaction mixture was kept at room temperature. Upon heating **27**, a new six-membered complex **28** was identified by ^{13}C NMR and ^{11}B NMR. ^{15}N NMR (Figure 1.2) of the six-membered complex **28** confirmed there were two protons (9.32 and 10.59 ppm), clearly coordinated to two distinct nitrogen atoms (two doublets; 135 and 174 ppm) in **28** and provided additional support for elucidation of **28**. ^{15}N NMR of the crude reaction mixture was very clean showing only **28** and the protonated aniline **11** (a quartet; ~50 ppm). Solvolysis of **28** should lead to the desired *ortho*-acylated aniline **13**, and the six-

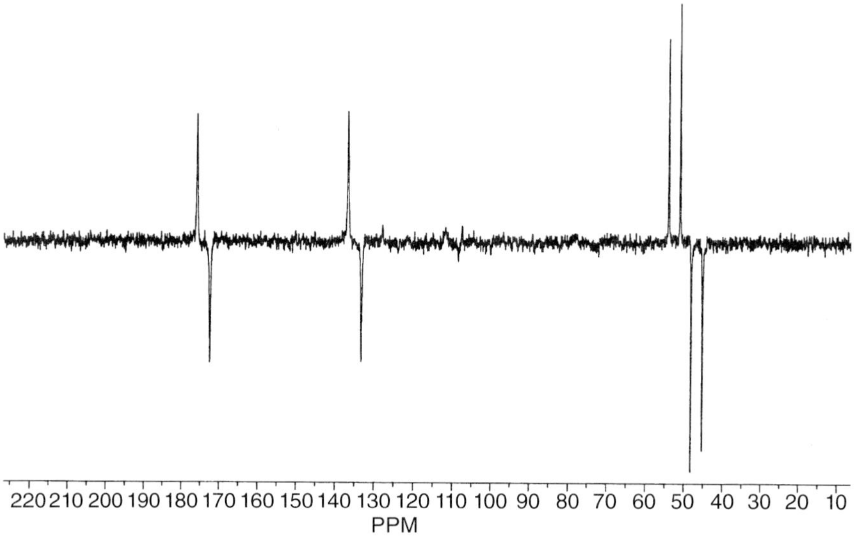

Figure 1.2 ^{15}N NMR (INEPT) of intermediate **28**. The quartet signal around 51 ppm is protonated 4-chloroaniline.

membered complex formation is the origin of the observed *ortho*-selectivity of the Sugasawa reaction.

However, is supercomplex 27 the true intermediate? As previously mentioned, Sugasawa reported that reaction did not proceed with *N,N*-dialkyl anilines. Do *N,N*-dialkylanilines form a similar supercomplex? We examined the following three anilines, $ArNH_2$, ArNHMe, and $ArNMe_2$, as shown in Figure 1.3. Under Sugasawa conditions at room temperature, formation of the corresponding supercomplex, respectively (**29**, **30**, and **31**) was confirmed, based on their NMR analyses (Complex **29** and **31** were derived from toluidine and complex **30** comes from aniline).

Upon heating and subsequent solvolysis, supercomplexes **29** and **30** provided the desired *ortho*-acylated anilines (**32** and **33**) in high yields. On the other hand, supercomplex **31** from *N,N*-dimethyl *p*-toluidine did not cyclize upon heating and only starting material was recovered, as Sugasawa reported. The failure elucidated that the supercomplex was not the true intermediate, at least in the case of *N,N*-dialkyl anilines. Since the structures of the supercomplex **27** and the six-membered complex **28** share common structural features, the true intermediate should also have a similar framework. The electron density of the aniline ring in the supercomplex should be very low since the aniline moiety is a part of an electron deficient cationic species. So, it would be reasonable to expect that electrophilic acylation on such an electron-poor ring would be prohibited. A proton should be eliminated from the supercomplex to form the true intermediate **34**, which is a neutral compound, prior to acylation to form the 6-membered complex, as shown in Scheme 1.10. Since there is no removable proton available in the supercomplex from *N,N*-dialkyl aniline such as **31**, the Sugasawa reaction could not proceed from *N,N*-dialkyl cases, as reported.

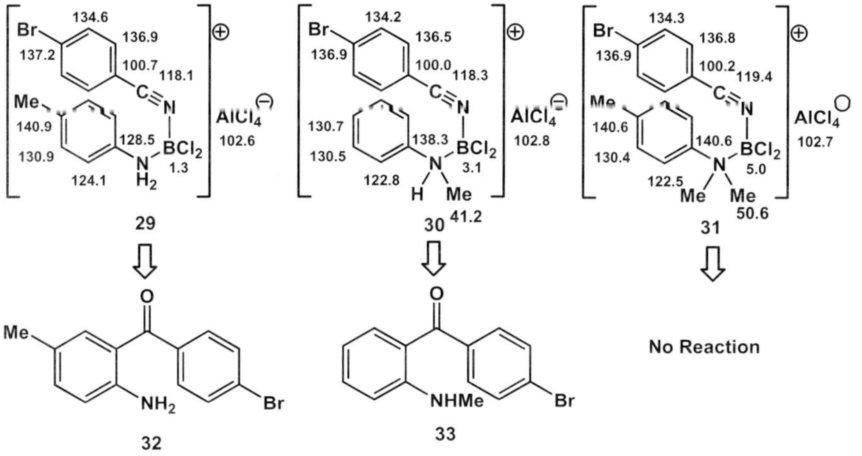

Figure 1.3 Structure of the "supercomplex" from NH_2, NHMe, and NMe_2 anilines.

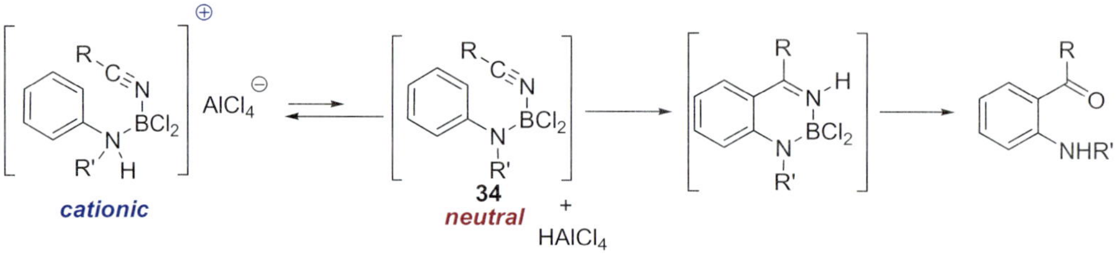

Scheme 1.10 What is the true intermediate?

1.1.2.1.2 Further Optimization of the Sugasawa Reaction Based on the Reaction Mechanism

Based on the elucidated mechanism, the role of an auxiliary Lewis acid has become clearer. The auxiliary Lewis acid bonds strongly with one chlorine of BCl_3. As a result, the boron can coordinate both nitrogen atoms in aniline and nitrile to form the supercomplex. The most chlorophilic Lewis acid is reported as gallium [14]. In fact, various Lewis acids were tested as an auxiliary, and formation of a supercomplex was confirmed in every case. Among them, $GaCl_3$ provided the best result, as shown in Table 1.1. Sugasawa reaction with $GaCl_3$ proceeded under milder conditions than with $AlCl_3$. When cyclopropyl nitrile was used, the product was isolated in 74% yield together with a cyclopropyl ring-opening product (~4%) with $GaCl_3$ as an auxiliary Lewis acid. However, the same reaction with $AlCl_3$ provided only 30–40% desired product, together with 15–20% ring-opening product. $GaCl_3$ appeared to be more effective, especially for electron deficient anilines. It is also noticed that BCl_3 is essential for this reaction and no reaction was found with an $AlCl_3$ and $GaCl_3$ combination. This is quite interesting since B, Al, Ga, Tl are in Group 13 in the Periodic Table.

This reaction generates 1 mole of $HAlCl_4$, which protonates anilines. Since protonated anilines could not coordinate with BCl_3, the reaction shuts down.

Table 1.1 Other auxiliary Lewis acids for the Sugasawa reaction.

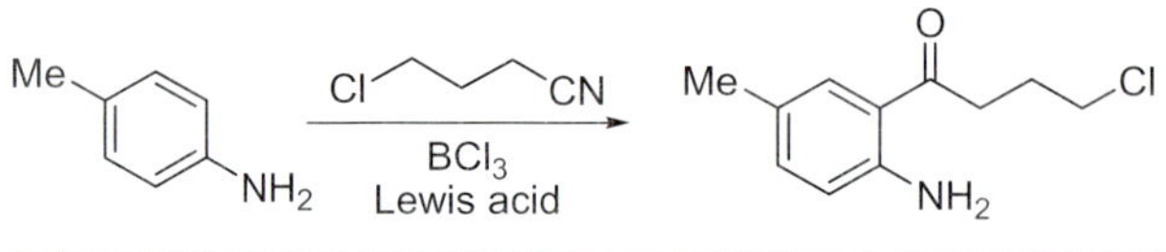

Lewis acid	$GaCl_3$	$InCl_3$	$AlCl_3$	$FeCl_3$	$SbCl_5$	AgOTf
Conditions	c, 26 h, 80 °C	c, 4 h, 132 °C	c, 4 h, 132 °C	t, 17 h, 96 °C	t, 5 h, 78 °C	c, 7 h, 100 °C
Yield (%)	72	63	45	44	26	24

c = chlorobenzene; t = toluene.

Therefore, addition of bases was studied. Gallium metal[2)] and amine bases were screened. However, the use of 2 equiv of aniline provided the best result.

1.1.2.2 **Asymmetric Addition of 2-Pyridinylacetylene Anion to Ketimine 5 and 17**

Asymmetric addition to ketimine in a reagent controlled manner has seldom been reported, even by 2008. When we investigated the potential for this asymmetric addition around 1992, there were no known examples. In 1990, Tomioka *et al.*, reported the first asymmetric addition of alkyl lithium to *N*-*p*-methoxyphenyl aldoimines in the presence of a chiral β-amino ether with 40–64% ee [8] (Scheme 1.11). In 1992, Katritzky reported the asymmetric addition of Et_2Zn to *in situ* prepared *N*-acyl imine in the presence of a chiral β-amino alcohol with 21–70% ee [15] (Scheme 1.12). In the same year, Soai *et al.*, reported the asymmetric addition of dialkylzinc to diphenylphosphinoyl imines in the presence of chiral β-amino alcohols with 85–87% ee [16] (Scheme 1.13). These three reports were, to the best of

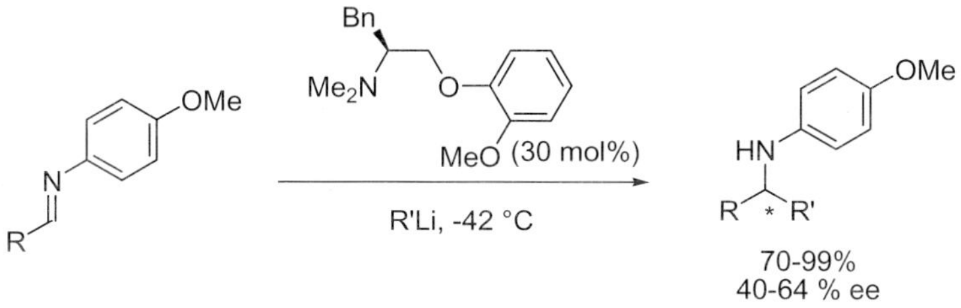

Scheme 1.11 Tomioka's report in 1990.

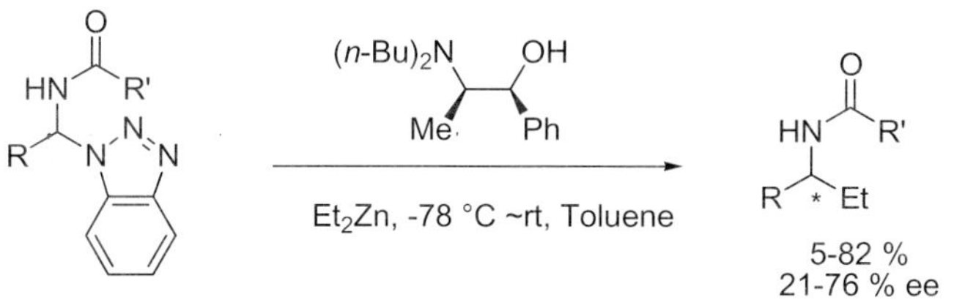

Scheme 1.12 Katritzky's report in 1992.

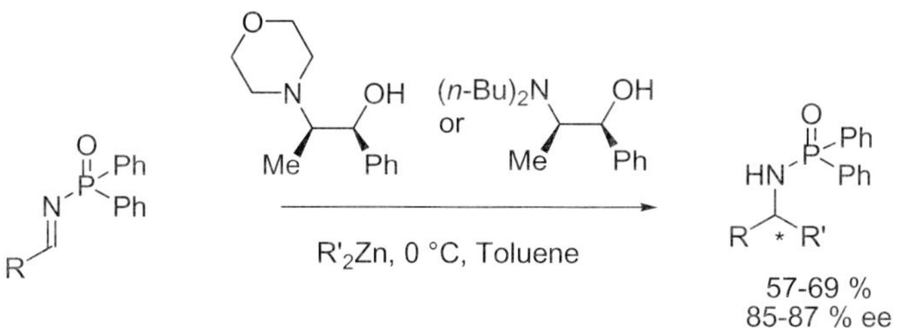

Scheme 1.13 Soai's report in 1992.

2) It was reported that Ga metal reacts with HCl to generate $GaCl_3$ and 1.5 equiv of H_2.

our knowledge, the only examples when we started our investigation on our ketimine **5**.

Based on these reports, we started investigation of the asymmetric addition of acetylide to *p*MB protected **5**, mainly in the presence of chiral β-amino alcohols. Many types of chiral amines were also screened (e.g., diamines, diethers), and it was soon found that addition of β-amino alkoxides effectively induced enantioselectivity on the addition. Since the best result was obtained with a stoichiometric amount of chiral amino alcohols, we focused our screen on readily available chiral β-amino alcohols and the results are summarized in Table 1.2.

While ephedrine derivatives showed some selectivity, the most promising results were obtained with cinchona alkaloids. Lithium alkoxides and lithium acetylides (*n*-BuLi or LiHMDS used to deprotonate both the acetylene and the alcohol) gave better results than the corresponding sodium or magnesium salts. Higher enantioselectivity was obtained in THF (homogeneous) than in toluene or diethyl ether (heterogeneous).

Both quinine and dihydroquinine favored the required (*S*)-enantiomer. A small ee difference of the product might be due to inconsistent purity of the naturally obtained cinchona alkaloids. It was noted that quinidine (the pseudo-enantiomer of quinine) gave the (*R*)-enantiomer with a similar 55% ee. Since quinine was

Table 1.2 Asymmetric addition of 2-pyridiylacetylide to *p*MB protected ketimine 5.

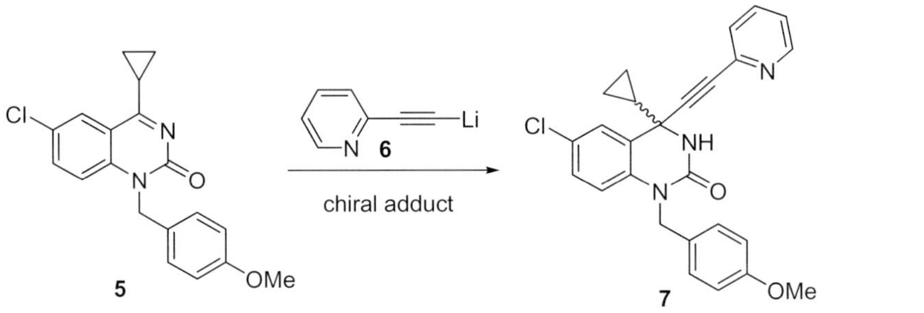

β-Amino alcohol	ee %	Configuration
(1*R*,2*S*)-Ephedrine	1	*S*
(1*R*,2*S*)-*N*-Methylephedrine	10	*R*
(*S*)-1-Methylpyrrolidine-2-methanol	0	
(*S*)-α,α-Diphenylpyrrolidine-2-methanol	0	
Quinine	59	*S*
Dihydroquinine	64	*S*
Cinchonidine	26	*S*
Quinidine	55	*R*
Dihydroquinidine	39	*R*
9-Epiquinine	28	*S*

Table 1.3 Optimization on protective group of ketimine.

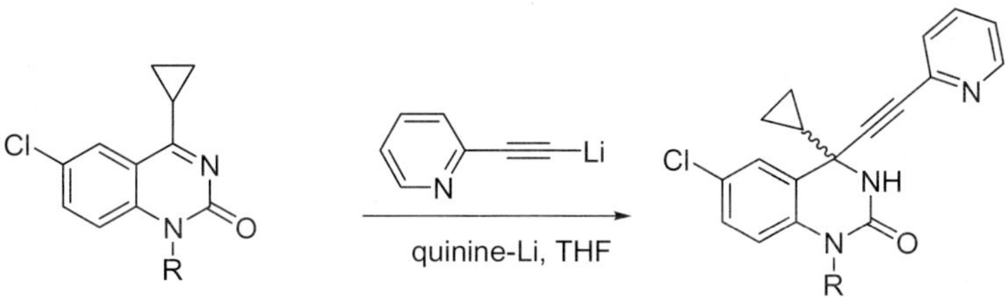

R	ee (%)
4-Methoxybenzyl (*p*MB)	64
Benzyl	56
4-Chlorobenzyl	37
Methyl	70
2,4,6-Trimethylbenzyl (TMB)	74
2,6-Dichlorobenzyl	80
9-Anthrylmethyl (ANM)	97

readily available and more affordable than dihydroquinine, we decided to optimize this asymmetric addition with quinine.

The effect of the protective group at the nitrogen was studied and the results are summarized in Table 1.3. Reaction conditions were optimized for each individual substrate.

There was a substantial electronic influence with electron-withdrawing substituents decreasing enantioselectivity. Interestingly, steric bulkiness at this remote part of the molecule was found to be highly effective for asymmetric induction. The bulky ANM group provided 97% ee with a high isolated yield.

Furthermore, it was important to note that this reaction system was very dynamic. There was a large temperature effect on ee and optimum temperature was dependent on the protective groups, as depicted in Figure 1.4. The best yields with *N*-ANM, *N*-trimethylbenzyl (TMB), and *N*-*p*MB were obtained at −25, −20, and −30 °C, respectively. Either higher or lower temperatures resulted in poor enantioselectivity. These phenomena might be a hint, suggesting that thermodynamic change of the anion species' aggregation stage played a key role in enantioselectivity. This was eventually confirmed during process development of Efavirenz®.

The scope and limitations were briefly studied. Unfortunately the scope of the reaction was rather narrow, as shown in Table 1.4. The limit of generality may originate from differences in aggregation of each individual lithium acetylide. For instance, changing 2-pyridyl to 3-pyridyl, the ee dropped to 36%. Furthermore, changing to 4-pyridyl, the ee further decreased to 13%. Fortunately, asymmetric addition of a TMS protected acetylide provided the desired adduct in 82% ee. Since

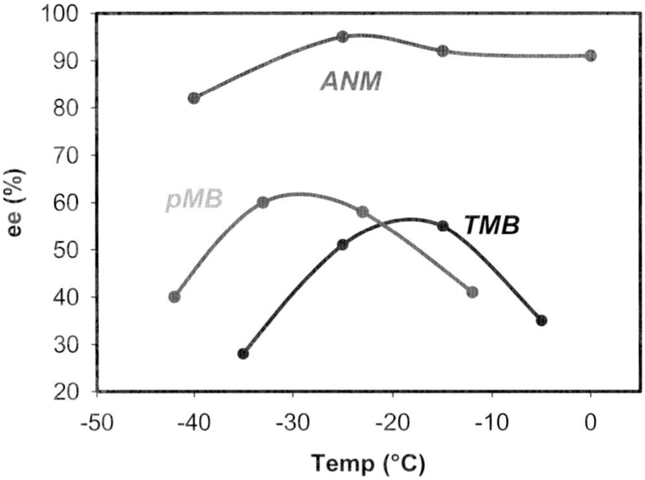

Figure 1.4 Temperature effect on asymmetric addition.

Table 1.4 Scope of acetylide.

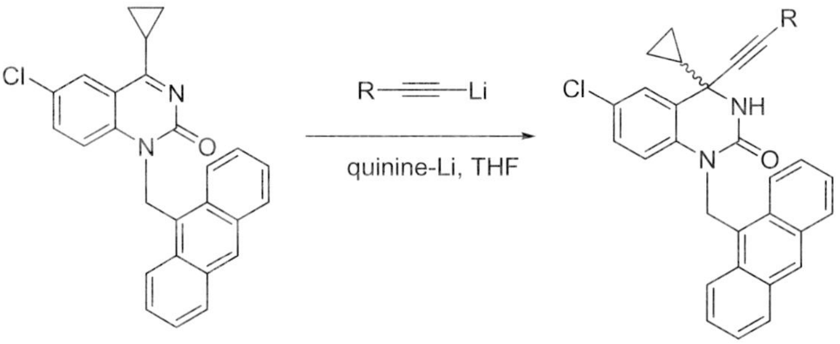

R	Temperature (°C)	ee %
2-Pyridyl	–25	94
2-Pyridyl	–15	92
3-Pyridyl	–25	22
3-Pyridyl	–15	36
4-Pyridyl	–25	6
4-Pyridyl	–15	13
4-MeO-Ph-	–25	86
Ph-	–15	65
4-Cl-Ph-	–15	58
Bu-	–25	77
TMS-	–25	82

the Sonogashira reaction allows any substitution on acetylene, this method became a general method, even though it required additional reaction steps.

Thus, we discovered the first asymmetric nucleophilic addition of acetylides to ketimines. The reaction mechanism was unfortunately not clear during this study but we felt that aggregation of lithium species might play an important role.

1.2 Efavirenz®

1.2.1 Project Development

1.2.1.1 Medicinal Route

Efavirenz® (**1**) was chosen over compound **2** as a developmental candidate in 1993 based on its better antivirus activities, especially against resistant strains [1, 17]. Efavirenz® is the first HIV non-nucleoside reverse transcriptase inhibitor (NNRTI) which was approved by the FDA on September 21, 1998. The original Medicinal Chemistry method to prepare Efavirenz® is depicted in Scheme 1.14.

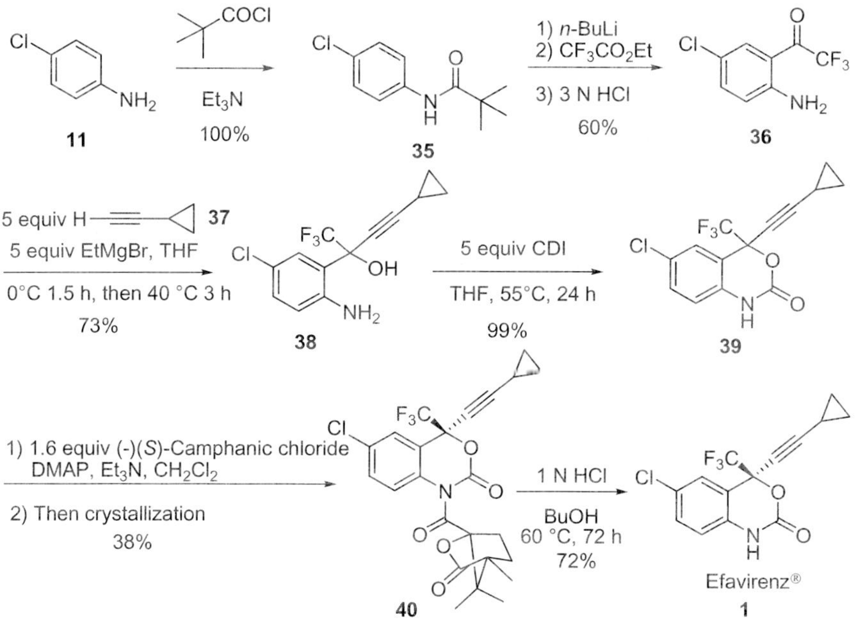

Scheme 1.14 Original Medicinal Chemistry route for Efavirenz® (**1**).

Efavirenz® (**1**) was prepared from 4-chloroaniline (**11**) rather straightforwardly in seven chemical steps in an overall yield of 12%. *Ortho*-Trifluoroacetylation of aniline **11** was carried out via a traditional three-step method yielding trifluorobenzophenone **36** [18] in 60% overall yield. First, the aniline nitrogen was protected as a pivalate **35** in quantitative yield. The dianion of **35**, generated by addition of *n*-BuLi, was reacted with ethyl trifluoroacetate to provide an *ortho*-acylated intermediate. Subsequent acidic solvolysis of the pivalate group gave the desired ketone **36**.

Addition of an acetylide to ketone **36** was sluggish and required 5 equiv of magnesium acetylide, even at 40 °C. This sluggishness may be due to reduction of electrophilicity of the carbonyl group by deprotonation of free aniline **36**. Nevertheless, the desired racemic *tert*-alcohol **38** was isolated in 73% yield by direct crystallization. When we started this project, cyclopropylacetylene (**37**) was rather limited in supply, and expensive. Therefore, the requirement of large excess amounts of **37** was one of the biggest issues in this project. After intensive research and efforts in the chemical industry, acetylene **37** is now one of the most affordable acetylenes due to its large demand for Efavirenz® production [19]. Racemic cyclic carbamate **39** was isolated in 99% yield after reacting alcohol **38** with carbonyldiimidazole (CDI). Racemic **39** was reacted with 1.6 equiv of (-)(*S*)-camphanyl chloride in the presence of triethylamine and a catalytic amount of *N*,*N*-dimethylaminopyridine (DMAP). The desired diastereomer **40** was isolated by simple crystallization in 38% yield. The undesired diastereomer is an oily compound and readily rejected into the mother liquor. Acidic solvolysis of **40** provided Efavirenz® in 72% yield as a crystalline compound.

1.2.1.1.1 Problems of the Original Route

The original Medicinal Chemistry route was straightforward but, from a process chemistry point of view [20], several problems were identified at the beginning of the project and some of them were quite similar to those for the previous development candidate:

1) A large excess of cyclopropylacetylene (**37**) was required. The compound was expensive and its supply was limited.
2) The target compound was obtained as a racemic mixture. Enantiomeric pure Efavirenz® had to be isolated via a classical chiral resolution of a diastereomixture of (-) camphanate imide.
3) (−)(*S*)-Camphanyl chloride is expensive and limited in supply. And the diastereomeric imide formation required 1.6 equiv of the reagent.

1.2.1.2 Process Development

All three previously mentioned issues associated with the Medicinal Chemistry route were rooted in cyclopropylacetylide (**37**) addition to the ketone **36**. Other steps in the Medicinal route are suitable for large scale preparation. Thus, our effort for this process development focused on asymmetric addition to ketone **36** with close to 1 equiv of **37** [21].

Naturally, we thought our novel asymmetric acetylide addition on ketenimine **5** (Scheme 1.6) could also be applicable in the preparation of Efavirenz®. The structure of **36** in Scheme 1.14 is somewhat misleading. We should expect that one of

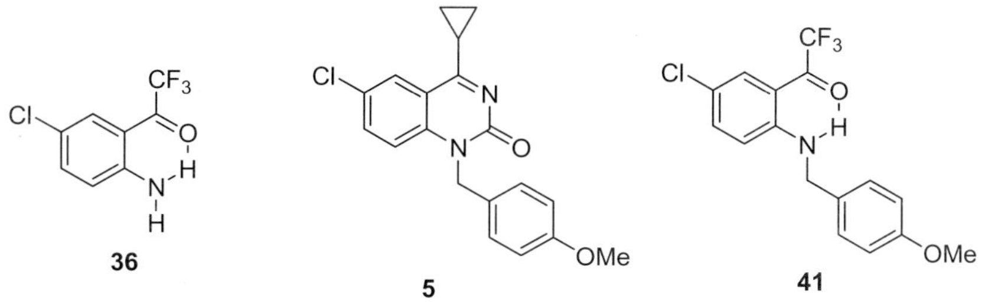

Figure 1.5 Structure resemblance between ketone aniline **36** and ketimine **5**.

the aniline hydrogens of **36** would hydrogen bond strongly to the ketone carbonyl, as shown in Figure 1.5. Therefore, ketone and aniline should consist of a six membered ring and the trifluoromethyl group should be located outside the ring. The other hydrogen in **36** should be protected to avoid deactivation of the ketone toward nucleophilic attack through *N*-anion formation. Once protected as a mono-*N*-*p*MB **41**, the special environment around the ketone of **41** would be quite similar to that of ketimine **5**. Thus, asymmetric addition of a lithium acetylide to **41**, mediated by the lithium alkoxide of cinchona alkaloids, should proceed similarly to the reaction with **5**. This working assumption was our starting point.

In the first half of this section for Efavirenz®, we will discuss the process development of the first and the current manufacturing route by going through each topic shown in the following list.

1) Selective mono-*N*-protection of **36**.
2) The first generation of asymmetric addition of lithium-cyclopropylacetylide to **41**.
   - Introduction
   - Preparation of the chiral modifier
   - Preparation of cyclopropylacetylene
   - Asymmetric addition of acetylide to the ketone
3) Preparation and isolation of Efavirenz® (first manufacturing route).
4) The second generation of asymmetric addition of zinc-cyclopropylacetylide to *N*-*p*MB ketone **41** (part of the current manufacturing route).

In the second half of this section, we will discuss the *mechanistic understanding of this chiral addition with lithium acetylide*, the cornerstone of the first manufacturing process. Based on the mechanism of asymmetric lithium acetylide addition, we will turn our attention toward *the novel highly efficient zincate chemistry*. This is an excellent example in which mechanistic studies paid off handsomely.

1.2.1.2.1 Preparation of Mono *N*-*p*-Methoxybenzyl Ketone 41

Initially, preparation of **41** was not an easy task and it very unexpectedly seems to be more difficult than the following key asymmetric acetylide addition. *N*-Mono alkylation of **36** with *p*MBCl **42** under various standard reaction conditions did not proceed as expected. It was found that the desired **41** was formed when **36** and chloride **42** were co-spotted on the TLC. So we turned our attention to reaction of

36 and **42** under acidic conditions. The reaction proceeded in the presence of silica gel, molecular sieves, or basic alumina in toluene, and among these, basic alumina worked the best. To the suspension of **36** and basic alumina in toluene was added chloride **42** and the reaction was complete in 3 h at room temperature with an assay yield of 85%. After filtering the alumina, the desired product **41** was isolated in 78% yield as a crystalline compound (Scheme 1.15).

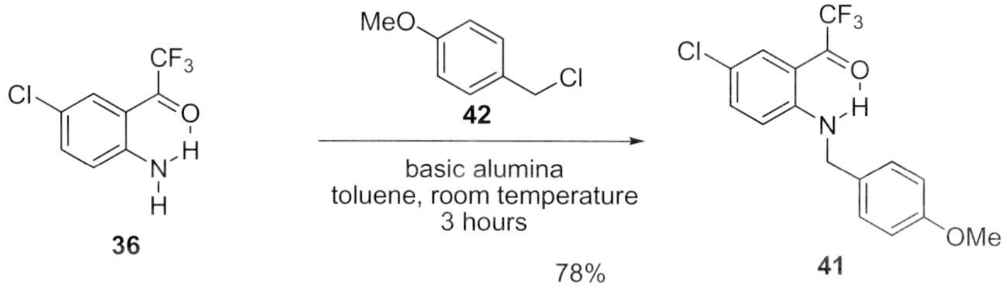

Scheme 1.15 Installation of *p*MB on **36**.

However, *p*MBCl **42** has a thermal stability issue and is expensive (Aldrich price: 25 g for \$69.90; the largest bottle). On the other hand, *p*MBOH **43** is stable and economically viable (Aldrich price; 500 g for \$84.90; the largest bottle). It was found that mono-*N*-alkylation of **36** proceeded well by slow addition (over 3 h) of **43** to a solution of **36** in acetonitrile in the presence of a catalytic amount of acid (*p*-TsOH) at 70 °C, as shown in Scheme 1.16. Slow addition of alcohol **43** minimized the self-condensation of **43** to form symmetrical ether **44**, which was an equally effective alkylating agent. The product **41** was then directly crystallized from the reaction mixture by addition of water and was isolated in 90% yield and in >99% purity. A toluene solution of **41** can be used for the next reaction without isolation but the yield and optical purity of the asymmetric addition product were more robust if isolated **41** was used. In general, the more complex the reaction, the purer the starting materials the better.

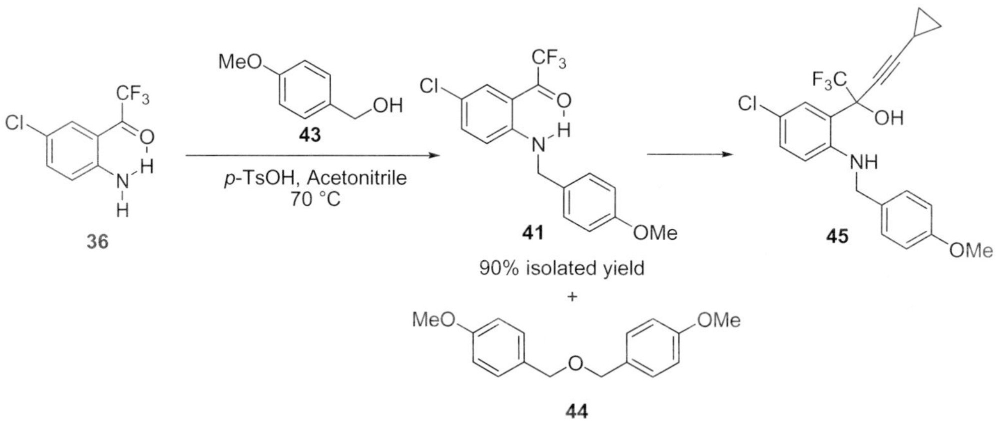

Scheme 1.16 Alternative installation of *p*MB on **36**, followed by acetylide addition.

1.2.1.2.2 The First Generation of Asymmetric Addition of Lithium Cyclopropylacetylide to the Ketone 41

Introduction Since we had already developed the novel asymmetric addition of lithium acetylide to ketimine **5**, we did not spend any time on investigating any chiral resolution methods for Efavirenz®. Our previous method was applied to **41**. In the presence of the lithium alkoxide of cinchona alkaloids, the reaction proceeded to afford the desired alcohol **45**, as expected, but the enantiomeric excess of **45** was only in the range 50–60%. After screening various readily accessible chiral amino alcohols, it was found that a derivative of ephedrine, (1*R*,2*S*) 1-phenyl-2-(1-pyrrolidinyl)propan-1-ol (**46**), provided the best enantiomeric excess of **45** (as high as 98%) with an excellent yield (*vide infra*). Prior to the development of asymmetric addition in detail, we had to prepare two additional reagents, the chiral modifier **46** and cyclopropylacetylene (**37**).

Preparation of the chiral modifier–(1*R*,2*S*)-1-phenyl-2-(1-pyrrolidiny)propan-1-ol (46) Our best chiral modifier **46** has been utilized in many asymmetric transformations by Mukaiyama [22] and Soai [23], and recently by Bolm [24]. The ligand **46** was prepared by heating norephedrine (**47**) with 1,4-dibromobutane (**48**) in the presence of K_2CO_3 in either EtOH or acetonitrile. The isolated yield by distillation was reported as only 33% [25]. It was found that $NaHCO_3$ was a better choice for the base, as shown in Scheme 1.17. A suspension of **47**, 1.1 equiv of **48**, and 2 equiv of $NaHCO_3$ in toluene was heated under reflux for 18–22 h. The solid was removed by simple filtration. The toluene solution could be used directly for the asymmetric addition reaction after washing with water and azeotropic drying. Free base can be isolated as a crystalline solid by switching the solvent to heptane at <0 °C. More conveniently, its HCl salt could be isolated by the addition of HCl in isopropyl alcohol in 90% yield [21, 26]. The HCl salt can be converted to the free base by neutralization.

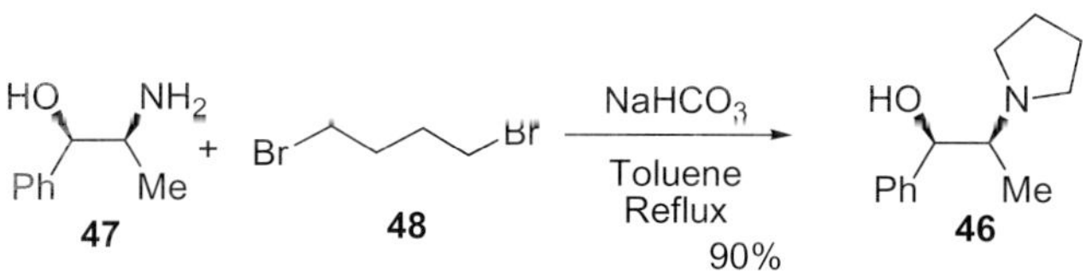

Scheme 1.17 Preparation of the chiral modifier **46**.

Preparation of cyclopropylacetylene (37) Cyclopropylacetylene (**37**) was a known compound and its synthetic method from vinylcyclopropane via dibromination had been reported [27] when we started our investigation. Large scale preparation of **37** was not an easy task. Actually, many chemists in our department worked on establishing the process for such a simple compound as **37** at the peak of the project.

Since the final product is a pharmaceutical, high purity of the product is definitely required. Furthermore, the amount of any impurities in the final product has to be rigorously regulated under ICH guidelines. Rejection of impurities related to cyclopropylacetylene (**37**) was difficult throughout this whole process [28]. Thus, not only the isolated yield but the impurity profile of **37** was critical.

The well documented synthetic method for **37** is chlorination of cyclopropylmethylketone followed by base treatment [29]. However, this method did not provide a suitable impurity profile. The most convenient and suitable method we found was the one-step synthesis from 5-chloro-1-pentyne (**49**) by addition of 2 equiv of base, as shown in Scheme 1.18 [21, 30]. Two major impurities, starting material **49** and reduced pentyne, had to be controlled below 0.2% each in the final bulk of **37**, to ensure the final purity of Efavirenz®. Acetylene **37** was isolated by distillation after standard work-up procedure.

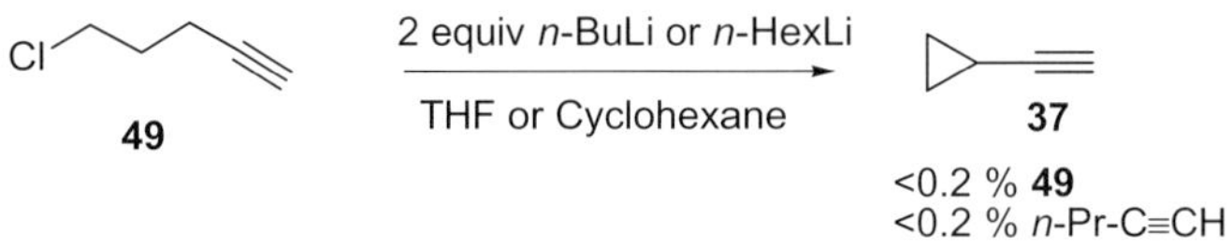

Scheme 1.18 Preparation of cyclopropylacetylene **37**.

Scientists at DuPont Merck Pharmaceuticals [31] had also developed a new process to prepare **37**, based on a modification of the Corey–Fuchs method, from cyclopropylaldehyde, prepared by thermal rearrangement of butadiene monoxide.

Asymmetric addition of acetylide to the ketone Having the two key reagents in hand, we optimized the asymmetric addition reaction on ketone **41**. First, chiral modifiers were screened from among readily accessible β-amino alcohols and the results are summarized in Table 1.5.

Among them, (1*R*,2*S*)-1-phenyl-2-(1-pyrrolidinyl)propan-1-ol (**46**) was selected as a chiral modifier for further optimization. It is interesting to point out that *N*-methyl ephedrine was not a suitable chiral modifier for ketimine **5** (only 10% ee as shown in Table 1.2), but in the case of ketone **41**, *N*-methyl ephedrine provided a respectable 53% ee, as shown in Table 1.5.

During optimization, a few important factors for this reaction were identified. First, 2 equiv of lithium acetylide and 2 equiv of the chiral modifier **46** were required for better chemical yield and high enantiomeric excess. Secondly, warming the mixture of lithium acetylide and chiral modifier to at least 0 °C prior to addition of ketone **41** is the key to ensuring consistently high selectivity and high yield. By doing so, the enantiomeric excess was improved from 82% to 96–98%. Thirdly, the reaction temperature had little effect, as summarized in Table 1.6, as long as the mixture of lithium acetylide and **46** was warmed prior to addition of ketone **41**.

Table 1.5 Amino alcohol-mediated addition to ketone **41**.

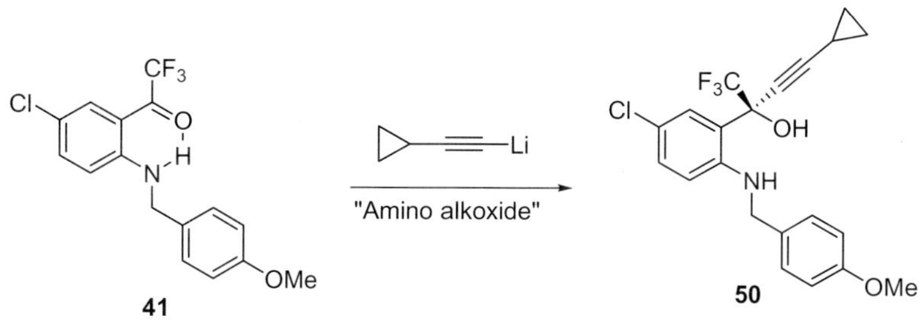

41 **50**

Amino alcohol	Structure	Product ee%
N-Methyl ephedrine	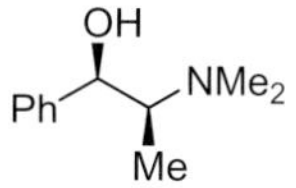	53
(1*R*,2*S*)-2-(*N*,*N*-Diethylamino)-1-phenylpropan-1-ol	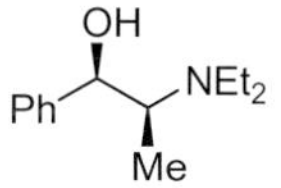	70
(1*R*,2*S*)-(*N*,*N*-Di-*n*-propylamino)-1-phenylpropan-1-ol	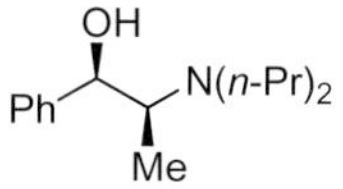	59
(1*R*,2*S*)-(*N*,*N*-Di-*n*-butylamino)-1-phenylpropan-1-ol	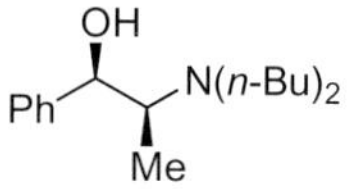	60
(1*R*,2*S*)-1-Phenyl-2-(1-pyrrolidinyl)propan-1-ol	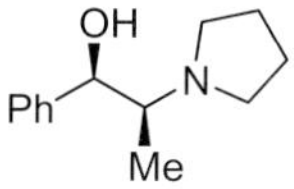 **46**	82
(1*R*,2*S*)-1-Phenyl-2-(1-piperdinyl)propan-1-ol	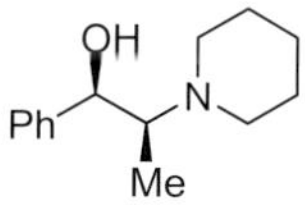	72
(1*R*,2*S*)-2-(1-Pyrrolidinyl)-1,2-diphenylethanol	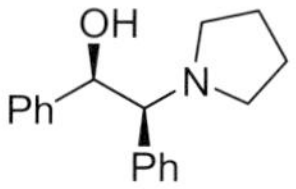	61
N-Methyl pseudoephedrine	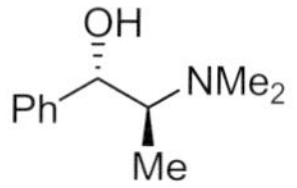	35

Table 1.6 Temperature effect.

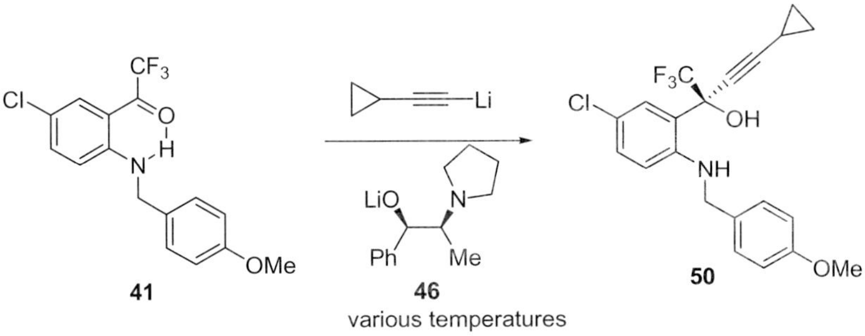

Reaction temperature (°C)	Enantiomeric excess (%)
−30	91
−40	95
−50	97–98
−60	99
−70	99

The effect of the nitrogen protective group in **37** was briefly studied and the results are summarized in Table 1.7. The *p*MB group provided a good selectivity. It is also noted that the reaction was sluggish and provided a lower enantiomeric excess (72%) if the nitrogen atom was not protected.

Experimentally, a chiral nucleophile was prepared by reaction of *n*-BuLi (or *n*-HexLi) with a mixture of chiral modifier **46** and cyclopropylacetylene **37** at −10 to

Table 1.7 Effect of the protecting group.

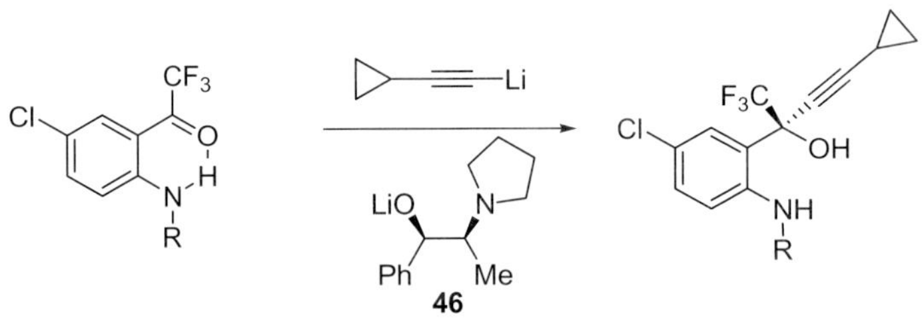

R =	Enantiomeric excess (%)
p-Methoxybenzyl (*p*MB)	96–98
H	72
3,4-Dimethoxybenzyl	99
Triphenylmethyl	90

0 °C in a THF–toluene–hexane mixture. After the mixture was cooled below −50 °C, ketone **41** was added. After ~60 min, the reaction was quenched with aqueous citric acid. The organic layer was then solvent switched into toluene, and the product **50** was crystallized by the addition of heptane (91–93% isolated yield, >99.5% ee). The chiral modifier **46** is easily recycled from the aqueous layer by basification with NaOH and extraction into toluene to recover **46** (>99% purity, 98% recovery yield). The modifier has been recycled up to nine times in subsequent chiral addition reactions without any problem.

This asymmetric addition method is robust and provides the desired chiral alcohol **50** with high ee % and good overall yield and it became a cornerstone for our first manufacturing route of Efavirenz®.

1.2.1.2.3 Preparation of Efavirenz® (1)

Obviously, there are two ways to prepare Efavirenz® from the *p*MB protected chiral amino alcohol **50**; (i) creation of the benzoxazinone first then removal of the *p*MB group; or (ii) removal of the *p*MB first then formation of benzoxazinone. Preparation of the benzoxazinone was demonstrated by Medicinal Chemistry from the amino-alcohol with CDI.

Initially, **50** was converted into the benzoxazinone **51** by reaction with phosgene in the presence of triethylamine and **51** was isolated in 95% yield upon crystallization from methanol. Deprotection of the *p*MB group from **51** was accomplished with ceric ammonium nitrate (CAN) in aqueous acetonitrile. Efavirenz® was isolated in 76% yield after crystallization from EtOAc-heptane (5 : 95), as shown in Scheme 1.19. There were two issues identified in this route. First, 1 equiv of anisaldehyde was generated in this reaction, which could not be cleanly rejected from product **1** by simple crystallization to an acceptable level under the ICH guideline. Anisaldehyde was removed from the organic extract as a bisulfite adduct by washing with aqueous $Na_2S_2O_5$ twice, prior to the crystallization of **1**. Secondly,

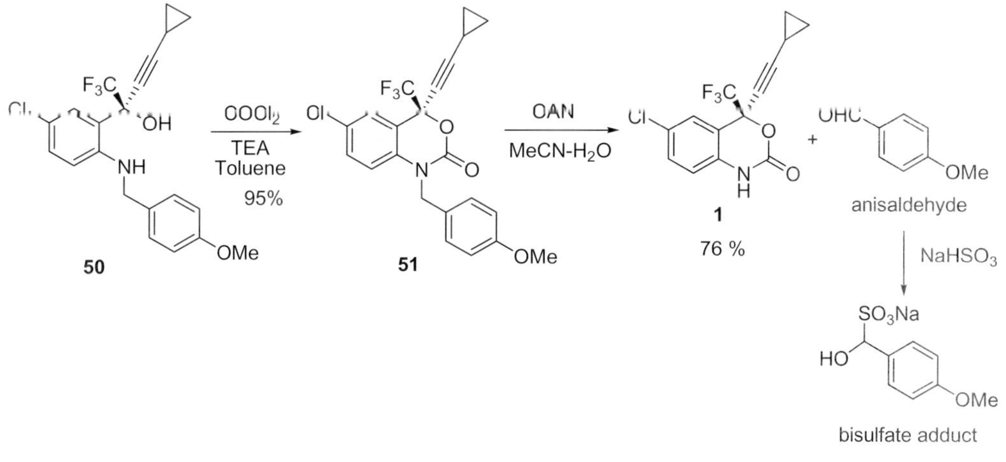

Scheme 1.19 Initial end game for Efavirenz®.

residual cerium salt in the water layer was difficult to recycle, thus the aqueous waste was an environmental issue which added an extra e-factor number to this process.

In order to overcome these two issues, we reversed the order of the reaction sequence, as summarized in Scheme 1.20. We took advantage of the alcohol functional group in **50**. Oxidation of *p*MB of **50** with DDQ proceeded smoothly to form cyclic aminal **52** (as a mixture of α and β = 11.5 : 1) in toluene at 0–10 °C. The resulting DDQH, which is insoluble in toluene, was filtered off, and isolated DDQH could be recycled as we demonstrated in the Proscar process (see p. 92) [32]. Thus, this process minimizes the impact to the environment from an oxidizing reagent. Cyclic aminal **52** was solvolyzed with NaOH in MeOH at 40 °C. The resulted anisaldehyde was reduced *in situ* to *p*MBOH **43** by addition of $NaBH_4$ and the desired amino alcohol **53** was isolated by direct crystallization from the reaction mixture, upon neutralization with acetic acid, in 94% yield and >99.9% ee after crystallization from toluene–heptane.

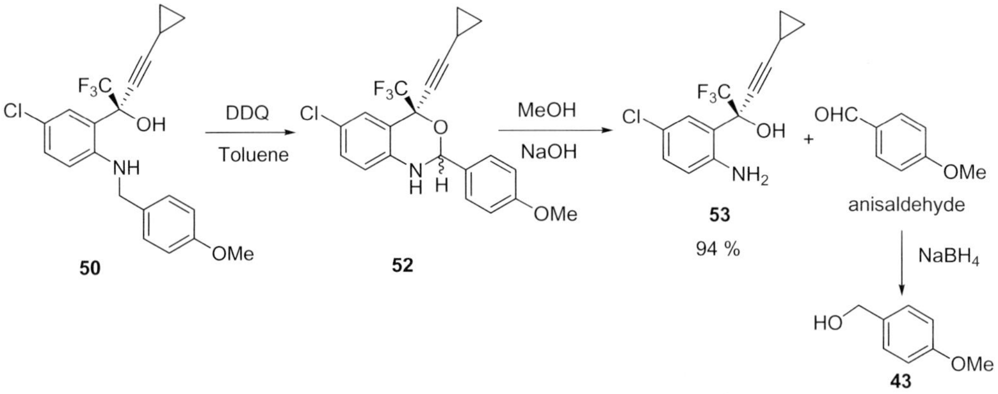

Scheme 1.20 Removal of *p*MB from **50**.

Conversion of the amino alcohol **53** to Efavirenz® (**1**) was readily accomplished by reaction with phosgene or phosgene equivalents. The most convenient and economically sound method is to react **53** with phosgene in the absence of base in THF–heptane at 0–25 °C. After aqueous work-up, Efavirenz® was crystallized from THF–heptane in excellent yield (93–95%) and purity (>99.5%, >99.5% ee).

Alternatively, two phosgene equivalents were studied, methyl chloroformate and *p*-nitrophenyl chloroformate. When methyl chloroformate was used for the end game, *N*-carbamate **54** was obtained smoothly but subsequent cyclization to benzoxazinone **1** was sluggish. Furthermore, removal of the unreacted intermediate methyl carbamate **54** from Efavirenz® was not trivial, thus we did not pursue this method. On the other hand, reaction of **53** and *p*-nitrophenyl chloroformate initially provided the corresponding *p*-nitrophenyl carbamate **55** under mild basic conditions ($KHCO_3$). Carbamate **55** was smoothly cyclized to **1** upon increasing the pH by addition of KOH, and **1** was isolated in 94% yield. When *p*-nitrophenyl chloroformate was added to amino alcohol **53** under stronger basic conditions (pH > 11) from the beginning of the reaction, the generated *p*-nitrophenol reacted with *p*-nitrophenyl chloroformate to form symmetric carbonate **56**. Thus, stepwise pH adjustment was critical for this reaction, as summarized in Scheme 1.21.

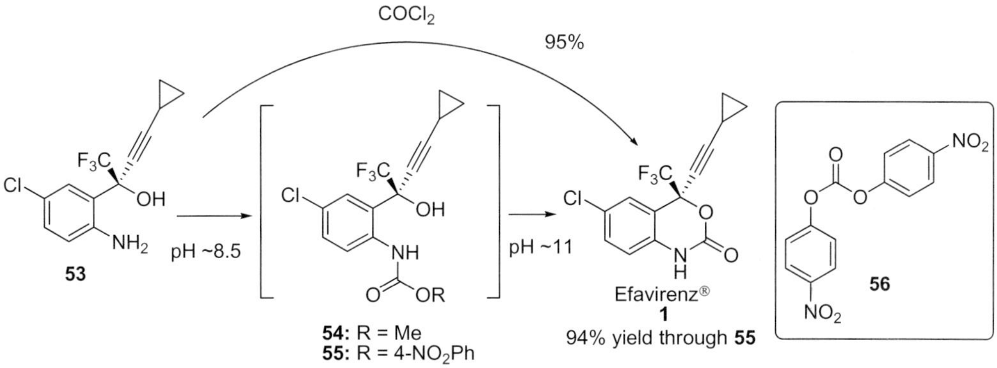

Scheme 1.21 Optimized end game for Efavirenz®.

1.2.1.2.4 **The Second Generation Asymmetric Addition of Zinc-Cyclopropylacetylide to 36 (Part of the Current Manufacturing Route)**

The overall process from amino ketone **36** to Efavirenz® (**1**) required four steps with an overall yield of 72% and quite high purity of the isolated **1**, as described above. This process supported initial marketing of Efavirenz® but there were a few drawbacks. The key asymmetric addition of acetylide required 2 equiv of precious cyclopropylacetylene (**37**). In addition, two steps out of the total four steps were protection with *p*MB and its deprotection.

It would be ideal if the asymmetric addition could be done without a protecting group for ketone **36** and if the required amount of acetylene **37** would be closer to 1 equiv. Lithium acetylide is too basic for using the non protected ketone **36**, we need to reduce the nucleophile's basicity to accommodate the acidity of aniline protons in **36**. At the same time, we started to understand the mechanism of lithium acetylide addition. As we will discuss in detail later, formation of the cubic dimer of the 1 : 1 complex of lithium cyclopropylacetylide and lithium alkoxide of the chiral modifier[3] was the reason for the high enantiomeric excess. However, due to the nature of the stable and rigid dimeric complex, 2 equiv of lithium acetylide and 2 equiv of the lithium salt of chiral modifier were required for the high enantiomeric excess. Therefore, our requirements for a suitable metal were to provide: (i) suitable nucleophilicity; (ii) weaker basicity, which would be

3) Many of the papers from Merck reported the 1 : 1 complex of lithium acetylide and lithium alkoxide of the chiral modifier as monomer and the dimer of the 1 : 1 complex as tetramer.

compatible with free aniline; and (iii) a favorable equilibrium between a monomer and a dimer to reduce the requirement of acetylene.

Kitamura and Noyori have reported mechanistic studies on the highly diastereomeric dialkylzinc addition to aryl aldehydes in the presence of (-)-3-*exo*-(dimethylamino)isoborneol (DAIB) [33]. They stated that DAIB (a chiral β-amino alcohol) formed a dimeric complex **57** with dialkylzinc. The dimeric complex is not reactive toward aldehydes but a monomeric complex **58**, which exists through equilibrium with the dimer **57**, reacts with aldehydes via bimetallic complex **59**. The initially formed adduct **60** is transformed into tetramer **61** by reaction with either dialkylzinc or aldehydes and regenerates active intermediates. The high enantiomeric excess is attributed to the facial selectivity achieved by clear steric differentiation of complex **59**, as shown in Scheme 1.22.

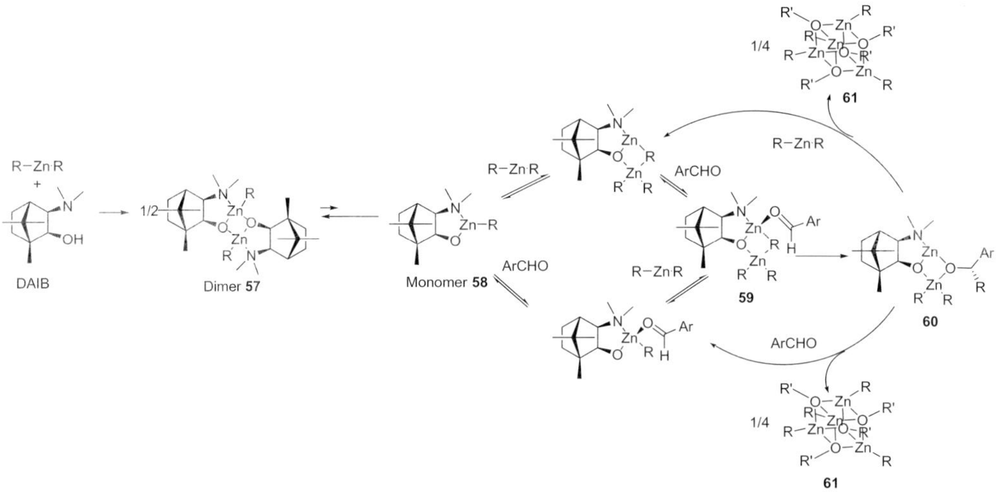

Scheme 1.22 Kitamura and Noyori's mechanism of the asymmetric addition of dialkyl zinc to aryl aldehydes.

These facts are perfectly matched with our above-mentioned desired requirements. In addition, alkyl zinc is known to be less basic and deprotonation of ketone-aniline **36** by zinc reagent is highly unlikely. However, one of the issues for this reaction was the requirement for two alkyl groups on the zinc metal since the product ends up as tetramer **61**, where the zinc atom still has one alkyl group, recalling that our cyclopropylacetylene (**37**) is not easy to obtain.

We came up with the idea of using a dummy ligand, as shown in Scheme 1.23 [34]. Reaction of dimethylzinc with our chiral modifier (amino-alcohol) **46** provided the methylzinc complex **62**, which was subsequently reacted with 1 equiv of MeOH, to form chiral zinc alkoxide **63**, generating a total of 2 moles of methane. Addition of lithium acetylide to **63** would generate an ate complex **64**. The ate complex **64** should exist in equilibrium with the monomeric zincate **65** and the dimer **66**. However, we expected that the monomer ate complex **64** and the mono-

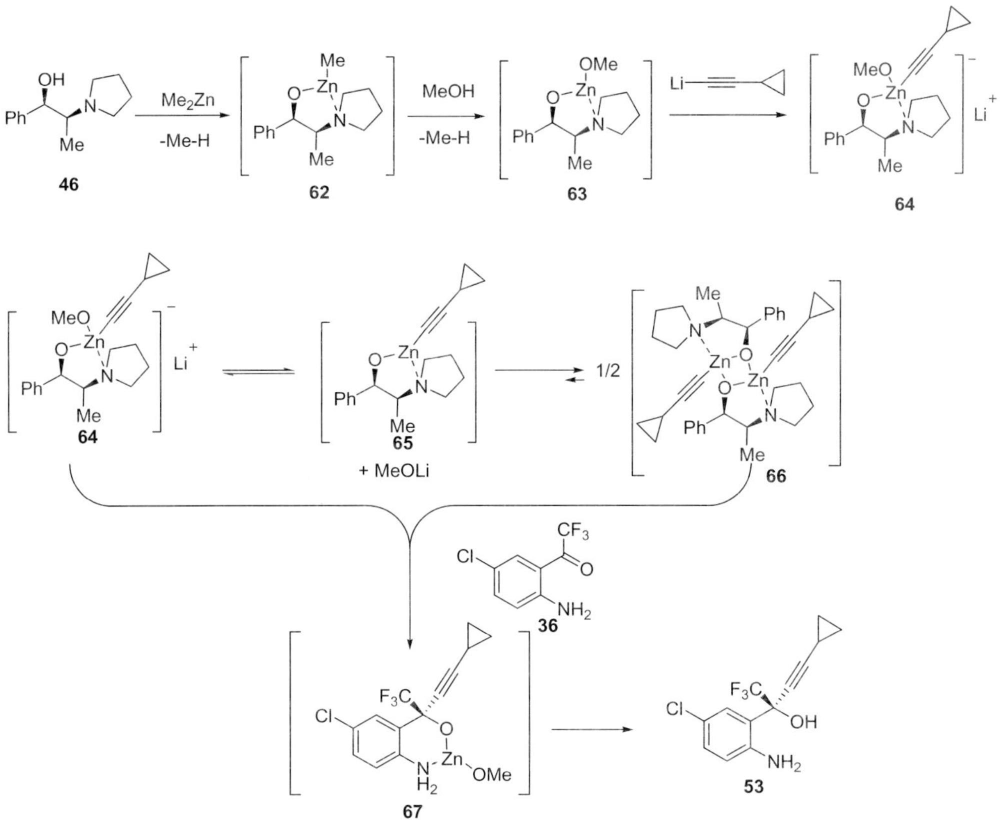

Scheme 1.23 Our initial thoughts on the organo-zinc reaction.

meric zincate **65** would react with the unprotected amino-ketone **36** to provide the desired non-protected amino alcohol **53**. Since the product **53** would remain as polymeric complexes of MeO–Zn–O–Product **67**, we expected only 1 equiv of cyclopropylacetylene to be needed for the completion of the reaction.

Chiral modifiers were screened in the zinc chemistry. Once again, in the case of aniline ketone **36**, chichona alkaloids, binaphthol, and tartaric acid derivatives gave very poor selectivity and ephedrine derivatives provided good selectivity. The results are summarized in Table 1.8.

The same chiral modifier used in the previous lithium chemistry also provided the best result in this case, with as high as 83% ee. Interestingly, the countercation also had a significant effect on the enantioselectivity. For example, with the chloromagnesium acetylide, the desired adduct **53** was obtained with 87% ee but only ~50% ee was obtained with the bromo- and iodomagnesium acetylide.

Furthermore, variation of the achiral adduct for formation of alkoxy zinc such as **63** (shown in Scheme 1.23) had a profound influence on the enantioselectivity of the alkynylation reaction; the results are summarized in Table 1.9.

Table 1.8 Effect of chiral modifiers on organo-zinc chemistry.

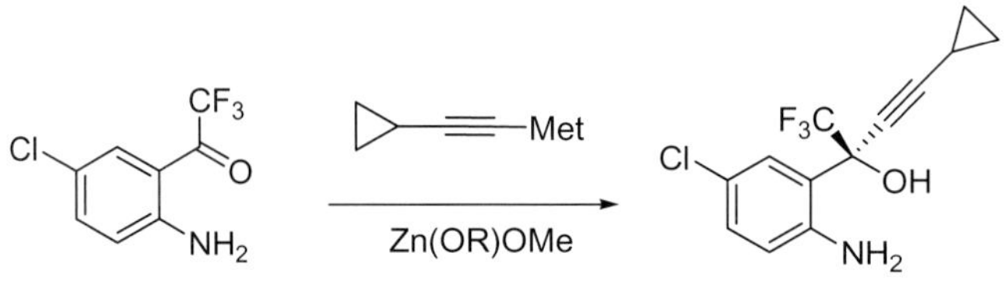

36 → **53**

Met	Ephedrine derivatives		ee% of 53
Li	Ephedrine	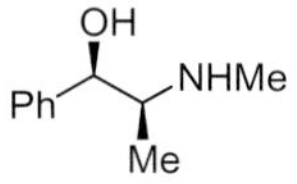	28
Li	Norephedrine	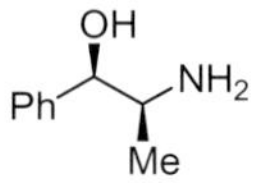	42
Li	*N*-Methyl ephedrine	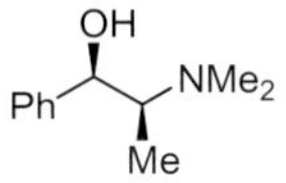	81
Li	(1*R*,2*S*)-1-Phenyl-2-(1-pyrrolidinyl)propan-1-ol	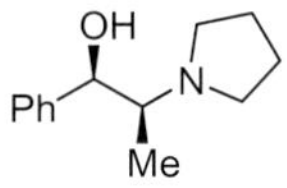 **46**	83
MgCl	(1*R*,2*S*)-1-Phenyl-2-(1-pyrrolidinyl)propan-1-ol	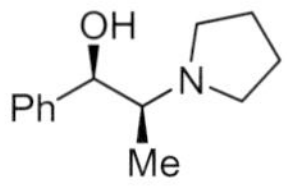 **46**	87
MgBr	(1*R*,2*S*)-1-Phenyl-2-(1-pyrrolidinyl)propan-1-ol	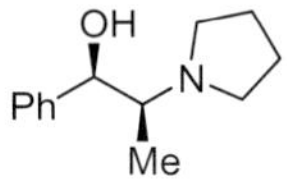 **46**	54
MgI	(1*R*,2*S*)-1-Phenyl-2-(1-pyrrolidinyl)propan-1-ol	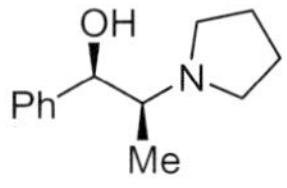 **46**	51

The use of ethanol as an achiral auxiliary gave the adduct **53** with 55% ee, while neopentyl alcohol and methanol gave 96 and 87% ee, respectively. These results suggested that the achiral alcohol might exert a steric effect on the stereoselectivity. However, the increase in enantioselectivity from 55% to about 96% when 2,2,2-trifluoroethanol (TFE) was used instead of ethanol indicates a possible significant inductive effect also. Good enantioselectivities were also obtained with carboxylic acids and phenols.

Table 1.9 Effect of achiral alcohol on organo-zinc chemistry.

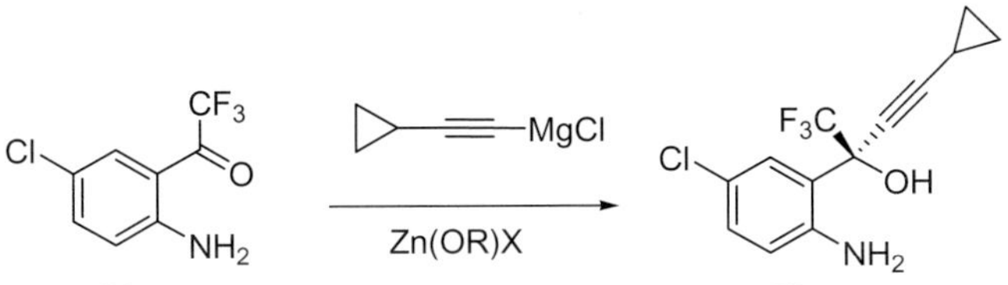

36 **53**

All reactions were carried out at 25 °C in THF/toluene with 1 equiv each of chiral modifier, achiral alcohol, dimethylzinc, and cyclopropylacetylide, and 0.83 equiv of **36**

Alcohol adduct	ee % of 53
MeOH	87
EtOH	55
$(CH_3)_3CCH_2OH$	96
$CH_2{=}CHCH_2OH$	90
$PhCH_2OH$	89
CF_3CH_2OH (TFE)	96
CF_3CO_2H	89
$(CH_3)_3CCO_2H$	72
4-NO_2-PhOH	89

Further optimization of this reaction was carried out with TFE as an achiral adduct, since reaction with TFE is much faster than that with neopentyl alcohol. We found that dimethyl- and diethylzinc were equally effective, and the chiral zinc reagent could be prepared by mixing the chiral modifier, the achiral alcohol and dialkylzinc reagent in any order without affecting the conversion and selectivity of the reaction. However, the ratio of chiral to achiral modifier does affect the efficiency of the reaction. Less than 1 equiv of the chiral modifier lowered the ee %. For example with 0.8 equiv of **46** the enantiomeric excess of **53** was only 58.8% but with 1 equiv of **46** it was increased to 95.6%. Reaction temperature has a little effect on the enantiomeric excess. Reactions with zinc alkoxide derived for **46** and TFE gave **53** with 99.2% ee at 0 °C and 94.0% ee at 40 °C.

Reaction procedure After optimization, the reaction was run as follows: diethylzinc (1.2 equiv in toluene) was slowly added to a solution of TFE (0.9 equiv) and **46** (1.5 equiv) in THF below 30 °C. To the solution was added a solution of chloromagnesium cyclopropylacetylide (1.2 equiv), prepared from cyclopropylacetylene and *n*-butylmagnesium chloride in THF. To the mixture was added a solution of **36** (1 equiv) in THF at 0 °C and then the mixture was aged for 15 h at room temperature. The solution was quenched by addition of aqueous K_2CO_3. The resulting inorganic salts were removed by filtration. The filtrate and washings were combined and washed with citric acid. The aqueous layer was kept for the recovery of **46**. The pH of the aqueous solution was adjusted to pH 11; toluene extraction and solvent switch to heptane afforded a solution which crystallized at low temperature to recover **46** in 95% yield. The organic layer was washed with water and solvent-switched to heptane and **53** was isolated by crystallization from heptane at 0 °C in 95.3% isolated yield with 99.2% ee.

With this novel zinc chemistry, the protection and deprotection sequence were eliminated, the requirement of expensive cyclopropylacetylene was reduced from 2.2 to 1.2 equiv and the previously required cryogenic temperature was eliminated. Finally, the overall yield was improved to 87% (in two steps) from 72% (in four steps).

The overall process for Efavirenz® is summarized in Scheme 1.24.

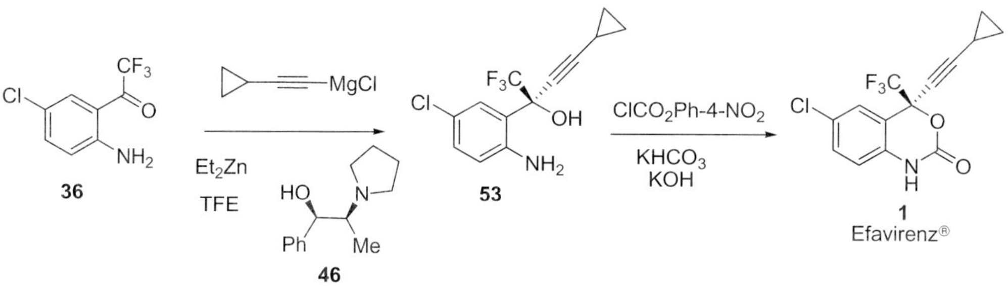

Scheme 1.24 Overall Efavirenz® synthesis.

1.2.2
Chemistry Development

When we worked on asymmetric addition to the ketimine **5**, we could not figure out the mechanism of this asymmetric addition. One of the authors still remembers his supevisor, Dr. Ed Grabowski, coming to his office just a few weeks before the final step of the large scale preparation of **2** and he did not ask about the preparation schedule but asked about the mechanism, especially the kinetics. Unfortunately, kinetic studies of the asymmetric addition to ketimine **5** were not fruitful, partially because the reaction was not totally homogeneous at low temperature. The only thing we were clear about was that the aggregation status of some lithium species would be important for this excellent enantiomeric excess based on the very unique temperature effect (Figure 1.4).

On the other hand, asymmetric addition of lithium acetylide in the presence of the ephedrine derivative **46** is a homogeneous reaction and reveals great detail about the reaction mechanism.

Here, we will discuss the reaction mechanism of the asymmetric lithium acetylide addition to *p*MB protected amino ketone **41**. Then we will discuss some speculation about the asymmetric addition via the novel zinc acetylide addition.

1.2.2.1 Reaction Mechanism for the Lithium Acetylide Addition to *p*MB Protected Amino Ketone 41

1.2.2.1.1 Circumstantial Evidence for the Reaction Mechanism

Before starting to describe detailed studies on the mechanism, we would like to summarize what we know about the reaction so far:

1) Two equiv of cyclopropylacetylene and two equiv of norephedrine derivative **46** are required to obtain good conversion and high enantiomeric excess.
2) Aging a mixture of lithium acetylide and the lithium alkoxide of **46** at higher temperature (−10 to 0 °C) prior to addition of ketone **41** is needed to obtain constantly high enantiomeric excess.

For the ketimine **5** case, the enantiomeric excess of adduct was dependent on the reaction temperature (there was an optimum temperature, lower or higher than that temperature gave lower enantiomeric excess). Thus, we assume the aggregation of the lithium complex with 2-ethynylpyridine and quinine dynamically changes with temperature. However, in this amino ketone **41** case, the suitable aggregate consisting of **46** and cyclopropylacetylene (**37**) seems to be stable once it is formed at higher temperature. Thus, lower temperature gave better enantiomeric excess with the pre-formed aggregate.

A few questions come to mind. What is the structure of the aggregate and why are 2 equiv of each reagent essential? Is it due to the acidic proton (N–H) in **41**? Before going into the detail of the mechanism, let us assemble more circumstantial evidence on this reaction.

First, we found a strong nonlinear effect on the adduct's enantiomeric excess, as indicated in Figure 1.6. The nonlinear effect strongly suggested there would be

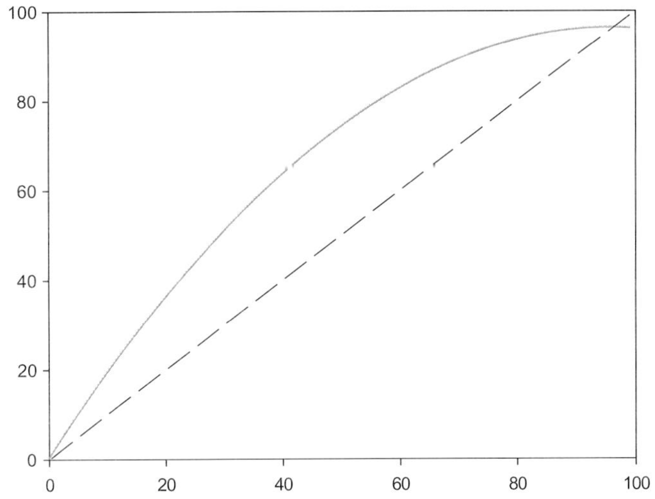

Figure 1.6 Nonlinearity of asymmetric acetylide addition.

polymeric species (including dimer), which participated in the rate-determining step.

When 1.2 equiv and 1.5 equiv of both lithium acetylide and chiral modifier **46** were used, the adduct **50** was obtained with high ee but the isolated yield was 59%. If deprotonation of the N–H of **41** by lithium acetylide is facile, 1.2 equiv and 1.5 equiv of reagents should afford a yield of 20%. However just 0.5 equiv of reagents gave adduct **50**. This experiment indicated that deprotonation of the N–H of **41** might not happen and only half of the reagent could be reacted with aminoketone **41**. Substituting deuterium in **41** (N–D) had no effect on the course of the reaction. Initial NMR and ReactIR studies eventually confirmed that no proton was abstracted from **41** under the reaction conditions (*vide infra*).

Thus, three additional pieces of circumstantial evidence are added to the list.

1) A strong nonlinearity relationship was observed between ee % of chiral modifier **46** and the adduct **50**.
2) Only half of the molar equivalents of the reagents are utilized.
3) No deprotonation of N–H in **41** was observed.

1.2.2.1.2 Structure Elucidation for Reaction Intermediates and Product by NMR Studies

In collaboration with Professor Collum and coworkers, ^{6}Li NMR (including ^{13}C-labeled acetylene **37** and ^{15}N-labeled chiral modifier **46** experiments) and Li aggregation studies were implemented to assist in the understanding of some of the factors responsible for the stereoselective nature of this chemistry [35].

All labeled compounds including *n*-Bu^{6}Li (from ^{6}Li ingot) were prepared by us. When *n*-Bu^{6}Li was added to a solution of cyclopropylacetylene (**37**) and chiral modifier **46** (1 : 1 ratio) in THF–pentane, the ^{6}Li NMR at −125 °C (**A**) is shown in Figure 1.7. A few sets of aggregates could be identified. (See Ref [35a] for full assignment).

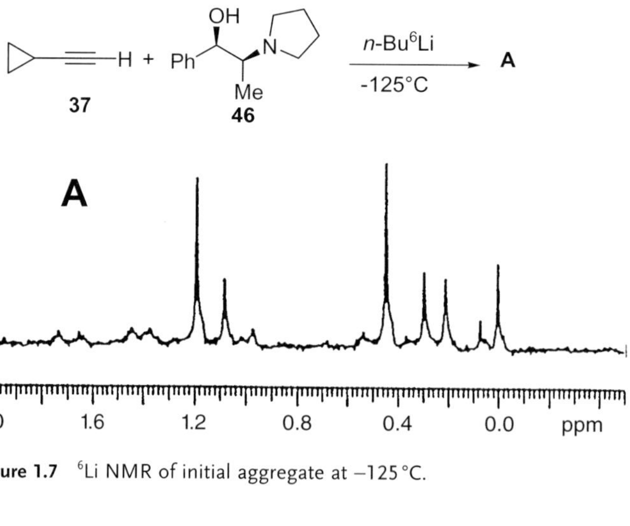

Figure 1.7 ^{6}Li NMR of initial aggregate at −125 °C.

The solution was warmed to 0 °C then cooled to −125 °C (**B**), the ^{6}Li NMR is much simpler than the original spectrum **A**, as shown in Figure 1.8. There are two equal intensity sets of lithium species (major and minor). This mixture is stable at various temperatures once formed. Generation of the stable set of aggregates provides a good correlation with our experimental data (the need to warm the lithium complex prior to addition to ensure high ee). The minor species was later assigned as a cubic aggregate from **37** and **46** (1 : 3).

The structure of the major aggregate was identified by labeling studies. Since the major set has two equal intensity ^{6}Li signals, these signals could be assigned as a 1 : 1 complex **68** of lithium acetylide and lithium alkoxide or a dimer (such as **69**) of the 1 : 1 complex **68** shown in Figure 1.9. Both structures have two different Li species. In order to discriminate between **68** and **69**, a terminal acetylene carbon of **37** was labeled with ^{13}C. In the case of **68**, both lithium signals will be a doublet

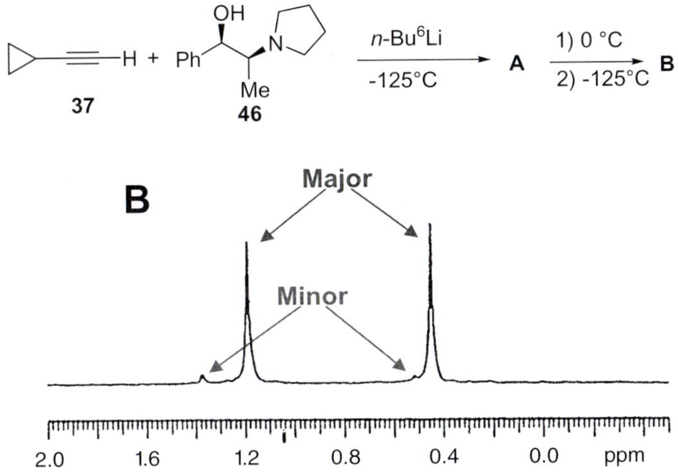

Figure 1.8 ^{6}Li NMR of Li aggregate after aging at higher temperature.

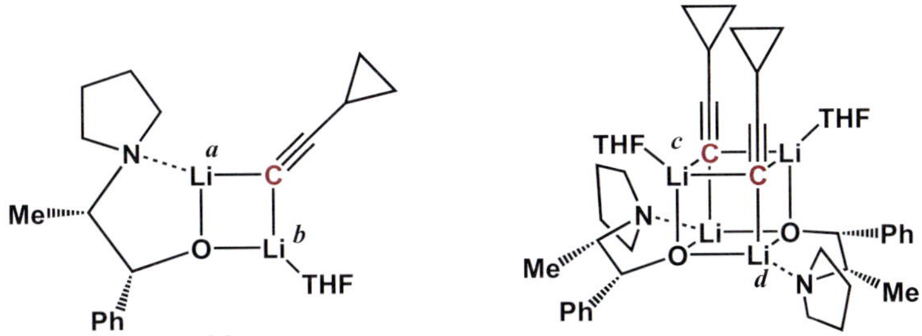

68 69

Figure 1.9 Proposed structures of a 1 : 1 complex and a 2 : 2 complex.

because both Li-***a*** and Li-***b*** coordinate to one ^{13}C atom. On the other hand, **69** will show one set of doublets due to Li-***d*** (coordinates with only one ^{13}C) and one set of triplets since Li-***c*** coordinates two ^{13}C.

^{6}Li NMR data from ^{13}C labeled cyclopropylacetylene (**37**) are shown in Figure 1.10. This spectrum is the definitive evidence that the aggregate is not **68**, as also proven by our experimental results such as nonlinearity of ee. Based on the coupling, the triplet signal at 1.2 ppm is assigned to Li-*c*, and the doublet signal at 0.42 ppm is assigned to Li-***d***.

Of course, there are two possible dimeric structures of the 1 : 1 complex **68**, as shown in Figure 1.11, namely **69** and **70**. Both dimers of **68** should behave in similar fashion in ^{6}Li NMR to the previous experiment. To differentiate those two structures, the nitrogen atom in the chiral modifier **46** was labeled with ^{15}N. Li-***d*** (~0.42 ppm) would be a doublet if the intermediate is **69**. However, Li-***c*** (~1.2 ppm) would be a doublet if it is **70**.

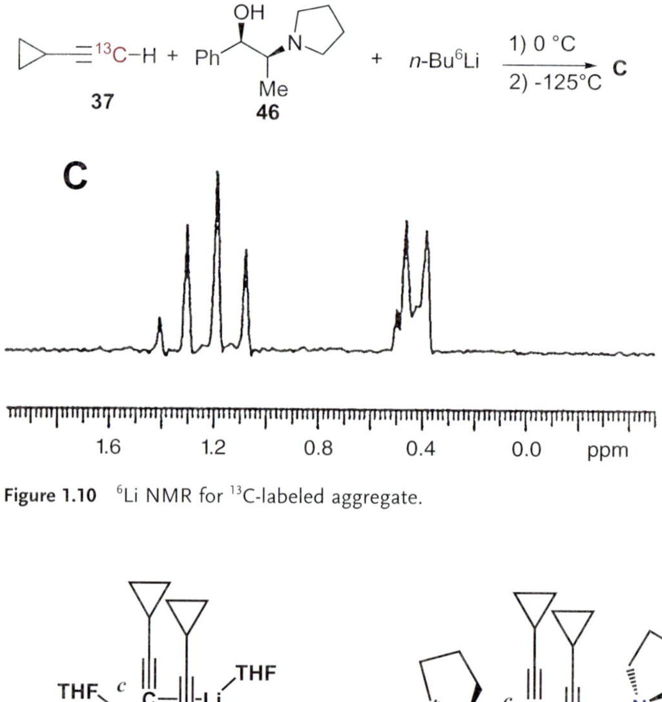

Figure 1.10 ^{6}Li NMR for ^{13}C-labeled aggregate.

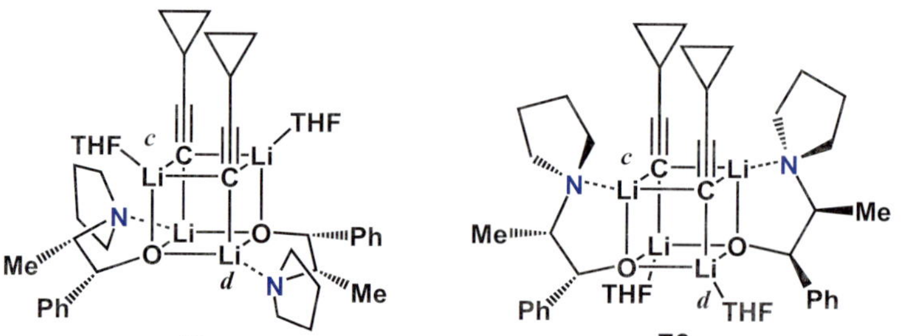

69 **70**

Figure 1.11 Two potential dimeric lithium aggregates of the 1 : 1 complex **68**.

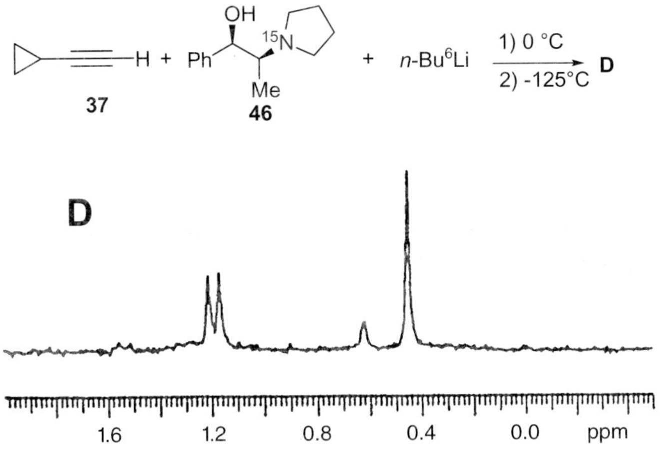

Figure 1.12 ^{6}Li NMR of ^{15}N-labeled aggregate.

The ^{6}Li NMR with ^{15}N labeled **46** is shown in Figure 1.12. Therefore, the aggregate **70** is the true intermediate for this asymmetric addition.

Next, we investigated the structure of the product by NMR, as shown in Figure 1.13 (0.5 equiv of **41** was added to **70**). The asymmetric addition to the dimer **70** proceeded almost instantaneously at −90 °C. Generation of cyclopropylacetylene was not observed by NMR. Following React-IR at the reaction temperature, no-C=O absorbance of **41** at 1660 cm^{-1} is observed until 0.5 equiv of the amino-ketone **41** was added to **70**. Absorbance of **41** was observed after more than 0.5 equiv of **41** was added. This is also consistent with our conclusion that no deprotonation of **41** occurs during the reaction.

When Li-NMR was measured with ^{13}C labeled cyclopropylacetylide (0.5 equiv of Li-acetylide), there was a major set of four singlets with equal intensity (Li-***a***, ***b***, ***c***, ***d***, assignments are depicted in Figure 1.14) as shown in Figure 1.13a under ^{13}C-decoupling conditions. When Li-NMR of the same sample was taken under ^{13}C-coupling conditions (Figure 1.13b), one of the singlets remained as a singlet but the other three singlets became doublets. Therefore, one of the Li atoms (***d***) in the product does not connect to ^{13}C and the other three Li atoms (***a***, ***b***, ***c***) connect to one of the ^{13}C. This is consistent with the proposed cubic aggregate **71**, assembled from two molecules of alkoxide of **46**, one molecule of cyclopropyl acetylide, and one molecule of product, as shown in Figure 1.14.

The aggregate **71** is not reactive toward amino-ketone **41** at low temperature, where the reaction runs typically. The loss of the reactivity of **71** may be attributed to the reduced Lewis acidity of lithium atoms.

These stereochemistry outcomes would be easily predicted based on the assumption that the carbonyl oxygen is coordinated to the lithium atom such as ***d*** in **70**. The larger aryl function will locate in the less sterically hindered side (left-hand side in **70**), providing the desired stereoselectivity. Semiempirical (MENO) computational methods were applied and the results supported our conclusion.

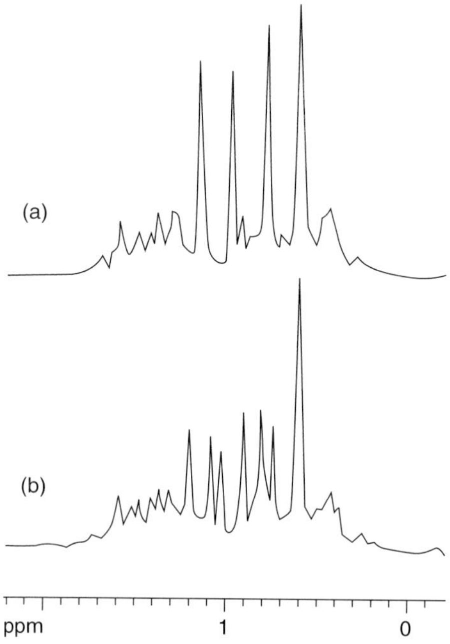

Figure 1.13 ^{6}Li NMRs of the product with ^{13}C-labeled acetylide (0.5 equiv of **41**). (a) ^{13}C-deoupled, (b) ^{13}C-coupled.

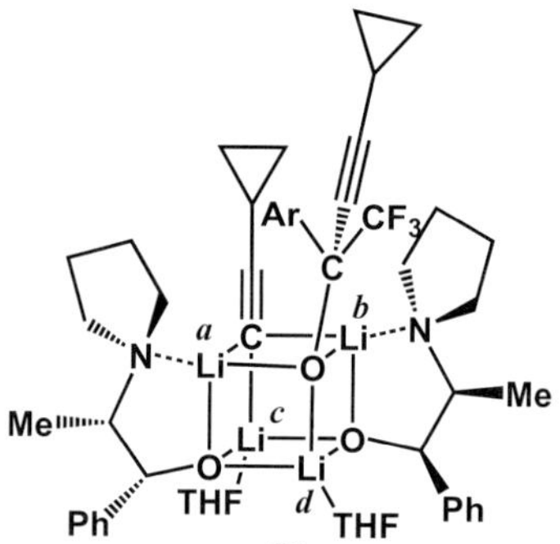

Figure 1.14 Proposed structure of the product.

1.2.2.2 **Reaction Mechanism for the Zinc Acetylide Addition to Amino Ketone 36**

Nonlinearity was also found for this asymmetric organozinc addition, for example, using 50% ee of chiral modifier **46** resulted in 80% ee of adduct **53**. The enantioselectivity is also dependent on the reaction concentration; >98% ee was obtained at 0.1–0.5 M but only 74% ee at 0.005 M. Kitamura and Noyori's work strongly suggested that heterodimer **72** might be more thermally stable than the homodimer

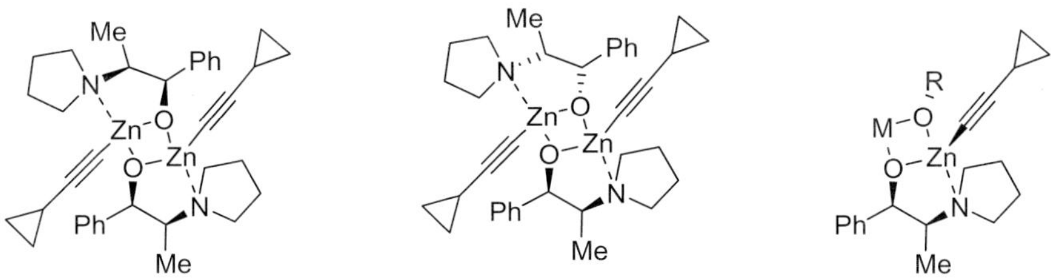

66 72 73

Figure 1.15 Other organo-zinc species.

66, thus asymmetric amplification of the reaction was observed. However, the thermodynamical equilibrium itself cannot explain a few things: (i) the effect of the counter cation of acetylide; (ii) the role of achiral alcohol; (iii) the effect of reaction concentration on the enantiomeric excess? Unfortunately, investigation of the zinc addition reaction with NMR and IR was so complex that these issues could not be resolved. However, intermediate **64** (Scheme 1.23) might exist as a mixed bimetallic species like **73**, which would be similar to the reactive intermediate **59** in Kitamura and Noyori's paper. The structure of **73** might offer some answers to these questions, but the structure of the key intermediate is still unknown (Figure 1.15).

1.3 Conclusion

A highly efficient manufacturing method for a non-nucleoside reverse transcriptase inhibitor, Efavirenz®, was devised and implemented. The final manufacturing method was crafted based on our chemistry knowledge accumulated from a previous drug candidate (first asymmetric acetylide addition to the ketimine), and a clear understanding of the asymmetric addition of lithium acetylide to the ketone for Efavirenz®, through many collaborations not only within our department but also with academic colleagues. It is also important to note that such accumulation of chemistry knowledge is not only applicable in this project but should be applicable to other projects. A good example is the Sugasawa reaction, which was first studied in carbapenem projects, and was then successfully applied in other projects including the first non-nucleoside reverse transcriptase inhibitor.

Acknowledgments

The authors would like to thank all colleagues who worked on this project, whose names are listed in the references. The authors would also like to thank Dr. James McNamara for his careful proofreading and helpful suggestions.

References

1 (a) Young, S.D. (1993) *Perspect. Drug Discov. Design*, **1**, 181–192. (b) Young, S.D., Britcher, S.F., Tran, L.O., Payne, L.S., Lumma, W.C., Lyle, T.A., Huff, J.R., Anderson, P.S., Olsen, D.B., Carroll, S.S., Pettibone, D.J., O'Brien, J.A., Ball, R.G., Balani, S.K., Lin, J.H., Chen, I.-W., Schleif, W.A., Sardana, V.V., Long, W.J., Byrnes, V.W., and Emini, E.A. (1995) *Antimicrob. Agents Chemother.*, **39**, 2602–2605.

2 Tucker, T.J., Lyle, T.A., Wiscount, C.M., Britcher, S.F., Young, S.D., Sanders, W.M., Lumma, W.C., Goldman, M.E., O'Brien, J.A., Ball, R.G., Homnick, C.F., Schleif, W.A., Emini, E.A., Huff, J.R., and Anderson, P.S. (1994) *J. Med. Chem.*, **37**, 2437–2444.

3 Houpis, I.N., Molina, A., Douglas, A.W., Xavier, L., Lynch, J., Volante, R.P., and Reider, P.J. (1994) *Tetrahedron Lett.*, **35**, 6811–6814.

4 Sugasawa, T., Toyoda, T., Adachi, M., and Sasakura, K. (1978) *J. Am. Chem. Soc.*, **100**, 4842–4852.

5 Yasuda, N., DeCamp, A.E., and Grabowski, E.J.J. (1995) US Patent 5,457,201.

6 Recent reviews: (a) Riant, O., and Hannedouche, J. (2007) *Org. Biomol. Chem.*, **5**, 873–888. (b) Friestad, G.K., and Mathies, A.K. (2007) *Tetrahedron*, **63**, 2541–2569. (c) Wu, G., and Huang, M. (2006) *Chem. Rev.*, **106**, 2596–2616. (d) Enders, D., and Reinhold, U. (1997) *Tetrahedron Asymmetry*, **8**, 1895–1946.

7 Recently asymmetric Strecker reaction with ketone is reported; (a) Vachal, P., and Jacobsen, E.N. (2002) *J. Am. Chem. Soc.*, **124**, 10012–10014. (b) Masumoto, S., Usuda, H., Suzuki, M., Kanai, M., and Shibasaki, M. (2003) *J. Am. Chem. Soc.*, **125**, 5634–5635.

8 Tomioka, K., Inoue, I., Shindo, M., and Koga, K. (1991) *Tetrahedron Lett.*, **32**, 3095–3098.

9 Denmark reported asymmetric addition to C=N in the presence of Box ligands or sparteine; Denmark, S.E., Nakajima, N., and Nicaise, O.J.-C. (1994) *J. Am. Chem. Soc.*, **116**, 8797–8798.

10 Huffman, M.A., Yasuda, N., DeCamp, A.E., and Grabowski, E.J.J. (1995) *J. Org. Chem.*, **60**, 1590–1594.

11 March, J. (1985) *Advanced Organic Chemistry*, 3rd edn, John Wiley & Sons, Inc., p. 485.

12 Kobayashi, S., Komoto, I., and Matsuo, J.-I. (2001) *Adv. Synth. Catal.*, **343**, 71–74.

13 Douglas, A.W., Abramson, N.L., Houpis, I.N., Molina, A., Xavier, L.C., and Yasuda, N. (1994) *Tetrahedron Lett.*, **35**, 6807–6810.

14 Baaz, M., and Gutman, V. (1963) Lewis acid catalysts in non-aqueous solutions, in *Friedel-Crafts and Related Reactions*, vol. **1** (ed. G.A. Olah), Interscience, New York, Ch. 5, pp. 367–397.

15 Katritzky, A.R., and Harris, P.A. (1992) *Tetrahedron Asym.*, **3**, 437–442.

16 Soai, K., Hatanaka, T., and Miyazawa, T. (1992) *J. Chem. Soc., Chem. Commun.*, 1097–1098.

17 Yong, S., Tran, L.O., Britcher, S.F., Lumma, W.C., Jr., and Payne, L.S. (1994) EP 0582455.

18 Another synthetic method was reported as follows; Jiang, B., Wang, Q.-F., Yang, C.-G., and Xu, M. (2001) *Tetrahedron Lett.*, **42**, 4083–4085.

19 There are many contributions for the preparation of cyclopropylacetylene. At one time, development for a method of manufacture for cyclopropylacetylene demanded the biggest manpower in the Merck Process Research. For example; Corley, E.G., Thompson, A.S., and Huntington, M. (2000) *Org. Synth.*, **77**, 231–235.

20 Some optimization of the original Medicinal route was reported from the DuPont Merck Pharmaceutical Company; Radesca, L.A., Lo, Y.S., Moore, J.R., and Pierce, M.E. (1997) *Synth. Commun.*, **27**, 4373–4384.

21 (a) Thompson, A.S., Corley, E.G., Huntington, M.F., and Grabowski, E.J.J. (1995) *Tetrahedron Lett.*, **36**, 8937–8940. (b) Pierce, M.E., Parsons, R.L., Jr., Radesca, L.A., Lo, Y.S., Silverman, S., Moore, J.R., Islam, Q., Choudhury, A., Fortunak, J.M.D., Nguyen, D., Luo, C.,

Morgan, S.J., Davis, W.P., Confalone, P.N., Chen, C.-y., Tillyer, R.D., Frey, L., Tan, L., Xu, F., Zhao, D., Thompson, A.S., Corley, E.G., Grabowski, E.J.J., Reamer, R., and Reider, P.J. (1998) *J. Org. Chem.*, **63**, 8536–8543.

22 (a) Mukaiyama, T., Suzuki, K., Soai, K., and Sato, T. (1979) *Chem. Lett.*, 447–448. (b) Mukaiyama, T., and Suzuki, K. (1980) *Chem. Lett.*, 255–256.

23 Niwa, S., and Soai, K. (1990) *J. Chem. Soc., Perkin Trans. I*, 937–943.

24 Zani, L., Eichhorn, T., and Bolm, C. (2007) *Chem. Eur. J.*, **13**, 2587–2600.

25 Soai, K., Yokoyama, S., and Hayasaka, T. (1991) *J. Org. Chem.*, **56**, 4264–4268.

26 Zhao, D., Chen, C.-y., Xu, F., Tan, L., Tillyer, R., Pierce, M.E., and Moore, J.R. (2000) *Org. Synth.*, **77**, 556–560.

27 Slobodin, Y.M., and Egenburg, I.Z. (1969) *Zh. Org. Khim.*, **5**, 1315.

28 Tillyer, R.D., and Grabowski, E.J.J. (1998) *Curr. Opin. Drug Discov. Devel.*, **1**, 349–357.

29 A new improved process has been reported but the isolated yield is mediocre. Schmidt, S.E., Salvatore, R.N., Jung, K.W., and Kwon, T. (1999) *Synlett*, 1948–1950.

30 Application of magnesium amide instead of alkyl lithium for the cyclopropanation formation from 5-chloropentyne was reported in the patent application: Stickley, K.R., and Wiley, D.B. (1999) US 5,952,537.

31 (a) Wang, Z., Yin, J., Campagna, S., Pesti, J.A., and Fortunak, J.M. (1999) *J. Org. Chem.*, **64**, 6918–6920. (b) Wang, Z., Campagna, S., Yang, K., Xu, G., Pierce, M.E., Fortunak, J.M., and Confalone, P.N. (2000) *J. Org. Chem.*, **65**, 1889–1891. (c) Wang, Z., Campagna, S., Xu, G., Pierce, M.F., Fortunak, J.M., and Confalone, P.N. (2000) *Tetrahedron Lett.*, **41**, 4007–4009.

32 Bhattacharya, A., DiMichele, L.M., Dolling, U.-H., Douglas, A.W., and Grabowski, E.J.J. (1988) *J. Am. Chem. Soc.*, **110**, 3318–3319.

33 Kitamura, M., Okada, S., Suga, S., and Noyori, R. (1989) *J. Am. Chem. Soc.*, **111**, 4028–4036.

34 (a) Tan, L., Chen, C.-y., Tillyer, R.D., Grabowski, E.J.J., and Reider, P.J. (1999) *Angew. Chem. Int. Ed.*, **38**, 711–713. (b) Chen, C.-y., and Tan, L. (1999) *Enantiomer*, **4**, 599–608.

35 (a) Thompson, A., Corley, E.G., Huntington, M.F., Grabowski, E.J.J., Remenar, J.F., and Collum, D.B. (1998) *J. Am. Chem. Soc.*, **120**, 2028–2038. (b) Xu, F., Reamer, R.A., Tillyer, R., Cummins, J.M., Grabowski, E.J.J., Reider, P.J., Collum, D.A., and Huffman, J.C. (2000) *J. Am. Chem. Soc.*, **122**, 11212–11218.

2
CCR5 Receptor Antagonist

Nobuyoshi Yasuda

A CCR5 antagonist drug candidate **1** (Figure 2.1) was discovered at Merck Research Laboratories in Rahway, NJ for treatment of HIV infectious diseases [1]. The CCR5 receptor plays a key role in the entry of HIV to T-cells. Thus, antagonists to CCR5 receptors should be expected to prevent HIV infections. On October 2005, Merck granted the non-profit group, the International Partnership for Microbicides (IPM), a royalty-free license to develop, manufacture and distribute their compounds for use as microbicides in resource-poor countries. Development of **1** is in progress at IPM as CMPD 167 [2].

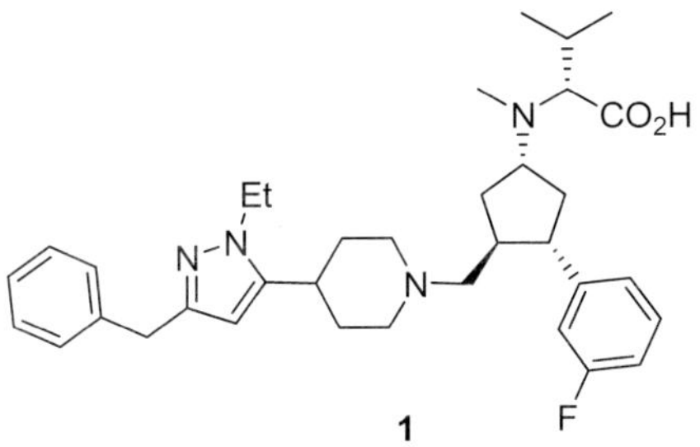

1

Figure 2.1 Structure of a CCR5 antagonist candidate.

2.1
Project Development

2.1.1
Medicinal Route

Originally, **1** was prepared by medicinal chemists from three key components, namely, a cyclopentanone moiety **2**, a pyrazole moiety **3**, and commercially available D-valine, as depicted in Scheme 2.1 [1]. These synthetic disconnections provided an applicable and convergent route to **1**, and consequently, were utilized in our strategy to develop an efficient and scalable synthesis of **1**.

The Art of Process Chemistry. Edited by Nobuyoshi Yasuda

ISBN: 978-3-527-32470-5

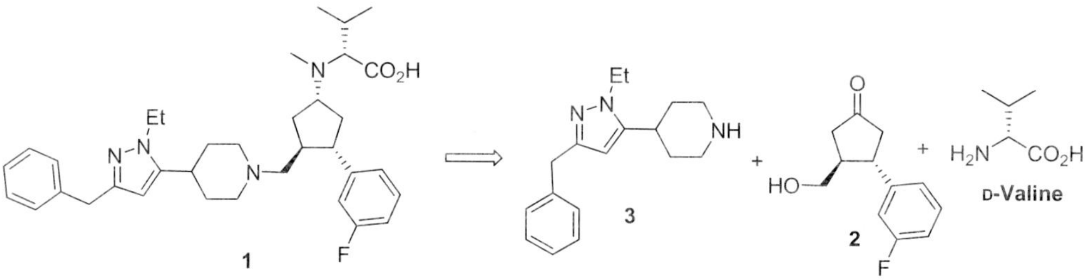

Scheme 2.1 Three-components coupling strategy for **1** by medicinal chemists.

Since preparation of pyrazole **3** was deemed straightforward (*vide infra*) and D-valine *tert*-butyl ester was commercially available, our efforts focused on developing a synthesis of the more challenging cyclopentanone **2** [3]. The original synthetic method for **2** by medicinal chemists is depicted in Scheme 2.2.

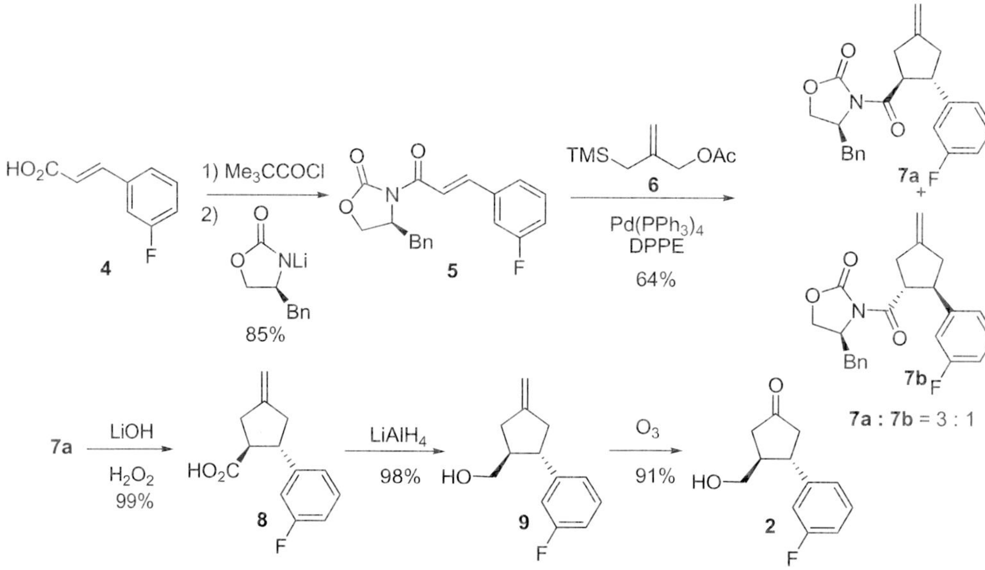

Scheme 2.2 Original preparation method for cyclopentanone **2**.

Construction of the cyclopentane ring was accomplished by utilization of Trost's Pd-mediated diastereoselective [3+2] trimethylenemethane (TMM) cycloaddition [4] on the cinnamate **5** having an Evans type chiral auxiliary [4b]. The resulting diastereomeric mixture (3 : 1 at best) of **7a** and **7b** was separated by careful silica gel column chromatography (**7a** is less polar than **7b** under normal phase). Purified **7a** was converted to the desired cyclopentanone **2** via solvolysis of the chiral auxiliary, reduction of the acid, and ozonolysis of the *exo*-methylene. The overall yield was good given the length of the synthesis, however, there were several significant issues that render the scale-up of this route difficult.

After preliminary evaluation of the original route, we identified several problems in the preparation of cyclopentenone **2**:

1) The chiral auxiliary method was not commercially ideal for large scale preparation because of the required introduction of two additional steps (protection and deprotection) and because of its poor atom economy.
2) The TMM chemistry with **5** was precarious and nonreproducible, and consequently, would be extremely risky to perform on any large scale setting.
3) The TMM reagent **6** was not commercially available in large quantities, and thus added additional synthetic steps.
4) The diastereoselectivity of the TMM chemistry with **5** was mediocre at best, thus the diastereomeric product mixture would require a chromatography step.
5) Absence of on-site large scale ozonolysis technology (This may not be an issue for other companies).

Therefore, we decided to abandon the original route for **2**.

2.1.2
Process Development

The major issue for the large scale preparation of our target **1** was the preparation of the cyclopentanone **2**. With the exception of **2**, we felt the original Medicinal route was suitable for the large scale preparation with standard development optimization. Thus, the outline of this section is:

1) Route selection for cyclopentenone **2**.
   - Diels–Alder/Dieckmann route as the initial route.
   - Asymmetric nucleophilic addition of π-allyl molybdenum complex as the final route.
2) Optimization of the selected route for **2**.
3) Optimization of preparation of the pyrazole **3**.
4) Optimization of the assembly of **2**, **3**, and D-valine to finish the preparation of our target **1** (end game).

We will discuss each topic below. Later, we will discuss the key reaction, *asymmetric nucleophilic addition of a π-allyl Mo complex*, in great detail.

2.1.2.1 Route Selection for Cyclopentenone 2

Although cyclopentanone **2** is a rather simple looking small molecule, the 3,4-*trans*-substituted architecture in a cyclopentanone ring provides significant complexity to this molecule. We devised two alternative routes for the preparation of **2**.

2.1.2.1.1 Diels–Alder/Dieckmann Route

Our first approach to cyclopentanone **2** was a diastereoselective Diels–Alder reaction followed by ozonolysis and Dieckmann condensation, as summarized in Scheme 2.3 [5]. Boeckman reported a similar approach in 1980 [6]. The Diels–Alder reaction of **5**, which was the same intermediate in the original TMM chemistry, with butadiene provided the cyclohexene derivative **10** with very high diastereoselectivity but the isolated yield was only 52%. This low yield was attributed to the low reactivity of butadiene. The chiral auxiliary in **10** was removed by LiOOH and the resulting carboxylic acid was reduced with $LiAlH_4$ to the corresponding alcohol **11** in 76% yield. Ozonolysis of **11** in aqueous AcOH followed by acid treatment in MeOH gave lactone **12** in only 32% yield. Dieckmann condensation worked smoothly on **12** leading to cyclopentanone **13** in 64% yield. Acid mediated solvolysis of **13** cleanly afforded our target **2** in 95% yield. Although this route did provide **2**, the overall yield was poor. The chiral auxiliary was not atom-economical, and ozonolysis was not suitable for our scale-up equipment. Thus, we determined that this approach was not suitable for large scale preparation.

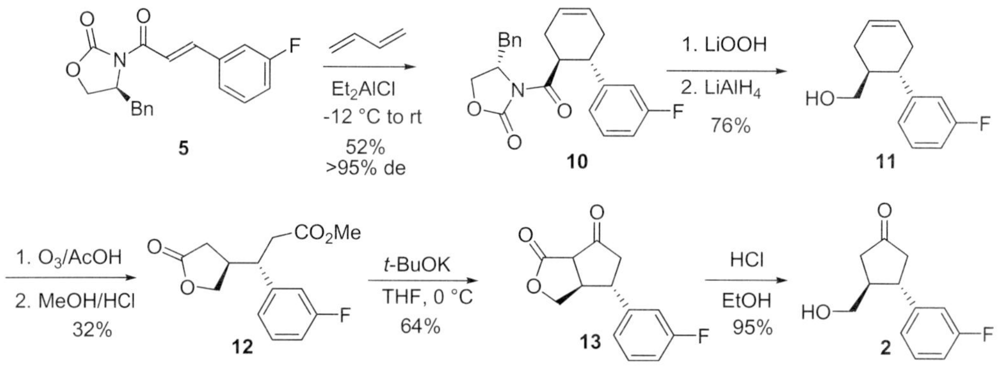

Scheme 2.3 Our first approach Diels–Alder/ozonolysis/Dieckmann.

2.1.2.1.2 Asymmetric Nucleophilic Addition of a π-Allyl Mo Complex route

A second route was devised using chiral *β*-keto ester **14**, which was identified as our precursor for **2** [7]. This idea was in analogy with the carbapenem chemistry [8], as depicted in Scheme 2.4, where Masamune reaction [9] for carbon elongation, diazo-transfer, and transition metal-mediated carbene insertion reaction [10] were employed as key steps sequentially.

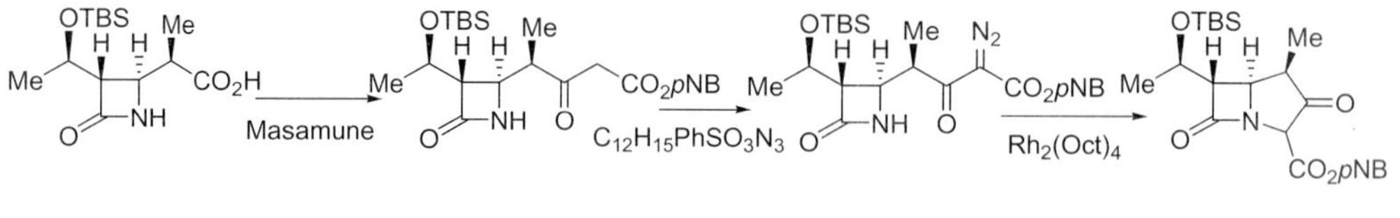

Scheme 2.4 Carbapenem projects.

We envisioned that compound **14** would be prepared as shown in Scheme 2.5. The chiral center would be installed from either linear carbamate **15** or branched carbamate **16** via the asymmetric addition of malonate anion to the π-allyl Mo complex reported by Trost *et al.* [11] to afford the branched chiral malonate derivative **17**. Decarboxylation of **17** should provide the mono-carboxylic acid **18**. Masamune homologation with **18** affords our common precursor **14**. Linear carbamate **15** was obtained from the corresponding cinnamic acid, and branched **16** was prepared in one pot from the corresponding aldehyde.

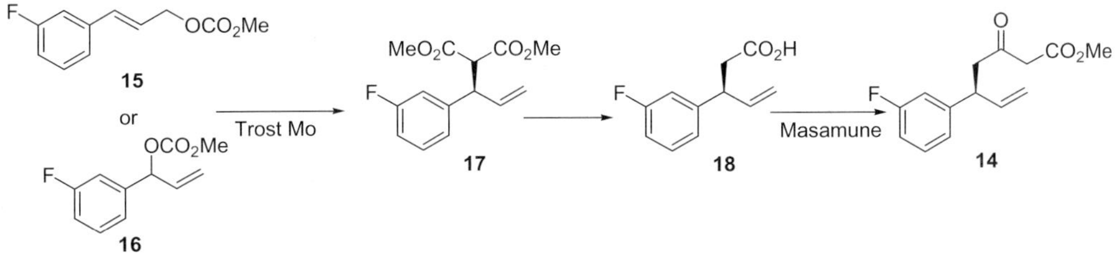

Scheme 2.5 Second approach; retrosynthetic analysis of common precursor **14**.

Three potential routes from **14** to **2**, shown in Scheme 2.6, were identified and evaluated. *Option A* was the original plan of preparation. Hydroboration of the carbon–carbon double bond in **14** followed by oxidation provided primary alcohol **19** (P=H). Beta-ketoester **19** was converted to the corresponding diazo compound

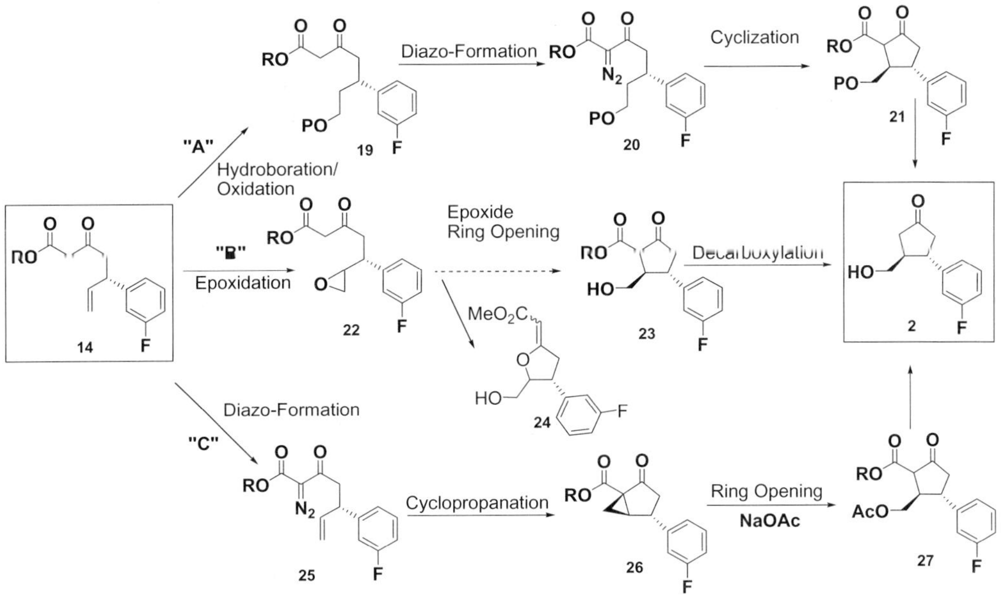

Scheme 2.6 Three potential routes to **2** from **14**.

20 with appropriate protection of the alcohol. Transition metal-mediated ring formations of **20** were studied. When the rhodium-catalyzed cyclization was attempted with the free alcohol **20** (P=H), only the seven-membered ether was observed [12]. When the alcohol in **20** was protected with TBDMS, cyclization in the presence of rhodium provided a mixture of the desired five-membered ring **21**, together with an undesired six-membered ring. This unusual regioselectivity might be due to stabilization of the α-carbocation of the oxygen [13]. Therefore development of this route was terminated.

Next, *option B* was examined. Oxidation of **14** with *m*CPBA proceeded well, leading to epoxide **22**. However, cyclization of the enolate of epoxide **22** did not provide the desired five-membered product **23**. The only isolated product was a tetrahydrofuran derivative **24**, which resulted from the *O*-attack of the enolate to the epoxide instead of the desired *C*-attack. Therefore, development of this route was also terminated.

Finally *option C* was suggested in consultation with Professor Barry M. Trost of Stanford University. He recommended the formation of the bicyclo[3.1.0] system **26** first via carbene insertion from the corresponding diazo compound **25**. The three-membered ring in the sterically strained bicyclo **26** would be easily cleaved by an oxygen nucleophile such as NaOAc, since two electron withdrawing groups were attached on the same bridge head carbon in **26**, and the reaction should yield the ring-opened cyclopentanone **27** [14]. Solvolysis of acetate and methyl ester in **27** promotes spontaneous decarboxylation and leads to our target **2**. This route worked well, and was used for the large scale preparation of **2**.

2.1.2.2 Process Optimization for Preparation of 2

2.1.2.2.1 Optimization of the Preparation of Allyl Carbonate 15 or 16

There were two potential starting materials for Trost's Mo chemistry (Scheme 2.7). The first approach utilized the commercially available 3-fluorocinnamic acid (**4**). However, reduction of **4** did not proceed well with various reducing agents and provided the desired allylic alcohol **28** in only mediocre yield. Alternatively, 3-fluorobenzaldehyde (**29**) was used as the starting material. Vinyl Grignard addi-

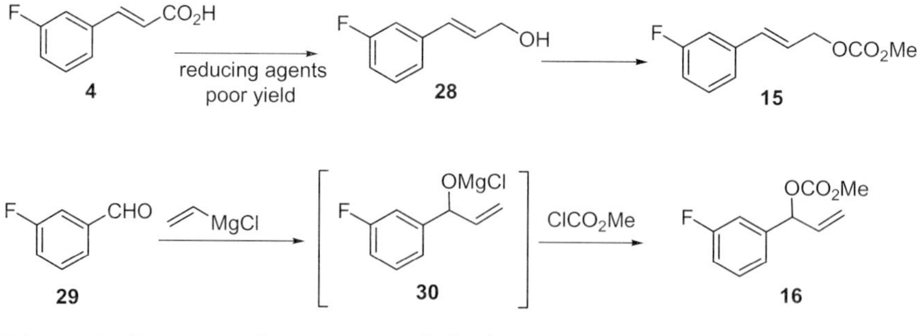

Scheme 2.7 Preparation of starting materials for the Mo chemistry.

tion to **29** followed by *in situ* trapping of the magnesium allyl alkoxide **30** gave the desired branched carbonate **16** in good yield in one pot. Since Trost reported similar results when either linear or branched carbonates were reacted under his reaction conditions, we selected **16** as our starting material.

2.1.2.2.2 Application of Trost's Mo Chemistry and Optimization

Initial attempt For initial attempts of the Mo chemistry with branched carbonate **16**, we used commercially available and crystalline $(C_7H_8)Mo(CO)_3$[1] as a molybdenum catalyst instead of the oily $(EtCN)_3Mo(CO)_3$ which was reported in the original paper and would have to be prepared [10]. The results from the initial attempts are summarized in Table 2.1. As reported, the reaction proceeded well when carried out in THF with 10 mol% of $(C_7H_8)Mo(CO)_3$ and 15 mol% of the chiral ligand **31** at 65 °C to give the desired chiral branched product **17** with high regioselectivity (**17** vs. **32**) in 90% ee. Reducing the amount of both $(C_7H_8)Mo(CO)_3$ and **31** from the original conditions (10 and 15 mol%, respectively) was feasible.

Based on these results, proof of concept was established for application of this chemistry. However, there were a few issues that needed to be addressed prior to scale up, namely preparation of the chiral ligand **31** and preparation of the Mo catalysts.

Table 2.1 Initial attempts to study Mo chemistry.

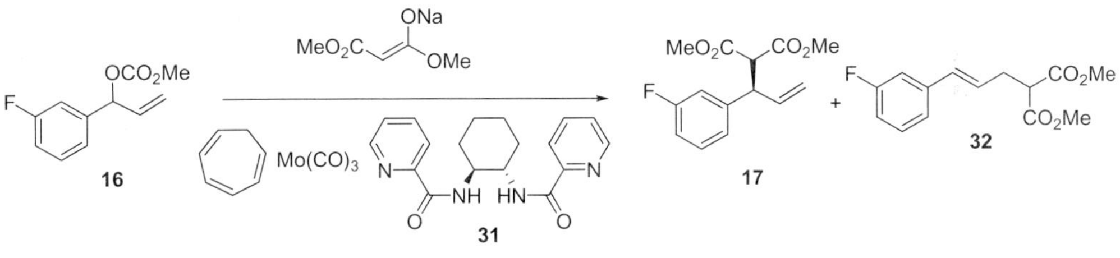

Entry	Equiv of $(C_7H_8)Mo(CO)_3$	Equiv of Ligand 31	Solvent	Temp (°C)	ee %	Regio (17/32)	Conversion (%) (isolated %)
1	0.067	0.081	THF	rt	89	11:1	62
2	0.076	0.095	THF	40	87	9:1	85
3	0.11	0.12	THF	65	88	8:1	>98
4	0.10	0.15	THF	65	90	7:1	>98 (75%)
5	0.10	0.15	DMF	65	24	6:1	>98
6	0.10	0.15	Toluene	65	–	10:1	6

1) Professor Trost recommended use of this catalyst since it was commercially available for small scale runs.

Preparation of ligand 31 Originally, chiral ligand **31** was prepared from (1*R*,2*R*)-1,2-diaminocyclohexane **33** based on the racemic synthesis reported by Barnes *et al.* in 1978 [15], where picolinic acid **34** was activated with $P(OPh)_3$ and then coupled with *trans*-1,2-diaminocyclohexane. The reported isolated yield in the case of racemate was only 47%. We optimized the preparation as shown in Scheme 2.8 [16]. Picolinic acid **34** was activated with CDI in THF. After confirmation of activation, chiral diamine **33** was added to the solution. When complete, the reaction was quenched via the addition of a small amount of water (to quench excess CDI). The reaction solvent was then switched from THF to EtOH, when the desired ligand **31** directly crystallized out. Ligand **31** was isolated in 87% yield by simple filtration of the reaction mixture in high purity. With a 22 litter flask, 1.25 kg of **31** was prepared in a single batch.

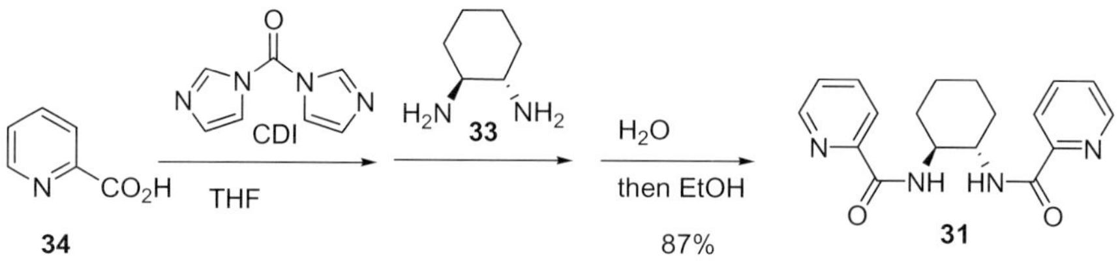

Scheme 2.8 Preparation of chiral ligand **31**.

Molybdenum catalyst Since $(EtCN)_3Mo(CO)_3$, which was reported in the original paper, was not commercially available and $(C_7H_8)Mo(CO)_3$, which was used for initial studies, was available but in very limited amount, we needed to devise an alternative Mo source. According to the literature, both Mo complexes were prepared [17] from $Mo(CO)_6$ with an excess amount of either propionitrile or cycloheptatriene in toluene under high temperature for approximately a day. $(EtCN)_3Mo(CO)_3$ was isolated by concentration and was not stable in air. On the other hand, $(C_7H_8)Mo(CO)_3$ was isolated by sublimation and was stable enough to be handled in air. Due to time constraints, we were about to start preparation of $(C_7H_8)Mo(CO)_3$ on a large scale and to look for a large scale sublimation apparatus.

Since this reaction with non-chiral Mo complexs with chiral ligand **31** provides very high enantioselectivity, the real active Mo catalyst should be coordinated by the chiral ligand **31**. The role of a weaker ligand, such as propionitrile or cycloheptatriene, should be to facilitate ligand exchange with the chiral **31**. The preparation method of $(C_7H_8)Mo(CO)_3$ and $(EtCN)_3Mo(CO)_3$ was evidence that weak ligands could exchange with three carbon monoxides from $Mo(CO)_6$. Therefore, we thought that we should be able to use air-stable and economical $Mo(CO)_6$, instead of the more sophisticated Mo complexes, with proper activation with chiral ligand **31**. Consequently, we attempted pre-heating $Mo(CO)_6$ and chiral ligand **31** prior to addition of carbonate **16** and sodium dimethyl malonate [18]. The results are summarized in Table 2.2.

Table 2.2 Activation with $Mo(CO)_6$.

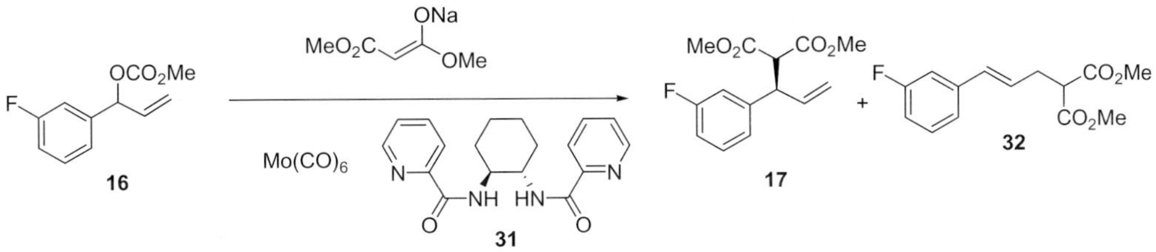

Entry	Solvent	A. time (h)	A. temp (°C)	ee%	17/32	Assay yield (%)
1	Toluene	0.75	85	95	95:5	77
2	Toluene	4	85	97	95:5	91 (84)
3	Toluene	15	85	89	92:8	91
4	THF	2	65	92	92:8	86
5	THF	4	65	92	89:11	83
6	THF	4	65–r.t.	96	84:16	33
7	DMF	4	85	87	86:14	55
8	DME	4	80	95	85:15	90
9	DCE	4	80	98	96:4	36

After 2–4 h preheating with chiral ligand **31**, $Mo(CO)_6$ was properly activated and the reaction proceeded well without using elaborate Mo catalysts in toluene, THF, DMF, DME and DCE. Longer heating (15 h) of $Mo(CO)_6$ with ligand **31** yielded lower selectivity due to degradation of the active catalyst (entry 3). Lowering the reaction temperature offered better enantioselectivity, but branch/linear selectivity did not change and the reaction was slower (entries 5 and 6). Among the solvents, toluene gave better results than THF, DMF and DCE. It was also noted that the reaction in toluene at lower temperature (65 °C, see Table 2.1 entry 6) was very slow.

With our optimized conditions in hand, the reaction was performed in a 100 litter flask on several occasions with good success. Even though we had React-IR data (p. 63, Figure 2.3) for monitoring the catalyst formation, the React-IR data did not provide additional information on the active catalyst. Therefore, a portion of activated catalyst solution was tested prior to addition of a whole catalyst solution to the real batch, to ensure success of the reaction. At the end of the reaction, the crude mixture was passed through a short silica gel pad to remove Mo and the mixture was used for the next reaction without further purification.

Decarboxylation, Masamune reaction, and diazotransfer Diazo **25** was prepared under optimized conditions, as summarized in Scheme 2.9. Decarboxylation of the malonate could be done under either acidic or basic conditions. Reaction of **17** under acidic conditions provided the desired mono-carboxylic acid **18** but lactone **35** was simultaneously formed (Figure 2.2). Under basic conditions,

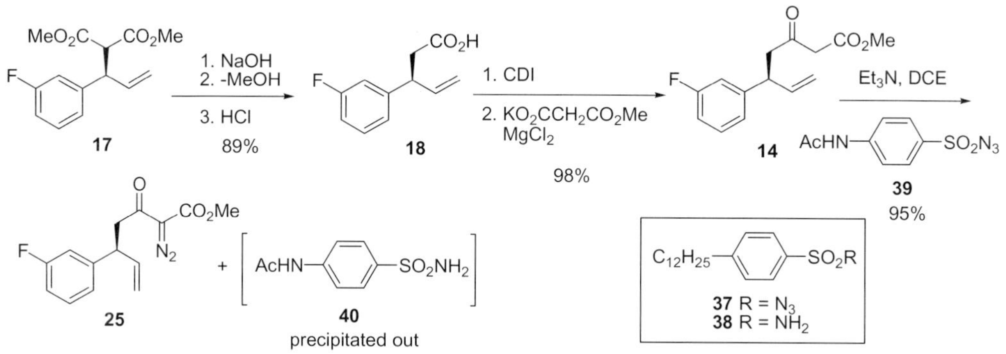

Scheme 2.9 Preparation of diazo **25**.

Figure 2.2 Potential impurities at solvolysis.

solvolysis of **17** proceeded well without any problem, but upon acidification of the crude reaction mixture, the reaction gave a mixture of product **18** and the corresponding methyl ester **36**. Thus, prior to acidification, MeOH was removed *in vacuo*. Carboxylic acid **18** was then isolated as the crystalline (+)-phenethylamine salt. Although the enantio-excess was not upgraded upon crystallization, the salt formation provided a convenient means of product isolation.

Masamune reaction of **18** using standard literature conditions went well without any problem, yielding β-keto ester **14** in high yield as expected. Historically we have successfully used dodecylbenzenesulfonyl azide (**37**) as a safe diazotransfer reagent, as previously demonstrated in Merck's carbapenem projects. The desired diazo compounds in carbapenem projects were isolated as crystalline compounds and the oily by product, dodecylbenzenesulfamide (**38**) was easily removed from the products by simple filtration. However, diazo **25** is not a crystalline compound. Thus removal of **38** from diazo **25** was not a simple operation. Furthermore, since we did not have enough safety data[2] on handling large amounts of **25**, it was desirable to keep **25** in solution with minimal operation. For this purpose, **37** was not a suitable reagent for diazo formation. 4-Acetamidebenzenesulfonyl azide (**39**), developed by Davies [19], was selected as the diazo-transfer reagent, since the byproduct, 4-acetamidebenzenesulfamide (**40**) crystallized out nicely from the rea-

2) Preliminarily, diazo **25** was evaluated to have a small amount of shock-sensitivity but is safe when handled in solution.

ction mixture in 1,2-dichloroethane. A solution of diazo **25** in 1,2-dichloroethane was isolated by simple filtration of the reaction mixture and used directly in the next reaction after washing with mild acid to ensure removal of triethylamine. Removal of triethylamine was required since it was found to be a catalyst poison for the next reaction. 1,2-Dichoroethane was selected based on better *trans/cis* selectivity in the subsequent cyclopropanation reaction, and the desire to avoid any solvent switching/concentration needs with the diazo intermediate. Thus, neither concentration nor solvent switch was required upon handing diazo **25**.

Cyclopropanation When we started this project, we expected, based on inspection of a molecular model, that the cyclopropanation would proceed with high *trans*-selectivity due to steric repulsion between the 3-fluorophenyl group and the forming three-membered ring. For the first reaction, we selected rhodium octanoate as catalyst, which was used for the Imipenem process. Surprisingly, this reaction gave a mixture of two compounds. More surprisingly, the major product was the *cis*-isomer **41**, based on its NMR. Several other rhodium catalysts were screened but almost all reactions screened gave the undesired *cis*-isomer **41** as a major product. Thus, we turned our focus to copper as a catalyst. The results are summarized in Table 2.3.

Generally speaking, copper-catalyzed cyclization gave the desired *trans*-**26** as the major product. The best selectivity obtained was an 85 : 15 mixture of diastereomers using CuOTf, which was prepared *in situ* from CuCl and AgOTf. $[(MeCN)_4Cu]PF_6$ gave comparable selectivity, providing 83 : 17 with high yield. For

Table 2.3 Cyclopropanation.

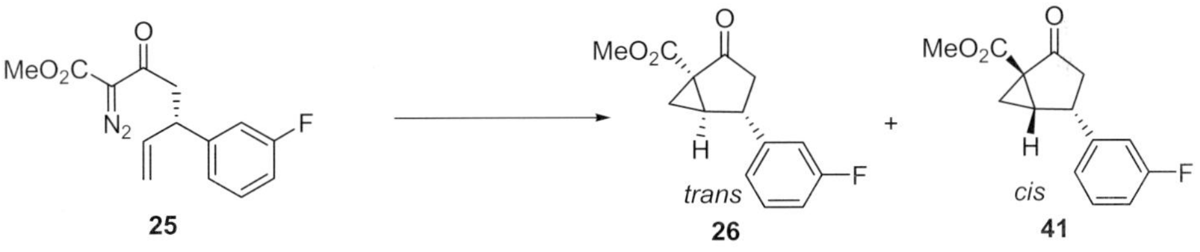

Entry	Catalyst	Solvent	Temp (°C)	Conv. (%)	*trans* 26 /*cis* 41
1	$Rh_2(OAc)_4$	CD_2Cl_2	rt	100 (90)	43 : 57
2	$Rh_2(O_2CC_7H_{15})_4$	CD_2Cl_2	rt	100 (57)	33 : 67
3	$Rh_2(cap)_4$	CD_2Cl_2	rt	95	33 : 67
4	CuCl	1,2-DCE	75	99	50 : 50
5	CuCl/AgOTf	1,2-DCE	75	100 (89)	85 : 15
6	$Cu(OTf)_2$	1,2-DCE	75	100 (92)	77 : 23
7	CuSCN	1,2-DCE	75	10	70 : 30
8	CuOAc	1,2-DCE	75	88	46 : 54
9	$Cu(acac)_2$	1,2-DCE	75	37	54 : 46
10	$[(MeCN)_4Cu]PF_6$	1,2-DCE	75	100 (98)	83 : 17

our first large scale preparation, we used both catalyst systems (CuCl/AgOTf and $[(MeCN)_4Cu]PF_6$) and both conditions gave us product mixtures of almost identical yield and selectivity. It is noted that diazo **25** solution was added slowly to the catalyst solution to minimize accumulation of potential thermally sensitive diazo **25** in the reaction mixture throughout the reaction.

Since our starting diazo compound **25** already has a chiral center, diastereo-matching and mismatching with chiral ligands was expected. Several chiral ligands (both enantiomers) were screened with both rhodium and copper catalysts but there was no influence on the *trans/cis* selectivity by changing ligands. It appears that the reaction site is so congested that the intermediate carbenoid would be almost ligand free. Thus, no diastereo-influence from chiral ligands would be observed. It is concluded that this cyclopropanation proceeded by substrate control not by catalyst control.

Later, two marginally related examples were reported. Both examples utilized Cu as catalyst and the *trans/cis* ratios were mediocre (Schemes 2.10 and 2.11).

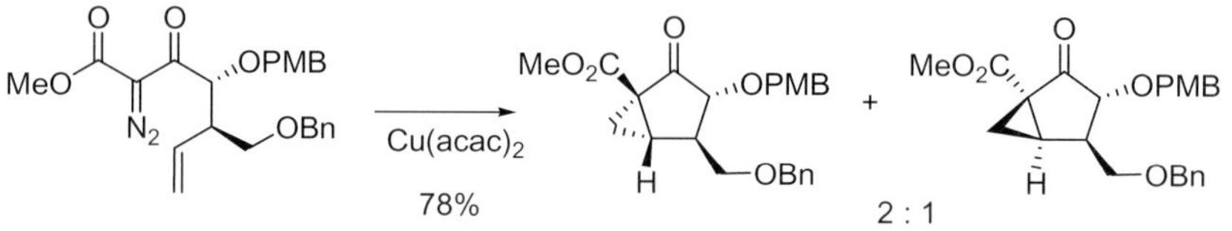

Scheme 2.10 Marquez's example [20].

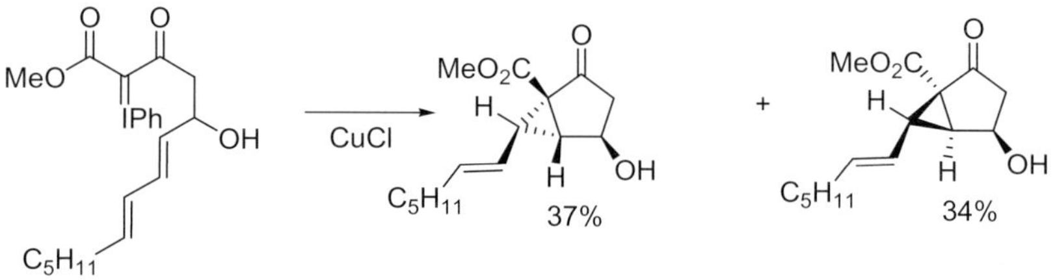

Scheme 2.11 Moriarty's example [21].

Ring-opening and isolation of our target 2 Ring opening of highly strained bicyclo[3.1.0]octane **26** was studied, and the results are summarized in Scheme 2.12. Heating a *trans/cis* mixture of **26** and **41** (9 : 1) in the presence of NaOAc in DMF and AcOH (DMF was added to improve the solubility of **26**) led to isolation of the desired **2** via **27** in 28% yield. The major by-product was a ring-opened compound **42** formed by nucleophilic attack of dimethylamine, which was generated by decomposition of DMF. Therefore, AcOH was identified as an alternative solvent to DMF and the solvent volume was increased to maintain the solubility of the reaction mixture. After ring-opening in a mixture of NaOAc and AcOH, ring-opened acetate **27** was observed as the major product. After AcOH was

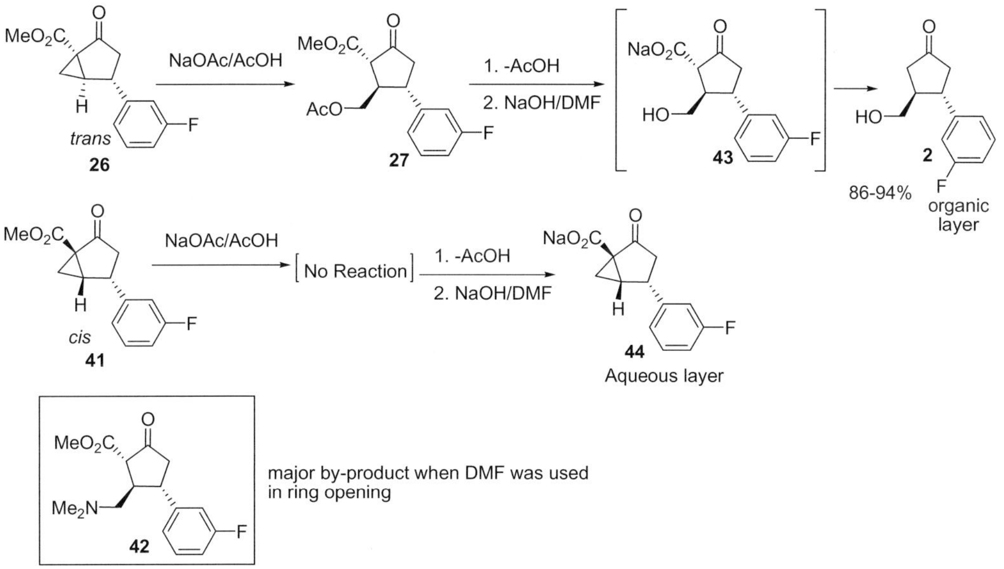

Scheme 2.12 Ring-opening and isolation of target **2**.

removed as much as possible by distillation, solvolysis of **27** was carried out by addition of DMF and NaOH. The resultant sodium salt of carboxylic acid **43** was spontaneously decarboxylated via a six-membered transition state to give our target **2** in good yield. Interestingly, ring-opening of the undesired *cis*-isomer **41** with NaOAc was extremely slow due to steric hindrance. The majority of **41** was unreacted under the ring-opening conditions. Unreacted **41** was then solvolyzed to give the corresponding bicyclo[3.1.0] carboxylic acid sodium salt of **44**. Salt **44** did not readily decarboxylate because **44** could not form the six-membered transition state required for decarboxylation. Thus, undesired *cis*-isomer **41** ended up in the aqueous layer as the sodium salt of **44**. On the other hand, the desired *trans*-isomer **26** was converted to **2**, and **2** was isolated in the organic layer in 86–94% yield based on an assay yield of **26** from the crude reaction mixture of cyclopropanation.

2.1.2.3 **Optimization of the Preparation of Pyrazole 3**

Synthesis of pyrazole **3** by the Medicinal Chemistry route was straightforward from *N*-Boc isonipecotic acid (**45**), so we utilized the route after some optimizations, as summarized in Table 2.4. The key 1,3-diketone intermediate **48** was prepared from **45** without issues. A minor problem in the original route was the exothermic nature of the Claisen condensation between methyl ketone **47** and methyl phenylacetate. Slow addition of 1.1 equiv of methyl phenylacetate to a mixture of **47**, 0.2 equiv of MeOH, and 2.5 equiv of NaH in THF at room temperature solved this exothermic issue and reduced the amount of self-condensation of

Table 2.4 Preparation of pyrazole.

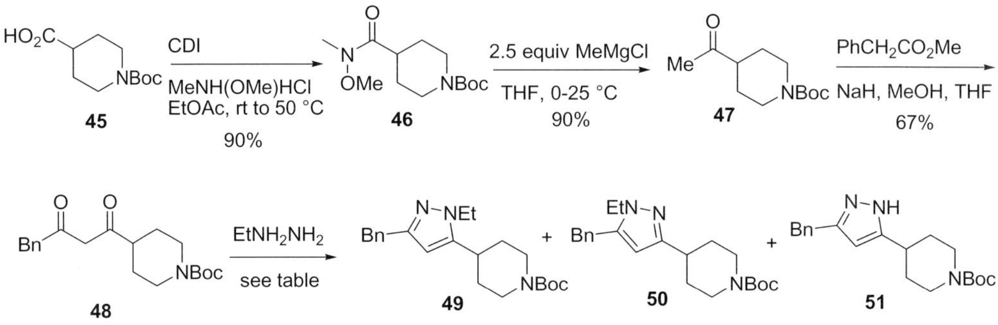

Entry	Solvent	Temp	49 : 50	Entry	Solvent	Temp	49 : 50
1	43% H_2O/MeCN	rt	4.0 : 1	9	MeOH	rt	3.4 : 1
2	38% H_2O/MeCN	rt	6.1 : 1	10	MeOH + $(CO_2H)_2$	rt	3.0 : 1
3	33% H_2O/MeCN	rt	6.0 : 1	11	CH_2Cl_2	rt	1.7 : 1
4	33% H_2O/MeCN	45 °C	4.0 : 1	12	Hexane	rt	1.6 : 1
5	32% H_2O/MeCN	3 °C	4.6 : 1	13	MTBE	rt	1.5 : 1
6	13% H_2O/MeOH	rt	3.8 : 1	14	THF	rt	1.4 : 1
7	9% H_2O/MeOH	rt	3.6 : 1	15	Toluene	rt	1.3 : 1
8	5% H_2O/MeOH	rt	3.6 : 1				

methyl phenylacetate. Based on this optimization, diketone **48** was able to be isolated as a crystalline compound directly from the reaction mixture. The major issue in the preparation of **3** was control of the regionselectivity (desired **49** vs. undesired **50**) in the reaction of diketone **48** and *N*-ethylhydrazine. Medicinal Chemistry used *N*-ethylhydrazine oxalate in MeOH for the pyrazole formation [1]. Since *N*-ethylhydrazine was available in large quantities as a 34 wt% aqueous solution, solvent effects on regioselectivity were carefully studied, mainly in an aqueous medium, and the results are summarized in Table 2.4. The mimicked original conditions (entry 10) gave mediocre selectivity (3.0 : 1). With less polar solvents, regioselectivity was generally poor. Protic polar solvents provided better selectivity. Acetonitrile was a good solvent and the amount of water in acetonitrile had a strong influence on regioselectivity. Aqueous acetonitrile (33–38%) was the best solvent choice. Reaction temperature was also screened and room temperature was determined to be optimal. Regioisomer **50** was relatively easy to remove via crystallization of **49** but *N*-unsubstituted impurity **51** was a little difficult to reject. Unsubstituted **51** was derived from **48** since hydrazine was present as an impurity in the aqueous solution of commercial *N*-ethylhydrazine. In addition, throughout the end game, all the corresponding impurities derived from **51** were very difficult to reject. Therefore, pyrazole **49** was recrystallized twice to reduce contamination from **51**, which negatively impacted the isolated yield of **49**.

Finally, the Boc group was removed by treatment with aqueous HCl. Free amine **3** was isolated by extraction with acetonitrile in the presence of NaCl after basification with NaOH. Free amine **3** crystallized upon standing at room temperature, as shown in Scheme 2.13. It was found that when **3** was left standing in CH_2Cl_2 for long period [22], which was the solvent for the next step, it reacted with solvent to form dimer **52**, together with formation of the HCl salt of **3**.

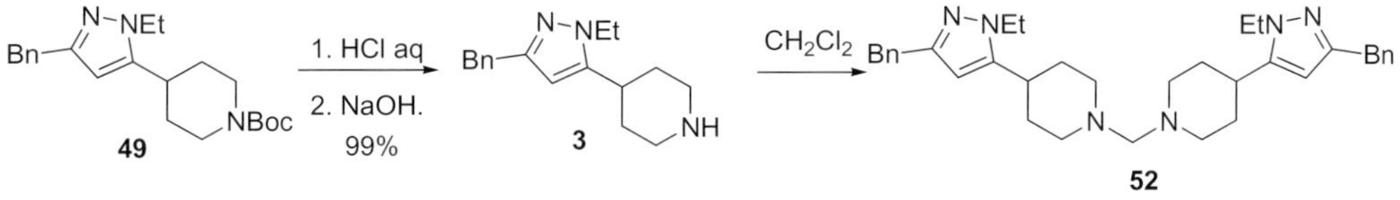

Scheme 2.13 Preparation of free pyrazole **3**.

2.1.2.4 Optimization of the Preparation of Our Target 1 (End Game)

To complete the preparation of drug candidate **1**, three components, **2**, **3**, and D-valine had to be assembled. First cyclopentanone **2** and *tert*-butyl D-valine (1.2 equiv) were coupled via reductive amination, resulting in the formation of an additional chiral center on the cyclopentane ring. Original conditions [$NaBH(OAc)_3$, CH_2Cl_2, room temperature] gave 1.9 : 1 selectivity (desired **53** vs. undesired **54**). Solvent screening revealed acetonitrile was a better solvent for this reaction giving 4.7 : 1 selectivity. Increasing the reaction temperature to 50 °C improved the selectivity further (7.4 : 1). Interestingly, the selectivity was improved to 8.2 : 1 by increasing the steric bulkiness of the reducing agent to $NaBH(OCOC_2H_5)_3$ at 50 °C. By further increasing the reaction temperature to 70 °C, the selectivity was finally optimized to 10 : 1. The mixture was subsequently treated with excess formaldehyde and $NaBH(OAc)_3$ to provide a diastereomeric mixture of *N*-methyl compounds (desired **55** and undesired **56**) in one pot (Scheme 2.14). Better selectivity was obtained when the reaction was run at higher temperature with bulkier reducing reagents with an excess of *tert*-butyl D-valine. In our first large scale preparation, the desired *N*-methyl **55** was purified by silica gel column chromatography. Later, we found that **55** can be directly crystallized from the reaction mixture as its HCl salt in 68% overall isolated yield with excellent purity.

Evans *et al.* reported highly diastereoselective reductive amination via hydride delivery controlled by a chelation from a hydroxymethyl group [23]. In our case, at the beginning of the reaction, the selectivity was very high with 1.7 equiv of *tert*-butyl D-valine, as Evans reported. Selectivity was as high as 100 : 1 after 50% conversion. However, as the reaction progressed, the selectivity deteriorated to 25 : 1. It is not clear why chelation control, as shown as **58** in Scheme 2.15 becomes less effective as the reaction progresses. Perhaps, borate exchange from the product borate **59** to the starting material ketone **2** and/or imine **57** may occur more frequently. The resulting borates **60** and/or **61** can no longer benefit from chelation of the free hydroxy group. This might be an explanation.

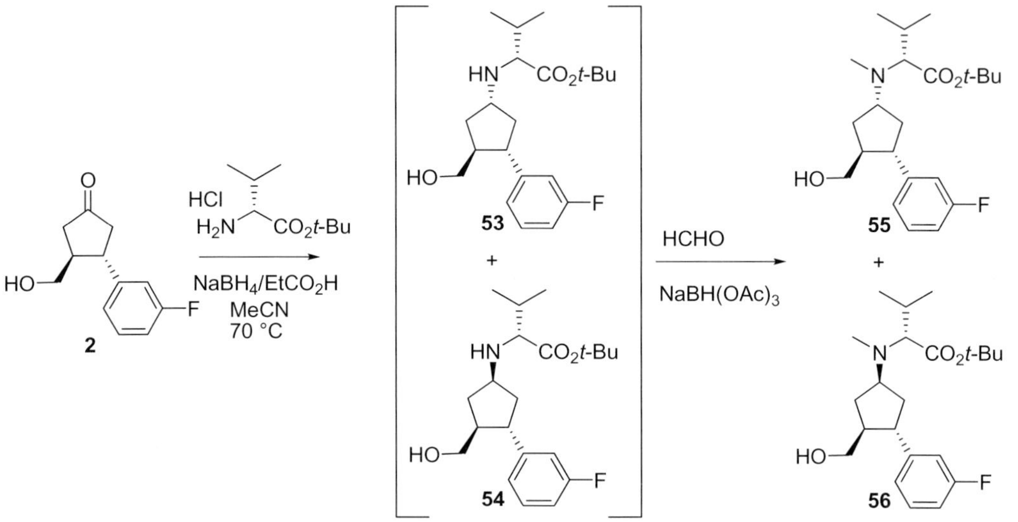

Scheme 2.14 Reductive amination.

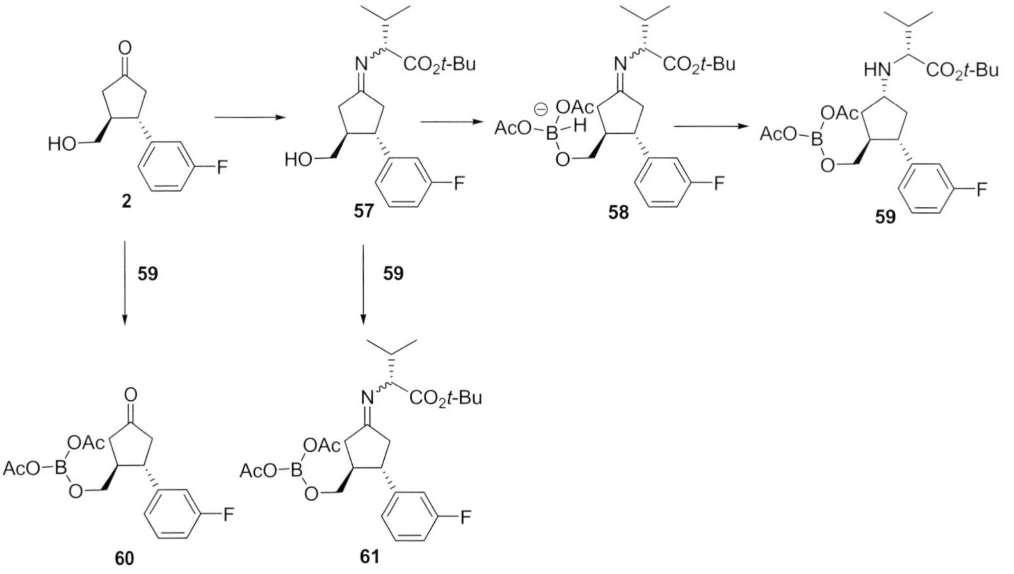

Scheme 2.15 Chelation control.

Target compound **1** was prepared from cyclopentane derivative **55** and pyrazole derivative **3**. The primary alcohol of **55** was activated as its mesylate, which was reacted with secondary amine **3**. However, it was found that the dialkylated impurity was formed as a significant by-product. Thus, we focused on a reductive amination method, as summarized in Scheme 2.16. Oxidation of **55** proceeded smoothly with DMSO, oxalyl chloride and TEA. Either free base or the HCl salt of **55** could be used to yield crystalline aldehyde **62** in almost quantitative yield. Epimerization of the α-position of the aldehyde was observed at 2~4% when the reaction mixture was quenched into water. This epimerization was totally prevented by quenching into phosphate buffer. Reductive amination was initially studied in acetonitrile with $NaBH(OAc)_3$. However, product **1** crystallized out from the reaction mixture. Therefore, the solvent was exchanged to CH_2Cl_2. One of the concerns with using CH_2Cl_2 was, as previously mentioned, the possible reaction between pyrazole **3** and CH_2Cl_2 (see Scheme 2.13). Based on control studies, we gained confidence that the rate of reductive amination is significantly faster than dimerization of **3**. Actually, the reaction proceeded well in CH_2Cl_2 and penultimate **63** was isolated as acetonitrile solvate in 99% yield. Heating of **63** in 3 M HCl at 50 °C for 3.5 h removed the *tert*-butyl group. Compound **1** could be isolated after neutralization. But **1** was not easily crystallized in the presence of NaCl. NaCl was removed by treatment with Amberchrome 161c (polystyrene resin). It was determined that crystallization of **1** required high temperatures. Consequently, the aqueous solution of **1** was heated at 60 °C with a small amount of seed for initiation of crystallization. Once the seed bed was formed, the mixture was slowly cooled down to room temperature for completion of crystallization. Compound **1** was isolated by filtration in 88% yield after drying.

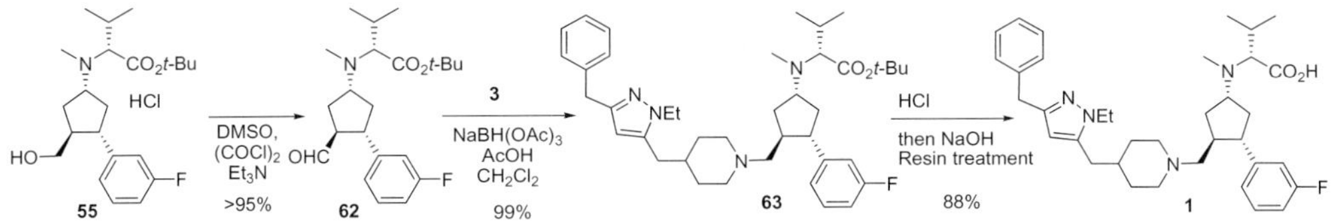

Scheme 2.16 End game.

2.1.2.5 **Overall Preparation Scheme**

Thus, we optimized our new process and the process improvements are summarized as follows:

1) Drug supply needs for the project were supported by providing **1** at short notice.
2) The overall yield of **1** was improved to 10% in 16 chemical steps with the longest linear sequence as ten steps.
3) A newly developed asymmetric nucleophilic addition of malonate to π-allyl Mo complex was the cornerstone for this preparative campaign.

- A better preparation method for the chiral ligand was used.
- Direct use of economically viable and stable $Mo(CO)_6$.

4) A better diastereoselectivity with Cu-catalyzed cyclopropanation was discovered.
5) Diastereoselective ring-opening of bicyclic compound led to simple isolation of the desired cyclopentanone.
6) Pyrazole synthesis was optimized, especially in terms of regioselectivity.
7) The end game was optimized, wherein diastereoselectivity of the reductive amination was further improved by using a more sterically bulky reducing reagent and by running the reaction at higher temperature.

2.2
Chemistry Development

When we used asymmetric nucleophilic addition of malonate to the Mo π-allyl complex in our first delivery, the Mo chemistry was not so clearly understood, and our application would be the first large scale example, to the best of our knowledge. Initially our contributions to Mo chemistry were two-fold; (i) replacement of non-commercially available $(EtCN)_3Mo(CO)_3$ or $(C_7H_8)Mo(CO)_3$ by more stable and inexpensive $Mo(CO)_6$ by incorporation of proper pre-activating time; (ii) simplified preparation of the chiral ligand. Even after we completed the project, we still had a strong interest in Mo chemistry.

When we activated the catalyst system on a large scale, we were unsure of whether the reaction would proceed. The only data for the catalyst activation available to us was *in situ* IR (React-IR) as shown in Figure 2.3. During activation of the catalyst, a single vibration frequency (~1980 cm^{-1}) of carbon monoxides in $Mo(CO)_6$ became five different frequencies of carbon monoxide in the catalyst solution. This IR data provided us some relief from the risk of running the large scale reaction but did not provide any clues on the structure of the true catalyst.

First, we examined whether we prepared the same active catalyst from $Mo(CO)_6$ as the original catalyst from $(EtCN)_3Mo(CO)_3$ or $(C_7H_8)Mo(CO)_3$. Thus, three different Mo catalysts were compared in two different solvent systems. The results are summarized in Table 2.5.

Entries 1–3 were run in toluene and entries 4–6 were run in THF with three different Mo catalyst sources. The activation time was longer when $Mo(CO)_6$ was used, but the results were dependent on the solvent not the catalyst source. Thus, it was confirmed that $Mo(CO)_6$ with the chiral ligand generated the same active catalyst as Trost reported.

Furthermore, we applied the π-allyl reaction with $Mo(CO)_6$ to other substrates and the results are summarized in Table 2.6.

The results with $Mo(CO)_6$ were similar to Trost's report. Reactions with *S,S*-ligand **31** yielded the *S*-products predominantly. It is interesting to point out that entry 1 and entry 2 should give the same result if this reaction was going through the same π-allyl Mo complex. However, branched carbonate (entry 1) gave slightly

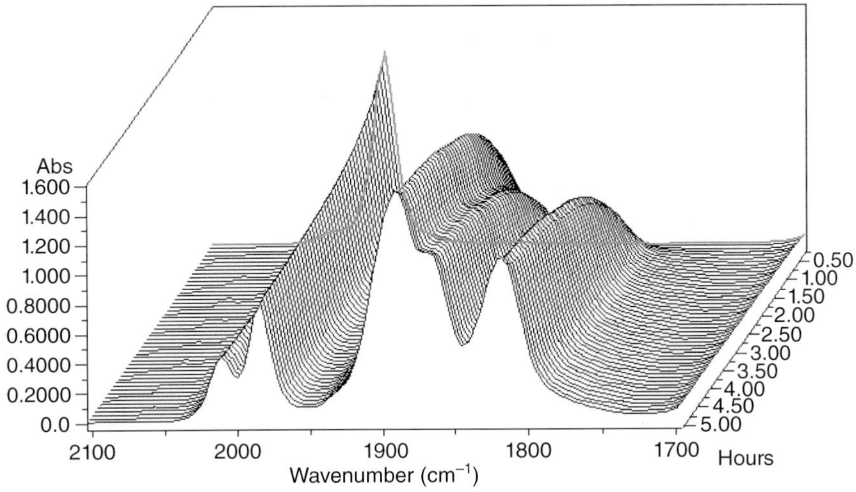

Figure 2.3 React-IR data of activation of $Mo(CO)_6$.

Table 2.5 Direct comparison between three Mo catalyst sources.

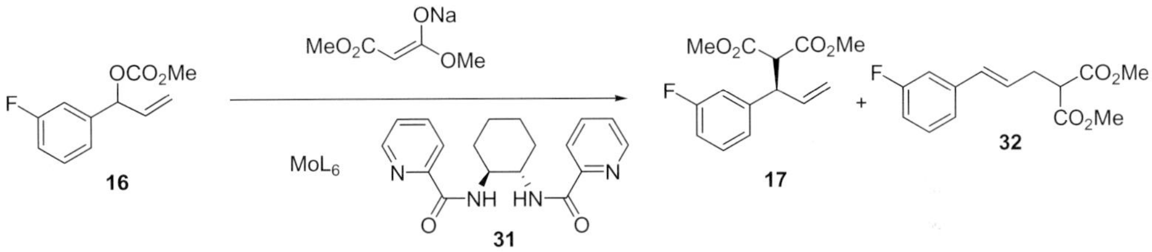

Entry	Mo catalyst	Solvent	A. time (h)	A. temp (°C)	ee %	17/32	A. yield (%)
1	$(EtCN)_3Mo(CO)_3$	Toluene	0.5	85	95	93:7	84
2	$(C_7H_8)Mo(CO)_3$	Toluene	0.5	85	96	96:4	97
3	$Mo(CO)_6$	Toluene	4	85	97	95:5	91
4	$(EtCN)_3Mo(CO)_3$	THF	0.5	65	91	90:10	88
5	$(C_7H_8)Mo(CO)_3$	THF	0.5	65	88	88:12	81
6	$Mo(CO)_6$	THF	4	65	91	89:11	83

reduced ee % (96 vs. 99% ee) and a little bit lower branch/linear ratio (93:7 vs. 95:5). When we reviewed the original paper, Trost also reported these subtle differences. We thought this reaction did not perfectly follow the Curtain–Hammett scheme. This was the beginning of our investigation into the Mo chemistry.

Table 2.6 $Mo(CO)_6$ method for other substrates.

Entry	Structure	ee %	Branched/linear	Isolated yield (%)
1	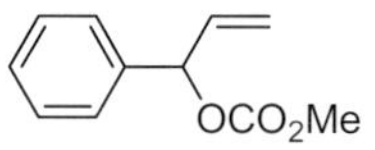	96	93:7	76
2	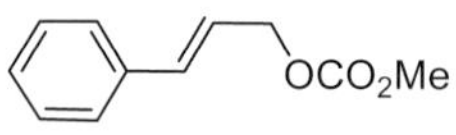	99	95:5	80
3	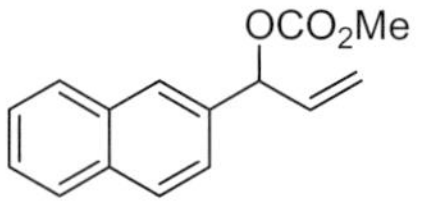	96	98:2	80
4	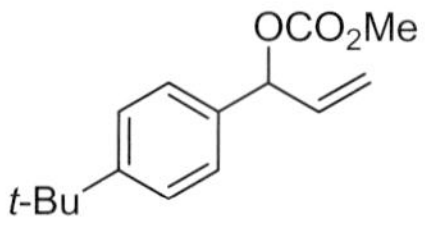	94	94:6	76

2.2.1 Kinetic Resolution

Based on the assumption that this reaction goes through π-allyl Mo intermediates (**A** and **B**), the result from either linear carbonate (**L-C**) or branched carbonate (**B-C**) should give exactly the same result if the equilibrium between **A** and **B** is much faster than nucleophilic addition of sodium dimethyl malonate to **A** or **B** (Curtain–Hammett) as shown in Scheme 2.17.

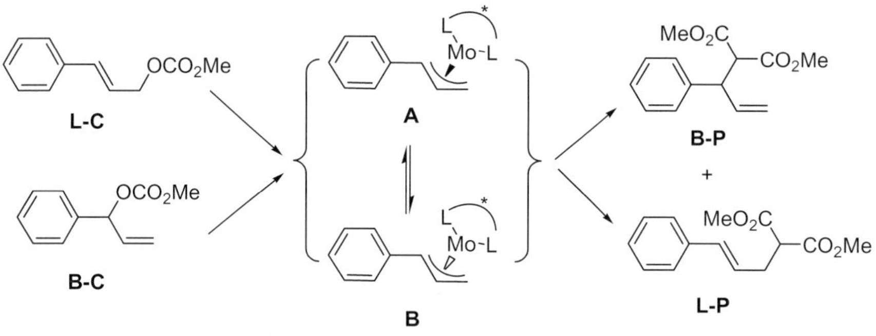

Scheme 2.17 Curtain–Hammett case.

As stated earlier, this reaction did not match perfectly with the Curtain–Hammett postulate. The chiral Mo complex can select the favored face (either **A** or **B**) from **L-C**. However, facial selection of **B-C** on formation of the π-complex (A or **B**) should be dictated by the orientation of the carbonate itself not by the chirality of the Mo complex. At the same time, we would expect the chiral Mo complex to show facial preference during formation of the π-complex. Since **B-C** is a racemic mixture (**B-C-S** and **B-C-R**), one of the enantiomers should be matched with the chiral Mo complex and the other should be mismatched. Thus, kinetic resolution of racemic **B-C** should be expected, as depicted in Scheme 2.18.

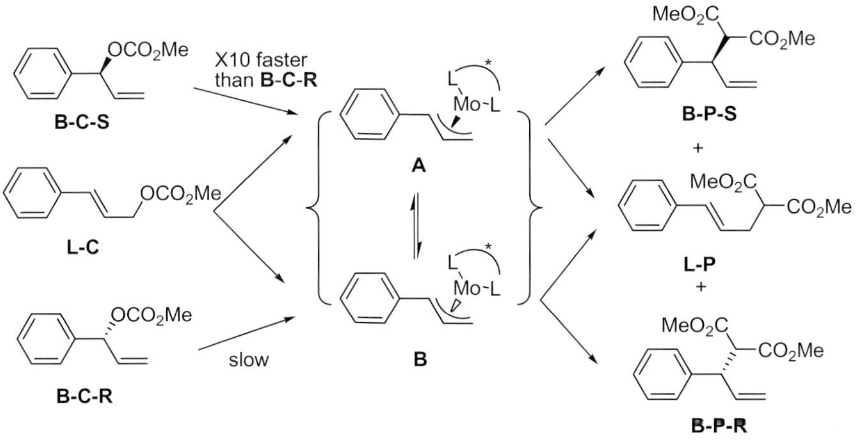

Scheme 2.18 Kinetic resolution with chiral ligand **31**.

Following the reaction with *S,S*-ligand **31**, it was found that the *S*-carbonate (**B-C-S**) reacted first and gave the desired **B-P** with a very high *S* selectivity. *R*-carbonate (**B-C-R**) reacted ten times slower than *S*. When the reaction was terminated around 60% conversion with *S,S*-ligand **31**, unreacted *R*-carbonate (**B-C-R**) was isolated from reaction mixture with >99% ee [24].

Let us assume that the π-allyl complex formation in the Mo reaction proceeds via retention and nucleophilic addition of sodium malonate goes via retention and the Mo chemistry gives the product in retention conformation in the matched case for sake of argument. (Later we will confirm the mechanism is retention and retention, not inversion and inversion like the corresponding Pd chemistry.) Under this assumption, matched branched carbonate **B-C-S** with *S,S*-ligand **31** initially forms the π-allyl Mo complex **A** with retention of configuration and **A** reacts with sodium dimethyl malonate proceeding with retention of configuration providing the desired **B-P-S** predominately, together with a very little linear product **L-P**. On the other hand, mismatched *R*-carbonate **B-C-R** with *S,S*-ligand **31** initially provides the π-allyl Mo complex **B**, which would be converted to complex **A** considerably faster than nucleophilic substitution to complex **B**. However, some nucleophilic substitution may have occurred from complex **B**, and nucleophilic substitution from complex **B** leads to the undesired product **B-P-R**, and more linear product **L-P** than that from complex **A**. For the case of linear carbonate **L-C**, *S,S*-ligand **31** selects a favored face to lead to the Mo complex **A** predominately.

Table 2.7 Experimental verification.

Predictions	Experimental Results
For "matched" carbonate (B-C-S)	
– increasing [malonate] will increase ee	– $[\text{malonate}]_0$ ~0.07 M: 92% ee – $[\text{malonate}]_0$ ~0.6 M: 97%ee
For "mismatched" carbonate (B-C-R)	
– decreasing [malonate] will increase ee	– All malonate present initially: 70% ee – Malonate added over six hours: 92% ee – ee higher in toluene than THF due to much lower solubility of Na-malonate in toluene

According to this equilibrium argument, the matched *S*-carbonate **B-C-S** should give a better branch to linear (B/L) ratio and enantiomeric excess if the nucleophilic substitution rate prior to π-allyl Mo conversion from complex **A** to **B** is increased. (see Table 2.7) For example, when the reaction was run at a higher concentration, $[\text{Malonate}]_0$ ~0.6 M rather than the typical ~0.07 M, the ee of the product increases to 97% from 92%.

In contrast, the mismatched *R*-carbonate **B-C-R** needs slower nucleophilic substitution for better selectivity, allowing the initially formed undesired π-allyl Mo complex **B** to convert to **A**, prior to substitution. For instance, when all the malonate was added at the beginning with the mismatched **B-C-R**, the ee was only 70%. On the other hand, when malonate was added to the reaction mixture over six hours, the ee was dramatically improved to 92%. Previously, we reported that the reaction in toluene gave better selectivity than in THF with branched carbonate as the starting material. We monitored the progress of the reaction in toluene and THF with chiral HPLC and the results are summarized in Figure 2.4.

In Figure 2.4, the *x*-axis is the percentage reaction conversion and the *y*-axis represents the ee% of the product (black square: toluene; black circle: THF). The first half of the reaction proceeded almost equally well in either toluene or THF. After 60% conversion, the ee % of the product remained at a similar level in toluene, but significantly deteriorated in THF. As previously mentioned, the matched **B-C-S** reacts ten times faster than the mismatched **B-C-R**. When the matched carbonate was reacting, the results were similar in toluene and in THF. After 60% conversion, almost all remaining carbonate was the mismatched **B-C-R**, which reacted in THF less selectively than in toluene. The reason is the solubility of sodium dimethyl malonate, which is freely soluble in THF but sparsely soluble in toluene. The reaction in toluene could be recognized as running under pseudo-high dilution conditions.

The reaction equilibrium issues have become clearer, but the mechanism of the reaction and the real active catalytic complex were unknown. Initially, we addressed these issues by measuring the reaction kinetics but the attempt did not lead us to a clear conclusion.

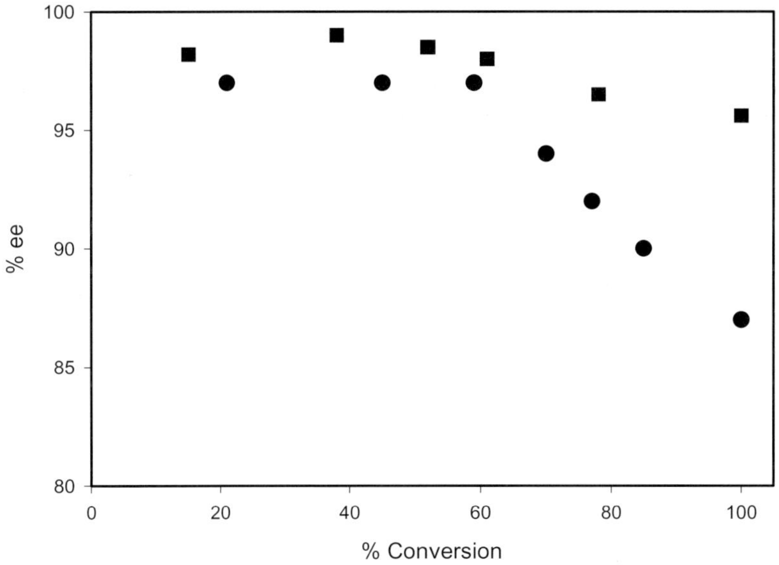

Figure 2.4 Progress of the reaction in THF vs. Toluene.

2.2.2
Modification of Ligands

Chiral ligand **31** has two C-2 symmetrical picolynyl amides. Initial kinetics taught us the reaction was 0.5 order in **31**. The kinetics indicated a dimeric nature of the active species. Therefore, systematic modification of the ligand was attempted and the results are summarized in Figure 2.5 [25].

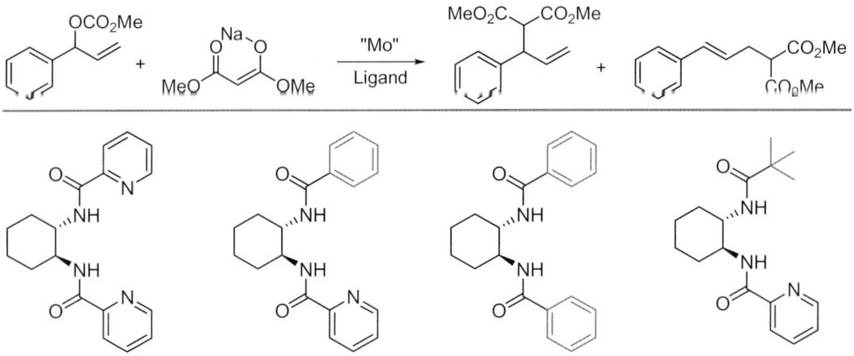

31
ee% 87%
B/L = 20
Reactivity = 1

64
ee% 92%
B/L = 35
Reactivity = 0.5

65
ee%24%
B/L = 1
Reactivity = 0.02

66
ee% 87%
B/L = 20
Reactivity = 0.25

Figure 2.5 Ligand modification.

As a baseline, Mo reaction with the standard chiral ligand **31** in THF gave the product in 87% ee with a B/L ratio of 20. Replacement of one of the picoline nitrogens with a carbon (ligand **64**) resulted in better ee % and improved B/L ratio even though the reactivity was reduced by half. Replacement of both picoline nitrogens with carbons (ligand **65**) virtually killed the reaction (reactivity was only 2% of the original conditions) but still provided modest 24 ee%. Replacement of one picoline amide with pivalate amide (ligand **66**) maintained the ee % and B/L ratio but somewhat reduced its reactivity.

These results strongly indicated that the minimum requirements for the ligand were two amides and one nitrogen atom on picoline. However, the real active Mo catalyst is not clear.

2.2.3
NMR Studies Revealed the Reaction Mechanism

In order to obtain insight into this reaction, we initiated detailed studies via NMR. $(C_7H_8)Mo(CO)_3$ was heated with ligand **31** in an NMR-tube. The result was complex, but spontaneous formation of $Mo(CO)_6$, which has a set of characteristic ^{13}C signals, was observed. Disproportionation between Mo and carbon monoxide was extremely facile in the presence of the ligand. From nOe studies, it was identified that Mo coordinated to a picolinyl amide of ligand **31** with both the nitrogen in pyridine and the oxygen in the amide. Either three or four carbon monoxides were coordinated to the Mo atom, as depicted as **67**. As a further complication, there is an equilibrium between the mono-Mo complex **67** and the bis-Mo complex **68**, as shown in Scheme 2.19. The NMR studies were profoundly complicated for further elucidation and we terminated our effort to identify the true catalyst with ligand **31**.

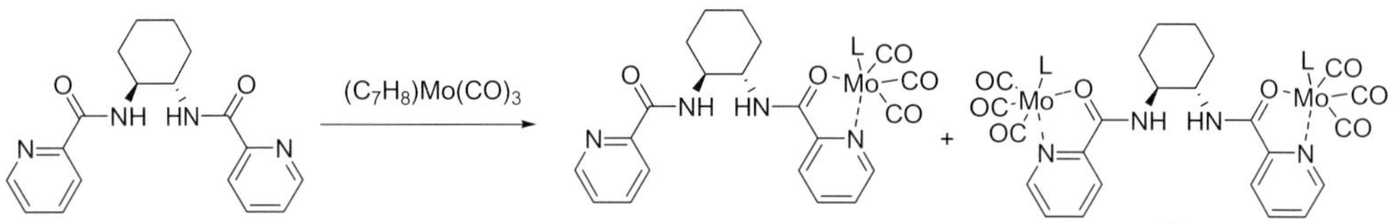

Scheme 2.19 Initial NMR studies with ligand **31**.

During our studies on ligands, we have found that mono-picolinyl mono-benzoyl amide ligand **64** is a better ligand than the original ligand **31** for this reaction, and we expected reaction with ligand **64** would provide simpler NMR spectra. Reaction of $(C_7H_8)Mo(CO)_3$ with ligand **64** gave a mixture of Mo complexes **69/70**, in which Mo was coordinated with a ligand **66** through a picolinyl amide via two atoms (pyridine nitrogen and amide oxygen) and with three or four carbon monoxides (Scheme 2.20). For further studies, dealing with an equilibrium mixture of three and four carbon monoxides coordinated to Mo, complex **69/70**, was not ideal.

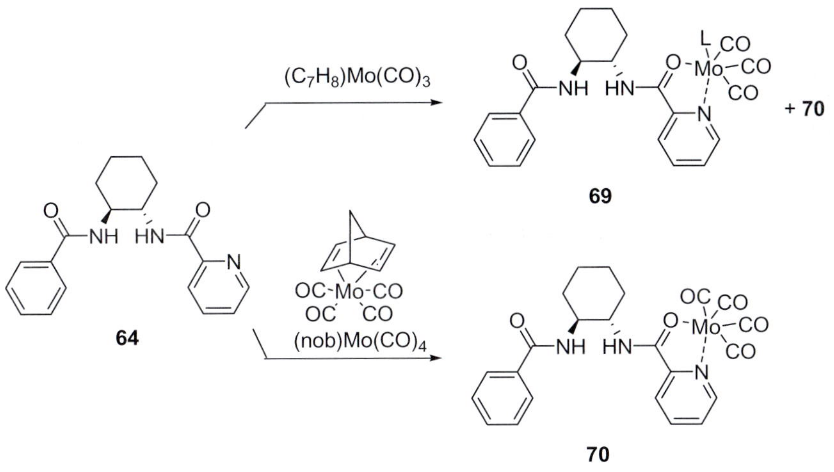

Scheme 2.20 NMR studies with simplified ligand **64**.

Thus, we prepared Mo complex **70** having four carbon monoxides from (nob) $Mo(CO)_4$ with ligand **64**.

Next we examined the reaction between the Mo complex **70** and linear carbonate **L-C** in an NMR tube [26]. The result was quite interesting, as summarized in Scheme 2.21.

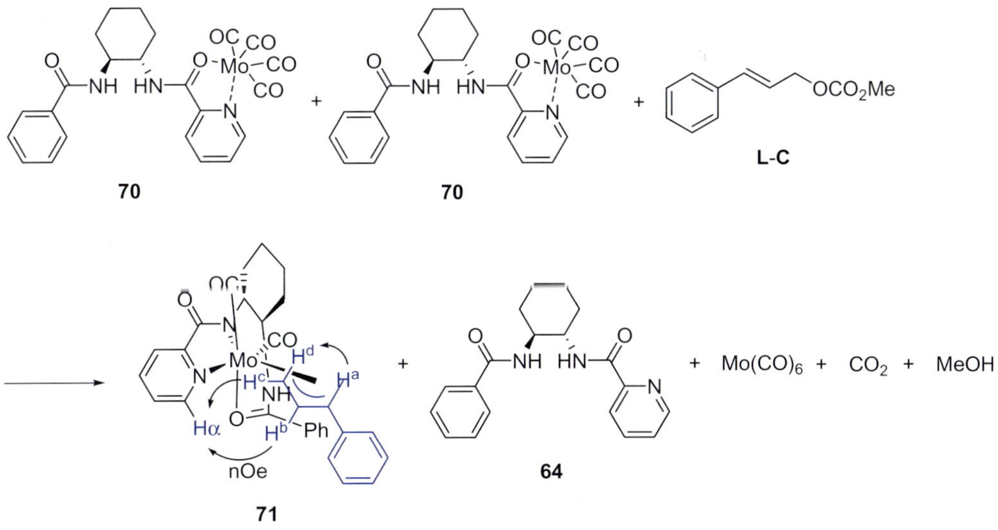

Scheme 2.21 Formation of π-allyl Mo complex.

Mo complex **70** (2 mole) were reacted with 1 mole of carbonate **L-C** to generate 1 mole of π-allyl Mo complex **71**, 1 mole of free ligand **64**, 1 mole of $Mo(CO)_6$,

1 mole of carbon dioxide and 1 mole of methanol. The structure of the π-allyl Mo complex **71** was initially elucidated by NMR nOe experiments and eventually confirmed by single crystal X-ray, as shown in Scheme 2.21. In complex **71**, Mo coordinated π-allyl, 2 mole of carbon monoxide (2 mole of carbon monoxide lost from **70**), pyridine nitrogen, the *N*-anion of the picolinyl amide (deprotonated by methoxide, which was generated upon formation of the π-allyl from carbonate), and oxygen at the benzoyl amide. The same complex **71** was also formed from 2 mole of Mo complex **70** and 1 mole of branched carbonate (either *S* or *R* carbonate or racemate **B-C**). *It is noted that nucleophilic attack of the Mo complex 71 from the cleanly open opposite side of the Mo atom should provide the R-adduct, instead of the experimentally observed S-adduct!*

More interestingly, reaction of the isolated crystalline π-allyl Mo complex 71 with sodium dimethylmalonate did not proceed at all. The result was shocking for us but at the same time complex **71** could be a resting intermediate for nucleophilic substitution since nucleophilic attack from the less sterically side of Mo should lead to the wrong stereoisomer (*vide supra*). (Professor Trost mentioned to us that the majority of isolated intermediate complexes would not be true active species, since those active species would be very difficult to isolate due to their reactivity). Subsequently, it was found that either carbon monoxide or $Mo(CO)_6$ was essential for the success of nucleophilic substitution of complex **71**. The reaction proceeded smoothly, as shown in Scheme 2.22. The desired product **B-P-S** was formed in 98% ee with high yield and a Mo ligand sodium salt **72**, which could be isolated as stable solid, was generated. It is noted that 4 mole of carbon monoxide coordinated to Mo in the sodium salt **72**, thus 2 mole of carbon monoxide was incorporated into the Mo complex upon going from **71** to **72**. Since nucleophilic substitution of complex **72** provided the *S*-isomer, the substitution occurred from the same side of the Mo atom (retention).

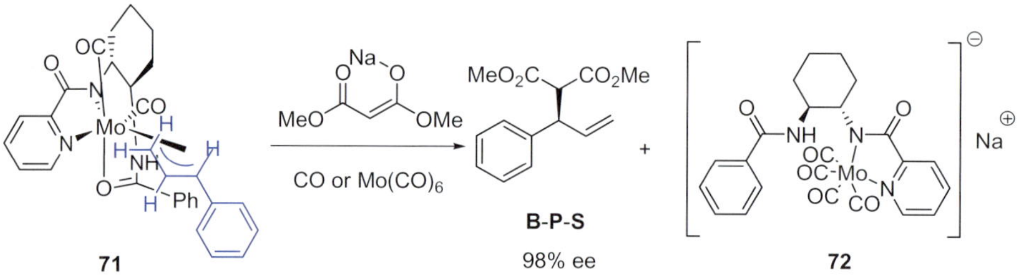

Scheme 2.22 Nucleophilic substitution in the presence of CO or $Mo(CO)_6$.

The isolated salt **72** was reacted with carbonate **L-C** to regenerate the π-allyl Mo complex **71**, releasing 1 mole of carbon dioxide, and sodium methoxide, and 2 mole of carbon monoxide (Scheme 2.23). Then, sodium dimethylmalonate reacts with the regenerated π-allyl Mo complex **71** in the presence of 2 mole of carbon monoxide.

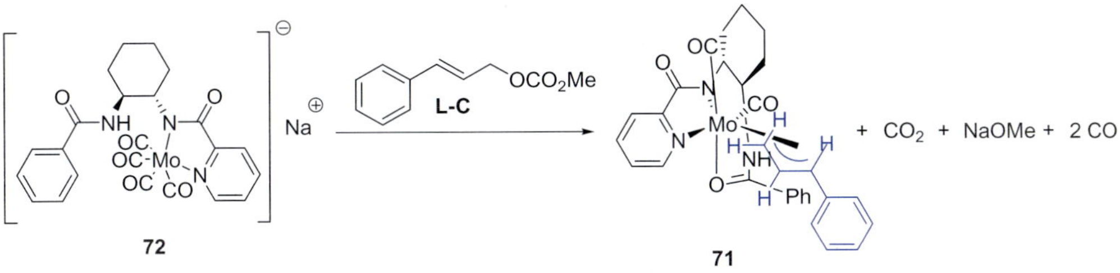

Scheme 2.23 Regeneration of π-allyl Mo complex.

The overall catalytic cycle is summarized in Scheme 2.24. The catalytic cycle from **71** to **72** is promoted by addition of 2 mole of carbon monoxide and the catalytic cycle from **72** to **71** releases 2 mole of carbon monoxide. Thus, carbon monoxide acts as the driver of this catalytic cycle.

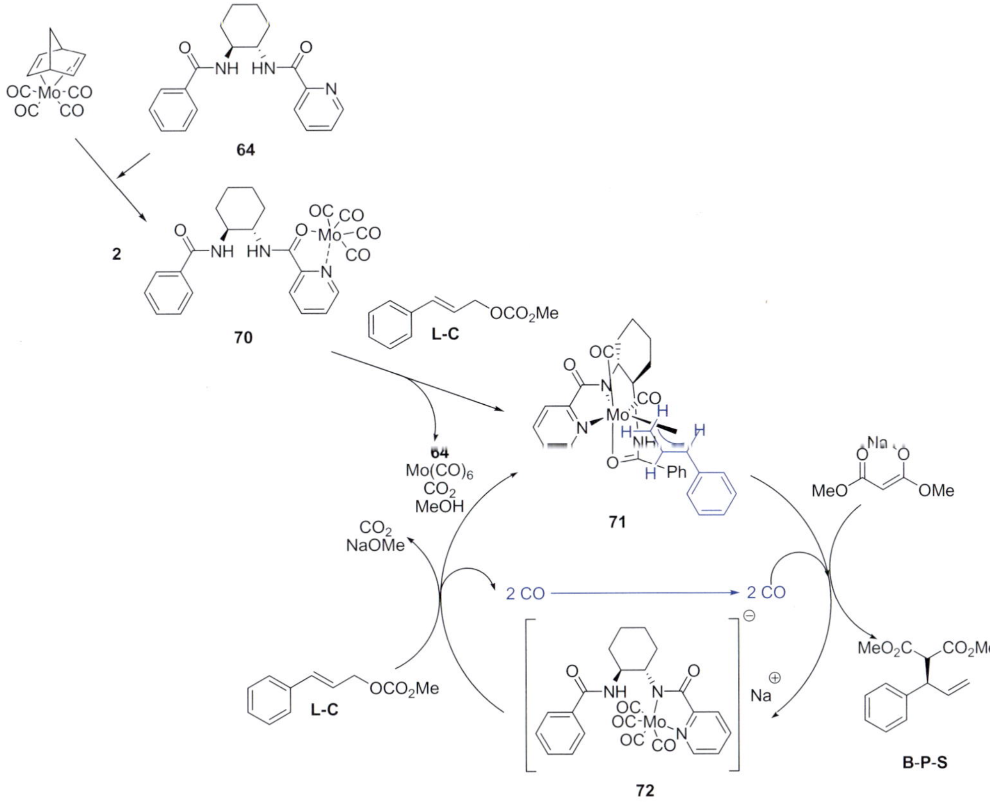

Scheme 2.24 Whole catalytic cycle of Trost's Mo-π-allyl nucleophilic reaction.

2.2.4 Additional Studies for Confirmation of the Retention–Retention Mechanism

As previously mentioned, the nucleophilic substitution on the Mo complex **71** most likely occurs with retention of configuration based on the stereochemistry outcomes of the product and **71**. The retention–retention mechanism was confirmed with labeling experiments in collaboration with Professor Lloyd-Jones, as shown in Scheme 2.25 [27].

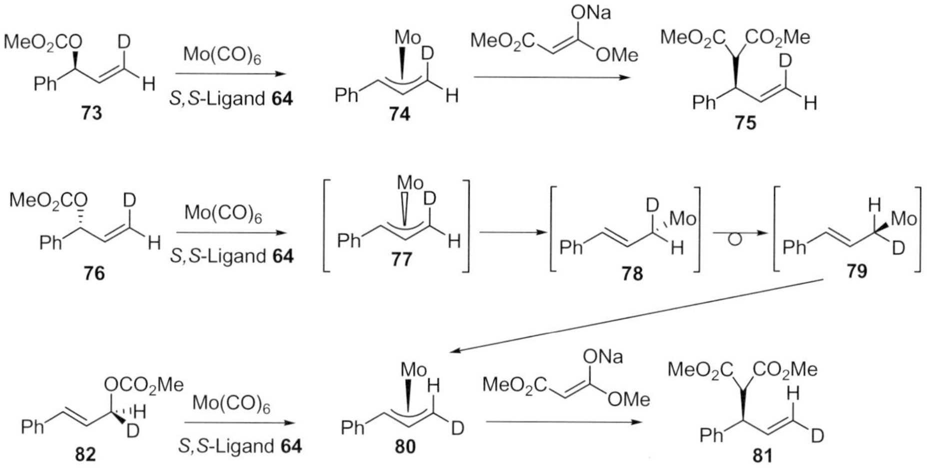

Scheme 2.25 Deuterium-labeling studies.

Stoichiometric reaction with matched *S*-carbamate having the D atom in the *Z*-position **73**[3] in the presence of *S,S*-ligand **64** without a nucleophile solely formed (no other isomer was observed by NMR) the Mo-complex **74** without transposition of the label. The structure of **74** was probed based on NMR studies by comparison with NMR studies and the X-ray structure of the protio complex **71**. Nucleophilic attack of sodium malonate on the Mo complex **74** provided the *S*-product **75**, where the D atom remained at the *Z*-position. On the other hand, stoichiometric reaction with mismatched *R*-carbamate having the D atom in the *Z*-position **76** without a nucleophile generated the Mo complex **80** as sole product, based on NMR studies. The structure of the complex **80** was elucidated by NMR. In **80**, Mo is located on the same face as in **74** but the D atom is transposed from the *Z* to the *E* position. The transposition could be explained as follows. Initially the π-allyl Mo-complex **77** (unobserved) must form with retention. Mo complex **77** is equilibrated into the more stable Mo complex **80**, where the D atom is moved

3) Actually, all deuterium labeled substrates are enantiomerically enriched but not 100% enatiomerically pure. However in this chapter, all discussion assumed them to be 100% pure to simplify the argument. For more precise discussions, please refer to the original papers [27].

to the *E*-position, presumably via σ-allyl Mo complexes (**78** and **79**) where the configuration of the carbon center adjacent to the Mo must be rotated by 180°. Nucleophilic addition of sodium malonate to the Mo complex **80** gave the *S*-adduct **81**, having the D atom at the *E*-position. Furthermore, enantiomerically D-labeled (*R*) linear carbonate **82** was subjected to the reaction conditions without a nucleophile and the only observed Mo complex was **80**, based on NMR data. Nucleophilic attack with sodium malonate provided the *S*-adduct **81**, having the D atom at the *E*-position. In the linear case, it seems that Mo attacks from the preferred face of the conformation (the carbonate group was facing up) where the carbonate group leaves from the same side as the Mo attack. It appears that all nucleophilic additions to the π-allyl Mo-complexes (**74** and **80**) occurred with retention.

There was still some room for uncertainty on this retention–retention mechanism. The argument was, if the unobserved π-allyl Mo complex (such as **77** or **B** in Scheme 2.18) was more highly reactive towards sodium malonate than experimentally observed π-allyl Mo complexes (such as **71**, **74**, and **80**), the reaction should proceed through inversion (since there is an equilibrium between the two π-allyl Mo complexes via the σ-allyl complex.) If so, when the isolated Mo-complex **71** was subjected to the reaction, **71** must be equilibrated to the enantiomer of **71** via the σ-allyl complex prior to reaction with a nucleophile. Therefore, reaction from the Mo complex **71** should proceed with less stereoselectivity than that from a mismatched branched carbonate. This hypothesis was examined, as shown in Scheme 2.26.

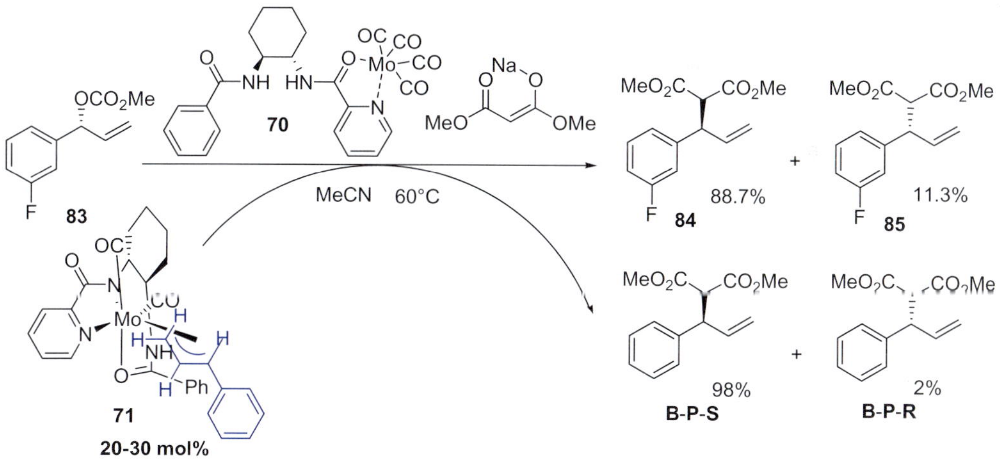

Scheme 2.26 Proof for retention–retention mechanism.

Reaction of mismatched 3-fluoro *R*-carbonate **83** with catalytic amounts of **70** (derived from *S*,*S*-ligand **64**) proceeded with sodium malonate in acetonitrile at 60 °C and the *S*-adduct **84** and the *R*-adduct **85** were obtained in a ratio of 88.7 : 11.3. Complex **71** (20–30 mol%) was added to the reaction mixture of **83**, **70**, and sodium malonate after the reaction was smoothly turning over (15–45% conversion). The result was totally opposite from the above-mentioned hypothesis and the ee % of the product from **71** was 96% (vs. 77% ee from **83**). Thus, this π-allyl Mo reaction was proved to proceed via a retention–retention mechanism.

Trost's original conditions required 10 mol% of Mo precatalyst and 15 mol% of chiral ligand **31**. Since 1 mole of $Mo(CO)_6$ and free ligand **31** would be generated but $Mo(CO)_6$ and **31** should regenerate the active catalyst like **71** with proper activating, the lower catalyst load was accomplished by running the reaction in refluxing toluene.

2.3 Conclusion

We have successfully prepared a CCR5 antagonist drug candidate, which has been licensed out to the International Partnership for Microbicides. A large scale preparation was developed in a very short time and this synthetic process is supporting current drug development. Based on a modification of Trost's asymmetric Mo π-allyl nucleophilic substitution, this reaction was found to proceed with highly effective kinetic resolution, and the reaction mechanism has become much clearer. Based on mechanistic considerations, the catalyst load was reduced to 1 mol% from the original 10 mol%. Thus, we contributed simultaneously to both project support and improvement of the chemical reaction based on deeper understanding of the reaction mechanism.

Acknowledgments

I would like to thank all colleagues who worked on this project, whose names are listed in the references. I would also like to thank Drs James McNamara and Michael Palucki for careful proofreading and their helpful suggestions.

References

1 (a) Finke, P.E., Hilfiker, K.A., Maccoss, M., Chapman, K.T., Loebach, J.L., Mills, S.G., Guthikonda, R.N., Shah, S.K., Kim, D., Shen, D.-M., and Oates, B. (2000) WO 2000076972 A1 20001221. (b) Kumar, S., Kwei, G.Y., Poon, G.K., Iliff, S.A., Wang, Y., Chen, Q., Franklin, R.B., Didolkar, V., Wang, R.W., Yamazaki, M., Chiu, S.-H.L., Lin, J.H., Pearson, P.G., and Baillie, T.A. (2003) *J. Pharmacol. Exp. Ther.*, **304**, 1161–1171.

2 IPM (2005) Annual report, http://www.ipm-microbicides.org/

3 An interesting route for **2** was reported; Zhang, W., Matla, A.S., and Romo, D. (2007) *Org. Lett.*, **9**, 2111–2114.

4 (a) Trost, B.M., and Chan, D.M.T. (1983) *J. Am. Chem. Soc.*, **105**, 2315–2325. (b) Trost, B.M., Yang, B., and Miller, M.L. (1989) *J. Am. Chem. Soc.*, **111**, 6482–6484. For some reviews, see: (c) Trost, B.M. (1986) *Angew. Chem. Int. Ed. Engl.*,

25, 1–20. (d) Lautens, M., Klute, W., and Tam, W. (1996) *Chem. Rev.*, **96**, 49–92. (e) Romero, J.M.L., Sapmaz, S., Fensterbank, L., and Malacria, M. (2001) *Eur. J. Org. Chem.*, 767–773. (f) Yamago, S., and Nakamura, E. (2002) *Org. React. (New York)*, **61**, 1–217.

5 Conlon, D.A., Jensen, M.S., Palucki, M., Yasuda, N., Um, J.M., Yang, C., Hartner, F.W., Tsay, F.-R., Hisao, Y., Pye, P., Rivera, N.R., and Hughes, D.L. (2005) *Chirality*, **17**, S149–S158.

6 (a) Boeckman, R.K., Jr., Naegely, P.C., and Arthur, S.D. (1980) *J. Org. Chem.*, **45**, 752–754. (b) Boeckman, R.K., Jr., Napier, J.J., Thomas, E.W., and Sato, R.I. (1983) *J. Org. Chem.*, **48**, 4152–4154.

7 Palucki, M., Um, J.M., Yasuda, N., Conlon, D.A., Tsay, F.-R., Hartner, F.W., Hisao, Y., Marcune, B., Karady, S., Hughes, D.L., Dormer, P.G., and Reider, P.J. (2002) *J. Org. Chem.*, **67**, 5508–5516.

8 Wildonger, K.J., Leanza, W.J., Ratcliffe, R.W., and Springer, J.P. (1995) *Heterocycles*, **41**, 1891–1900.

9 Brooks, D.W., Lu, L.D.-L., and Masamune, S. (1979) *Angew. Chem. Int. Ed. Engl.*, **18**, 72–74.

10 (a) Ratcliffe, R.W., Salzmann, T.N., and Christensen, B.G. (1980) *Tetrahedron Lett.*, **21**, 31–34. (b) The first example of a carbene insertion to a beta-lactam nitrogen atom; Cama, L.D., and Christensen, B.G. (1978) *Tetrahedron Lett.*, **19**, 4233–4236.

11 Trost, B.M., and Hachiya, I. (1998) *J. Am. Chem. Soc*, **120**, 1104–1105.

12 Heslin, J.C., Moody, C.J., Slawin, A.M.Z., and Williams, D.J. (1986) *Tetrahedron Lett.*, **27**, 1403–1406.

13 White, J.D., and Hrnciar, P. (1999) *J. Org. Chem.*, **64**, 7271–7273.

14 (a) Danishefsky, S. (1979) *Acc. Chem. Res.*, **12**, 66–72. (b) A similar transformation was reported; Tanimori, S., Tsubota, M., He, M., and Nakayama, M. (1995) *Biosci. Biotech. Biochem.*, **59**, 2091–2093.

15 Barnes, D.J., Chapman, R.I., Vagg, R.S., and Walton, E.C. (1978) *J. Chem. Eng. Data*, **23**, 549–550.

16 Conlon, D.A., and Yasuda, N. (2001) *Adv. Synth. Catal.*, **343**, 137–138.

17 For $(EtCN)_3Mo(CO)_3$: (a) Kubas, G.J., and Van der Sluys, L.S. (1990) *Inorg. Synth.*, **28**, 29–33. (b) for $(C_7H_8)Mo(CO)_3$: Cotton, F.A., McCleverty, J.A., and White, J.E. (1990) *Inorg. Synth.*, **28**, 45–47.

18 Palucki, M., Um, J.M., Conlon, D.A., Yasuda, N., Hughes, D.L., Mao, B., Wang, J., and Reider, P.J. (2001) *Adv. Synth. Catal.*, **343**, 46–50.

19 Baum, J.S., Shook, D.A., Davies, H.M.L., and Smith, H.D. (1987) *Synthetic Commun.*, **17**, 1709–1716.

20 Shin, K.J., Moon, H.R., George, C., and Marquez, V.E. (2000) *J. Org. Chem.*, **65**, 2172–2178.

21 Moriarty, R.M., May, E.J., Guo, L., and Prakash, O. (1998) *Tetrahedron Lett.*, **39**, 765–766.

22 A similar dimerization was reported; Mills, J.E., Maryanoff, C.A., McComsey, D.F., Stanzione, R.C., and Scott, L. (1987) *J. Org. Chem.*, **52**, 1857–1859.

23 Evans, D.A., Chapman, K.T., and Carreira, E.M. (1988) *J. Am. Chem. Soc.*, **110**, 3560–3578.

24 Hughes, D.L., Palucki, M., Yasuda, N., Reamer, R.A., and Reider, P.J. (2002) *J. Org. Chem.*, **67**, 2762–2768.

25 Trost, B.M., Dogra, K., Hachiya, I., Emura, T., Hughes, D.L., Krska, S., Reamer, R.A., Palucki, M., Yasuda, N., and Reider, P.J. (2002) *Angew. Chem. Int. Ed.*, **41**, 1929–1932.

26 (a) Krska, S.W., Hughes, D.L., Reamer, R.A., Mathre, D.J., Sun, Y., and Trost, B.M. (2002) *J. Am. Chem. Soc.*, **124**, 12656–12657. (b) Krska, S.W., Hughes, D.L., Reamer, R.A., Mathre, D.J., Palucki, M., Yasuda, N., Sun, Y., and Trost, B.M. (2004) *Pure Appl. Chem.*, **76**, 625–633.

27 (a) Llyod-Jones, G.C., Krska, S.W., Hughes, D.L., Gouriou, L., Bonnet, V.D., Jack, K., Sun, Y., and Reamer, R.A. (2004) *J. Am. Chem. Soc.*, **126**, 702–703. (b) Hughes, D.L., Lloyd-Jones, G.C., Krska, S.W., Gouriou, L., Bonnet, V.D., Jack, K., Sun, Y., Mathre, D.J., and Reamer, R.A. (2004) *PNAS*, **101**, 5378–5384.

3
5α-Reductase Inhibitors – The Finasteride Story

J. Michael Williams

5α-Reductase is the enzyme responsible for the conversion of testosterone into dihydro-testosterone in Man (Figure 3.1). It was proposed that regulation of this process through selective inhibition of 5α-reductase might provide a safe and effective therapy for androgen-dependent disorders such as benign prostatic hypertrophy and male-pattern baldness, both of which are a consequence of elevated levels of dihydro-testosterone [1]. Based on this premise, a program was initiated at Merck with the goal of identifying a potent, selective inhibitor of 5α-reductase. From this effort emerged the Δ^1-3-keto-4-azasteroids which are believed to serve as nonreducible structural mimics of the transition state in the enzymatic reduction of the natural substrate [2]. The enzyme proved relatively insensitive to the identity of the C17 side chain of the azasteroid. The C17 side chain was, however, found to be important in providing the desired biopharmaceutical properties and safety profile.

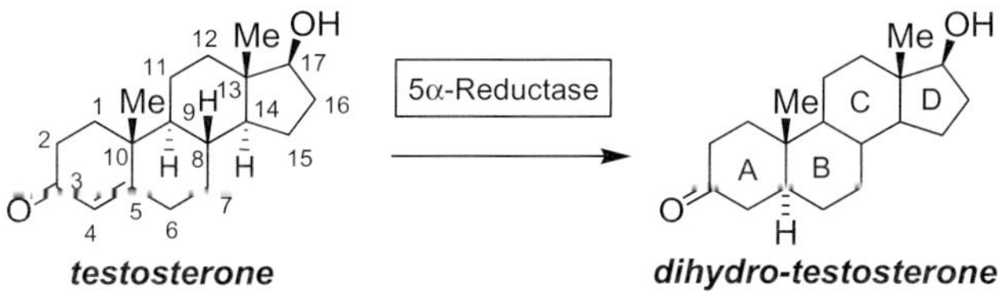

Figure 3.1 The biosynthesis of dihydro-testosterone.

Among the candidates considered for development in this series was the *t*-Bu amide which was subsequently given the name finasteride and became the active ingredient in both PROSCAR® and PROPECIA® (Figure 3.2). Section 3.1 will tell the story of the development of a manufacturing process for finasteride. As in most programs at Merck, drug candidates showing potential for improved performance over the lead compound were approved for development as the lead

The Art of Process Chemistry. Edited by Nobuyoshi Yasuda

ISBN: 978-3-527-32470-5

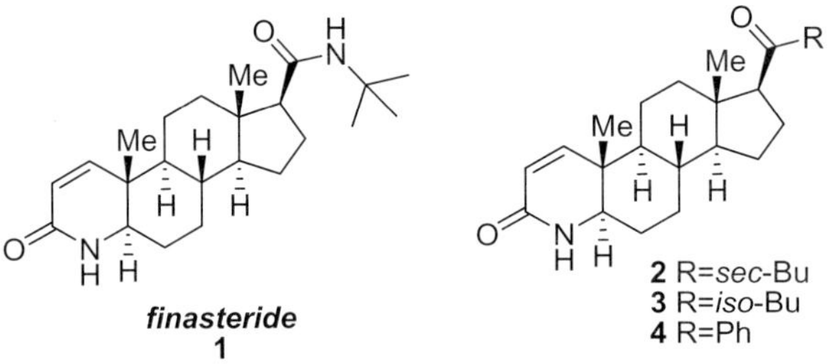

Figure 3.2 Finasteride and related back-up compounds.

progressed through safety assessment and clinical trials. In the second half of Section 3.1, we will focus on the development of processes for the kilogram-scale preparation of back-up candidates (**2**–**4**).

During development of the azasteroids, we discovered an efficient method for the conversion of a lactam to the corresponding α,β-unsaturated lactam using 2,3-dichloro-5,6-dicyanobenzoquinone (DDQ) and clearly elucidated the mechanism of this oxidation. Our understanding of the mechanistic details of this reaction guided optimization of the process. Although the second and third generation compounds, **2**, **3**, and **4**, did not become products for Merck, each provided opportunities for the development of new methods addressing difficult synthetic challenges, methods that have proven their value beyond azasteroid synthesis. We will discuss these discoveries in Section 3.2. In particular, a new general preparation of Weinreb amides from esters, which was discovered through this work, has become widely used [3].

3.1
Project Development

3.1.1
Finasteride

3.1.1.1 The Medicinal Chemistry Route

The Medicinal Chemistry synthesis of the Δ^1-3-keto-4-azasteroids, shown in Scheme 3.1, was designed to be divergent, thereby allowing installation of many different side-chains in the C17 position [2]. This strategy was key in establishing structure–activity relationships for this series. The thiopyridyl ester derivative **9** of 3-keto-4-azasteroid-17β-carboxylic acid **8** was the key intermediate in this strategy. Other means of activating carboxylic acid **8** were also used, including conversion of the acid to the acid chloride, the benzotriazole ester, and the acyl imidazolide, but these intermediates were not as broadly useful as the thiopyridyl ester. Fol-

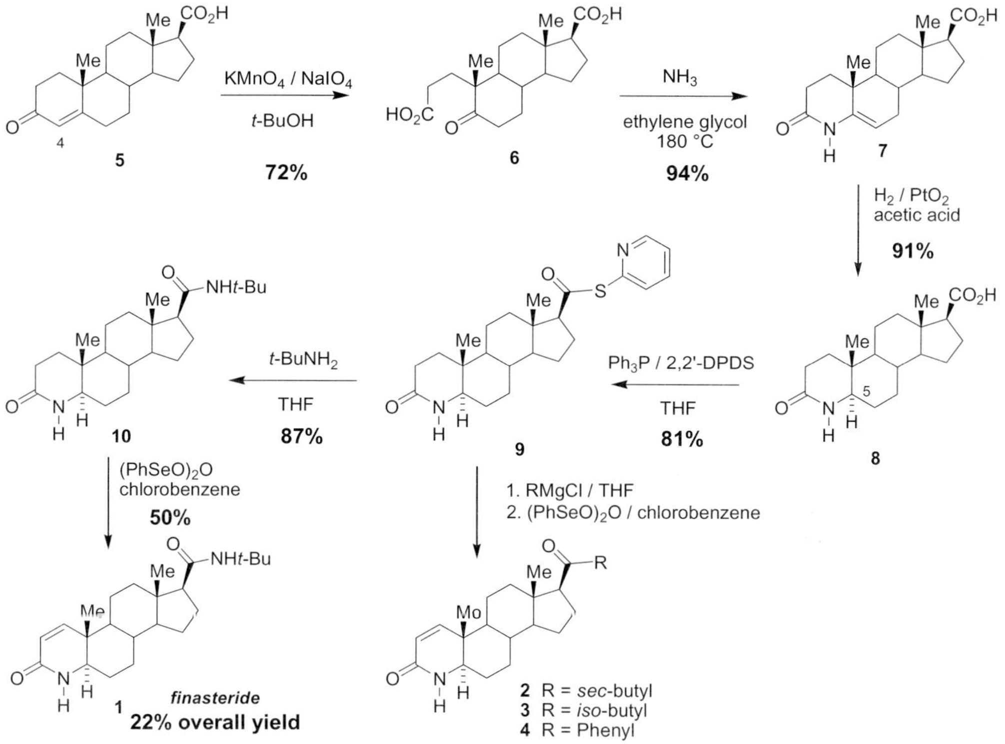

Scheme 3.1 The Medicinal Chemistry synthesis of Δ^1-3-keto-4-azasteroids.

lowing conversion of the acid to the desired functionality, the Δ^1 double bond was installed using $(PhSeO)_2O$. Attempts to introduce the double bond by this method at an earlier stage in the synthesis were not successful.

The Medicinal Chemistry synthesis of finasteride **1** began with a starting material, 4-androsten-3-one-17β-carboxylic acid (**5**), that was readily available on gram scale. The C4 carbon was oxidatively excised using KIO_4 with catalytic $KMnO_4$ giving the keto diacid **6**. Reaction with ammonia in ethylene glycol at 180 °C produced the Δ^5-aza-lactam **7** which was hydrogenated over PtO_2 setting the requisite stereochemistry at C5. The carboxylic acid **8** was then activated as the thiopyridyl ester using PPh_3 and 2,2′-dipyridyl disulfide (2,2′-DPDS) and the thioester **9** was purified by chromatography to remove by-products of the reaction. Subsequently, reaction of **9** with *t*-$BuNH_2$ gave the amide **10** and, finally, the double bond was introduced using $(PhSeO)_2O$ in refluxing chlorobenzene. Chromatographic purification was required to remove selenium-containing impurities. The overall yield for the six-step sequence to finasteride (**1**) was 22%. This scheme served well for preparation of the many Δ^1-3-keto-4-azasteroids (such as **2**, **3**, **4**) that were tested for 5α-reductase inhibition. For the preparation of kilogram quantities of the drug candidate, however, there were clearly a number of issues that needed to be addressed.

3.1.1.1.1 Problems of the Original Route

Following selection of the drug candidate, the key issues for process development below were identified.

1) The method used for introducing the double bond produced toxic selenium waste and impurities containing selenium that were difficult to remove.
2) Volume productivity was limited by the poor solubility of the azasteroid intermediates.
3) The permanganate reaction produced a heavy precipitate which limited reaction productivity and the resulting waste was a disposal issue. The yield proved to be variable on scale up.
4) Chromatographic purification was required in two steps.
5) One step required high temperature.
6) The platinum catalyst used in the hydrogenation was expensive and selectivity was variable.
7) A chlorinated solvent was used in one step.
8) The starting material was not available in the quality and quantity needed.

In addition to the challenges cited above, there are some special issues associated with steroid chemistry that should be noted. The steroidal impurities formed in the process are generally similar in structure to the desired product and, in some cases, co-crystallization with the product is a problem. It is, therefore, critical to limit the formation of steroidal impurities in the reactions. The structural similarity between product and impurities also creates challenges in developing assays for reaction monitoring and purity determination. Furthermore, the poor solubility of these compounds in the solvents typically used in a manufacturing process makes it very difficult to achieve practical volume productivity in process development.

3.1.1.2 Process Development

Early in development, the availability of drug for testing almost always limits the pace of development. In some cases, only minor modification is needed to scale up a Medicinal Chemistry synthesis making the drug available. In other cases, the existing synthesis is not suitable for scale-up and an entirely new approach is needed. The judgment of the process chemist based on very limited experience with the Medicinal Chemistry synthesis determines the course early in development. In this section, we will examine the strategy, decisions, and key discoveries that led from the early stages through development to demonstration of a manufacturing process for finasteride. The discussion is outlined below.

1) Early process development and modification of the Medicinal Chemistry synthesis for the first kilogram-scale delivery of finasteride
   – The strategy – introduce the *amide* early in the synthesis
2) A change in strategy – *the carboxylic acid* as a late-stage intermediate
   – Key discovery of a practical method for introduction of the Δ^1 double bond
3) The manufacturing process – targeting the *ester* as a late-stage intermediate

3.1.1.2.1 The First Kilogram-Scale Delivery of Finasteride 1

Following preliminary assessment of the Medicinal Chemistry synthesis, the decision was made to modify the synthesis for scale-up to address some of the issues identified but, without a ready alternative for the double bond introduction, $(PhSeO)_2O$ would be used in the first scale-up to kilogram scale.

Selection of the starting material Synthesis development and scale-up efforts were initiated early in 1985. At that time, sourcing acid **5** failed to identify an adequate supply of the quality needed for the first scale-up effort. Pregnenolone (**11**) was recognized as a suitable starting material that was available in multi-kilogram quantities. The process devised for the conversion of **11** to **5** is detailed in Scheme 3.2. The C17 methyl ketone was readily converted to the methyl ester **12** through the iodoform reaction. Oppenauer oxidation of the secondary alcohol **12** to the enone and hydrolysis of the ester **13** gave the acid **5** in 56% yield from **11**. For later development, commercial supplies of the acid **5** were identified.

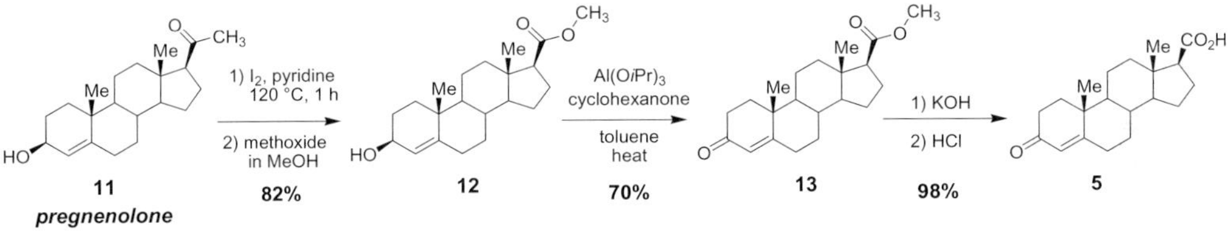

Scheme 3.2 A commercially available starting material.

Strategy for the first delivery of finasteride – the amide route At this point, a strategic decision was made to change the order of the steps in the original synthesis. Introducing the *t*-$BuNH_2$ early in the scheme could be effected through activation of the acid as the acid chloride, thus eliminating both the need to make the thiopyridyl ester and the associated chromatographic purification. Also, by introducing the amide at this stage, we believed that the solubility of intermediates would be better, thereby allowing improved volume productivity. The acid **5** was activated using oxalyl chloride and reaction of the acid chloride with *t*-$BuNH_2$ provided amide **14**. Oxidative cleavage of C4 to give keto-acid **15** followed by condensation with ammonia in ethylene glycol at 140 °C gave Δ^5 enamide **16** in 89% yield. Hydrogenation of **16** with PtO_2 set the C5 configuration and oxidation with $(PhSeO)_2O$ completed the first delivery in 30% overall yield (Scheme 3.3).

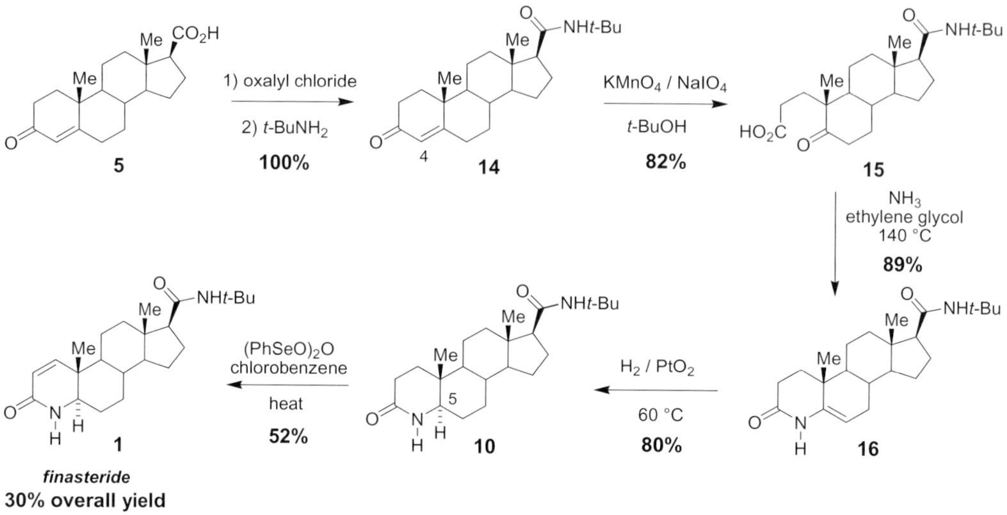

Scheme 3.3 The first scale-up route to finasteride.

3.1.1.2.2 **A Change in Strategy – the Carboxylic Acid as a Late-Stage Intermediate**
When second generation candidates differing from finasteride only at the C17 position were considered for development, a second team was tasked with defining the synthesis while the first delivery of finasteride was being completed. Three ketones were considered as potential back-up compounds, the *s*-Bu, *i*-Pr, and *i*-Bu ketones (**2**, **19**, **3** in Scheme 3.4). Ideally, the new route would allow divergence at a late stage of the synthesis to make both finasteride and the ketone selected for

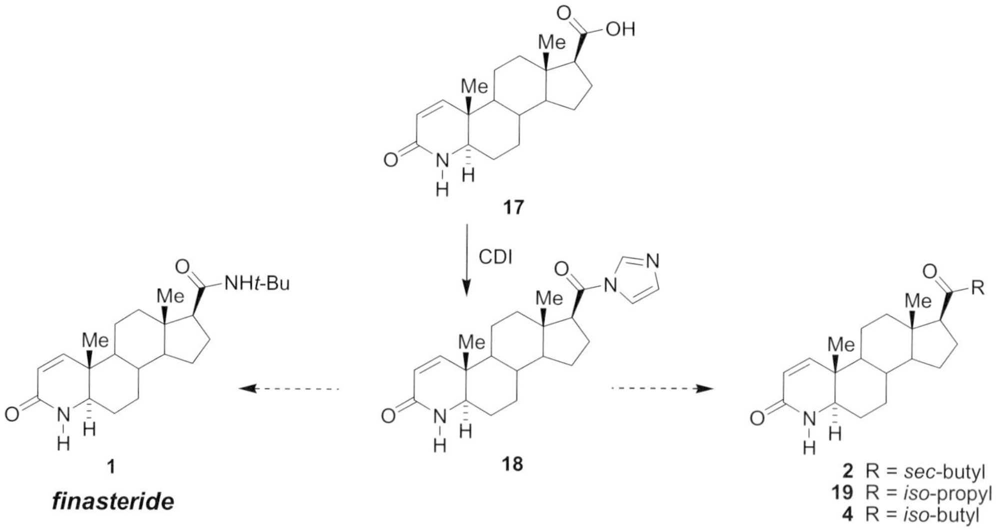

Scheme 3.4 A divergent strategy.

development. The plan was to return to the Medicinal Chemistry approach where the C17 carboxylic acid was activated later in the synthesis. It had been shown that the acid could be activated in the presence of the A-ring lactam using carbonyldi-imidazole (CDI) to make the acyl imidazolide.

We hoped to be able to introduce the Δ^1 double bond prior to installation of the C17 functionality. Medicinal Chemistry had reported that this would not be possible using $(PhSeO)_2O$. Furthermore, recognizing that the use of $(PhSeO)_2O$ would be restricted beyond the first delivery, there was an intense effort to identify a practical method for introduction of the Δ^1 double bond. Acyl imidazole **18**, prepared from Δ^1-4-aza-carboxylic acid **17**, became the target intermediate for preparation of both finasteride and the back-up ketone, as shown in Scheme 3.4. The azasteroid carboxylic acids were known to have low solubility in most organic solvents and volume productivity would be a challenge in making this approach practical.

Preparation of carboxylic acid 8 A commercial source of androst-4-en-3-one-17β-carboxylic acid **5** was found at this time and a number of key improvements were made in the conversion of **5** to 3-keto-4-aza-17β-carboxylic acid **8**, as shown in Schemes 3.5–3.7. Oxidative cleavage of the enone **5** could be effected in water using catalytic RuO_2 [4] with bleach as the stoichiometric oxidant [5] thereby eliminating the expense and waste associated with the permanganate/periodate cleavage reaction (Scheme 3.5). Assay yields as high as 92% were achieved on a small scale. A somewhat lower yield was observed initially on scale-up. Control of pH was critical. At pH above 8.5, over-oxidation of the ketone product was a problem so it was necessary to add the basic bleach over time to avoid exceeding pH 8. As the reaction progresses, the pH drops. At pH below 7.5, bleach is unstable. NaOH was added using a pump controlled by a pH meter to maintain the pH between 7.5 and 8.5. The catalyst loading could be reduced to 0.6 mol% by using 10% acetonitrile as cosolvent. Acetonitrile is believed to serve as a ligand for lower valent Ru species in the catalytic cycle thereby limiting the formation of insoluble carboxylate complexes [4]. Following completion of the reaction, the pH was adjusted with HCl to give the carboxylic acid form of the product **6** which was extracted into CH_2Cl_2. The product was not isolated; the solvent was switched to AcOH by distillation in preparation for the next step. On pilot scale, the reaction consistently produced 85–89% assay yield after extraction, a 10% improvement over the original conditions.

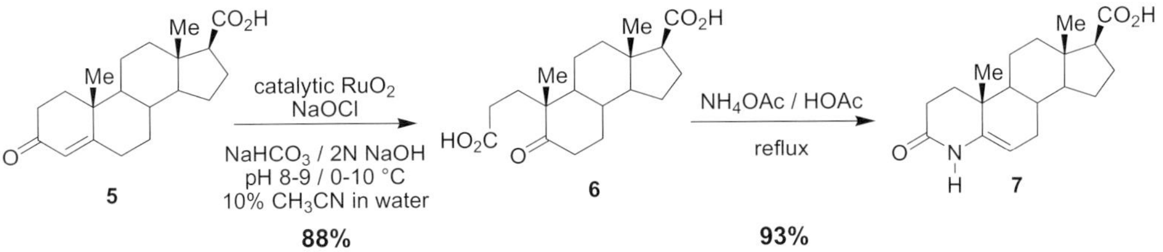

Scheme 3.5 Oxidative cleavage and ene-lactam formation.

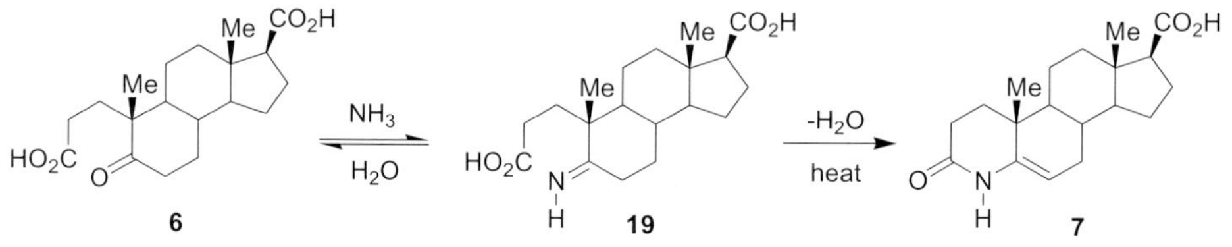

Scheme 3.6 Sequence of events in formation of the ene-lactam.

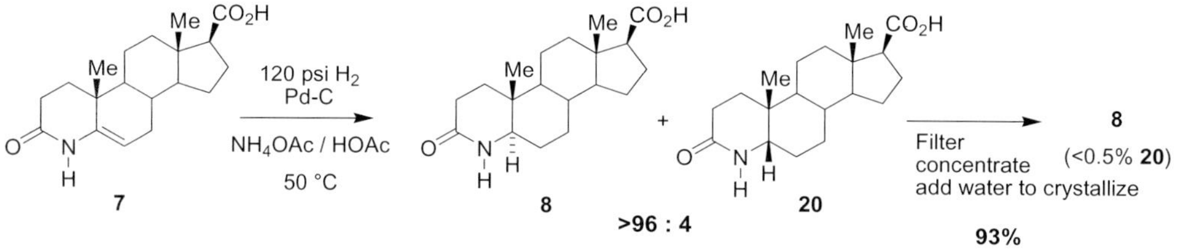

Scheme 3.7 Ene-lactam hydrogenation.

In the initial attempt to scale up the reaction using ammonia in ethylene glycol at 180 °C to form lactam **7**, the reaction, depicted in Scheme 3.6, did not go to completion, most likely due to the loss of ammonia during the longer time required to reach reaction temperature at this scale. On the gram scale, it was found that complete conversion could be achieved by heating at 140 °C in ethylene glycol and, on scale-up, improved performance was observed.

Although the reaction on kilogram scale could be completed by heating at 140 °C, this was not the best solution. It was subsequently found that the reaction could be performed in AcOH using NH_4OAc as the source of ammonia. Solubility in AcOH was good, allowing improved volume productivity, and the loss of ammonia was no longer a problem. In AcOH, heating to only 120 °C was sufficient to complete the reaction within 2 h. After concentrating the AcOH solution, the product **7** could be isolated simply by adding water to give a crystalline solid.

The Medicinal Chemistry synthesis had used PtO_2 as catalyst for the hydrogenation in AcOH to set the required configuration at C5. The catalyst was quite expensive and selectivity was found to be variable. Early in development, the switch was made to Pt/C but the selectivity remained variable. AcOH continued to be the best solvent for the reaction providing good solubility for the starting material and product. The reaction was promoted by the weak acid; no reaction was observed in alcohol or ester solvents. We found that strong acid accelerated the reaction and degraded the selectivity; the addition of trace sulfuric acid could produce up to 25% of the undesired isomer **20**.

When NH_4OAc was used in the hydrogenation, the reaction became reliable, consistently producing less than 4% of the undesired isomer. In addition to acting as a buffer, NH_4OAc served to limit hydrolysis of ene-lactam **7** and the yield increased as a result. Ultimately, it was possible to go to a cheaper Pd/C catalyst. Following completion of the reaction, the mixture was filtered to remove the catalyst and, after concentrating the AcOH solution, the product was crystallized by adding water. This process consistently gave **8** containing less than 0.5% of the undesired isomer **20** in 93% yield, as shown in Scheme 3.7.

Introduction of the Δ^1 double bond–the carboxylic acid as substrate At the time, there had been a very limited number of methods reported in the literature for the dehydrogenation of lactams and the most efficient used $(PhSeO)_2O$. All other methods required multiple steps producing only modest overall yields [6] and did not appear likely to provide the basis for a practical manufacturing process. This problem was seen as a great challenge and also a great opportunity. The groups working on the problem explored a wide range of ideas, but nearly all attempts produced disappointing results.

Two approaches that had been reported for the related dehydrogenation of ketones appeared to offer the promise of a more efficient method. These approaches relied on activation of a ketone as a trimethylsilyl (TMS) enol ether. A solubilized palladium catalyst had been reported to effect the dehydrogenation of TMS enol ethers [7]. DDQ had been shown to react with TMS enol ethers to give enones [8].[1] There was some concern, however, that the analogy between ketones and lactams was fatally flawed, as illustrated in Scheme 3.8. Silylation of the ketone **21** would give the silyl enol ether **22** leading to enone **23** on dehydrogenation. Silylation of a lactam such as **24** was expected to give the silylimidate **25** which would lead to the wrong dehydrogenation product **26**.

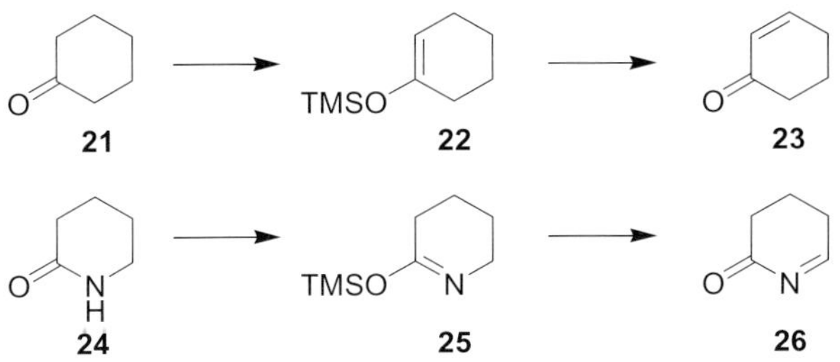

Scheme 3.8 The ketone / lactam analogy.

To overcome this problem, two approaches were investigated. In principle, lactam **24** could be deprotonated using strong base to give a dianion **27** that would react with a silylating reagent affording the bis-silyl lactam **28** which could then be dehydrogenated to give the product **30** with a double bond in the required position (path **A**, Scheme 3.9). Attempts to realize this approach with an azasteroid were not successful. Alternatively, a strong silylating agent might be capable of

1) DDQ had been used extensively in the dehydrogenation of 3-keto steroids under neutral or acidic conditions but the reactions generally gave a mixture of products in modest yield.

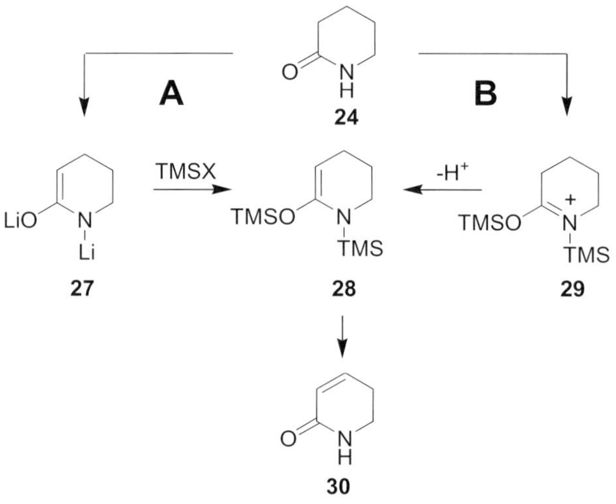

Scheme 3.9 Strategy for lactam activation.

doubly silylating the lactam **24** leading to **28** through **29**, also illustrated in Scheme 3.9 (path **B**).

All attempts to use Pd and other catalysts in a silyl-mediated dehydrogenation failed, but the reaction with DDQ showed promising results. The oxidation of acid **8** was performed using the strong silylating agent bis-(trimethylsilyl)trifluoroacetamide (BSTFA) with DDQ in hexamethyldisilazane (HMDS). Heating at 110 °C for 18 h produced enone **17** in a remarkable 89% assay yield. This was the lead that we had been looking for.

Repeating the reaction in HMDS without BSTFA gave 78% assay yield. The reaction could also be performed with BSTFA in other solvents. Of the solvents screened, dioxane gave the highest yields and cleanest reaction. The oxidation was also successful with amide **10** and ketone **31**, as shown in Scheme 3.10. Reactions

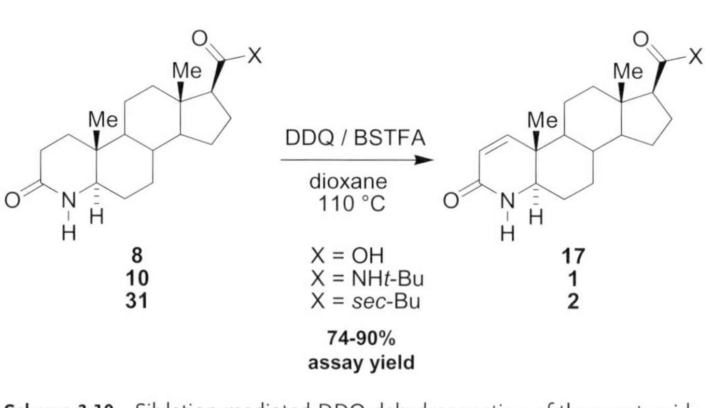

Scheme 3.10 Silylation-mediated DDQ dehydrogenation of the azasteroids.

with the *sec*-Bu ketone **31** showed only dehydrogenation of the lactam ring. Although we recognized that dioxane would not be suitable as a solvent for production, the reaction clearly showed promise as a practical solution to the Δ^1 double bond problem.

To complete the synthesis of finasteride, carboxylic acid **17** in 10 ml (gTHF)$^{-1}$ was activated with CDI (1.02 equiv) to form the acyl imidazolide **18** (Scheme 3.11). Without isolation, the acyl imidazolide was reacted with 4.6 equiv of *t*–BuNHMgBr heating to reflux in THF for 18 h to give finasteride in 98% yield [9].

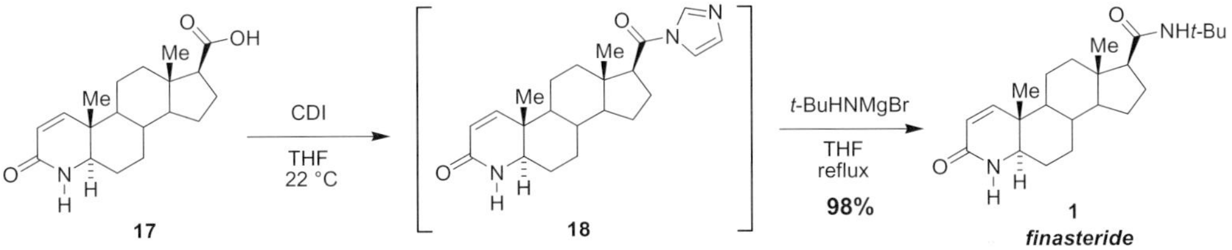

Scheme 3.11 Preparation of finasteride from carboxylic acid **16** via the acyl imidazolide.

3.1.1.2.3 The Manufacturing Process – Targeting the Ester as a Late-Stage Intermediate

Although we had made good progress toward a practical synthesis that could be used to make finasteride, solubility limitations for the carboxylic acid intermediates and the cost of CDI kept us thinking about how to further improve the process. Purification of the acid **17** by extraction following dehydrogenation was particularly challenging; we could not effectively use basic extractions to remove the acidic hydroquinone by-product without losing an unacceptable amount of the carboxylic acid product to the aqueous layer. The poor solubility of the acid limited productivity. We needed a simple way to mask the carboxylic acid early in the synthesis that would allow conversion to both amide and ketone in the last step. Esterification, as shown in Scheme 3.12, was considered the most practical means of masking the acid, but clean, efficient conversion of ester **33** to both amide **1** and ketone **2** was not assured.

Preparation of methyl ester 32 Esterification of the carboxylic acid **8** in MeOH does not proceed to completion using a catalytic amount of H_2SO_4. The A-ring of the azasteroid is opened to form the hydrogen sulfate salt of the dimethyl ester **34**, as illustrated in Scheme 3.13. Following addition of toluene, isopropanol, and water, the pH is adjusted to 8–9 by neutralizing first with NaOH, to avoid foaming, followed by the addition of $NaHCO_3$ to reproducibly give pH 8–9. Ring closure to give **32** takes about 1 h.

A key discovery for optimization of the dehydrogenation reaction In studying the dehydrogenation reaction, an observation led to a discovery that was key in

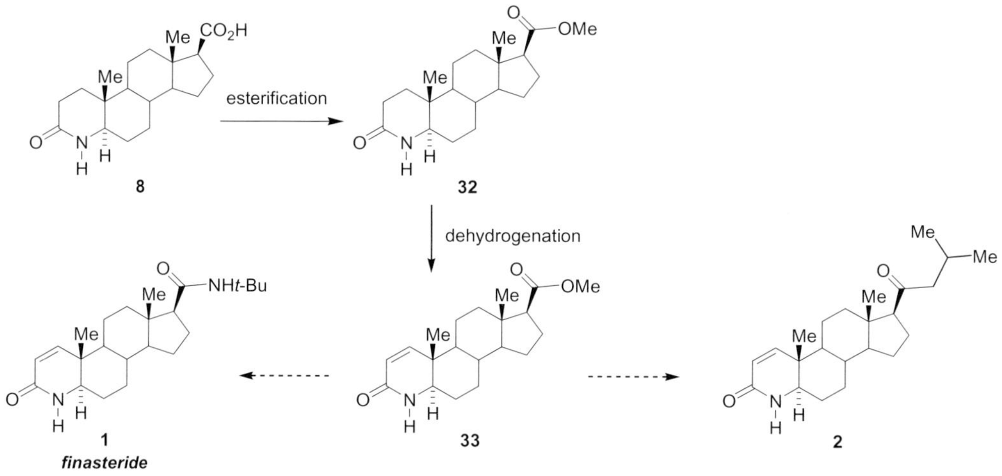

Scheme 3.12 A change in strategy and a new key intermediate.

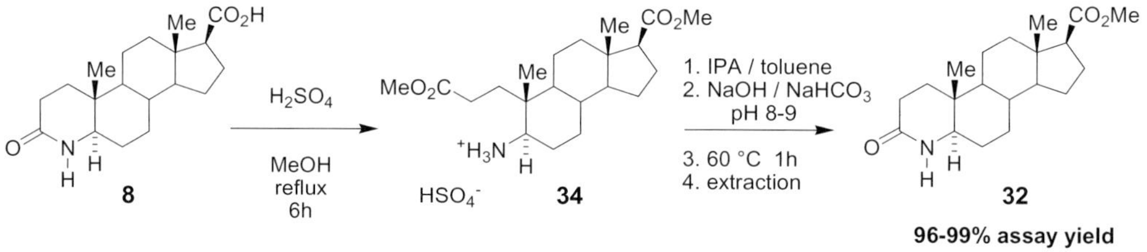

Scheme 3.13 Esterification.

developing this reaction. The reaction was originally performed by charging the reactants and immediately heating to 110 °C. When we ran the reaction at lower temperatures and monitored conversion by HPLC, we were surprised to find that the starting material **8** was consumed but the product **17** did not appear until the reaction was heated. NMR studies revealed that adducts were formed with DDQ (Scheme 3.14). This observation was key in changing our thinking about this reaction and how to make it work well.

Two types of adducts, *C–C* (**35**) and *C–O* (**36**) adducts, were identified and characterized by ^{1}H and ^{13}C NMR [10]. A pair of diastereomeric *C–C* adducts **35** were the predominant products formed. These adducts were converted to the desired double-bond-containing product with heating. This reaction most likely involves a pericyclic elimination (**38**) which is favored by formation of the aromatic hydroquinone **37**, as highlighted in Scheme 3.14.

A single *C–O* adduct **36** is formed in the reaction; this adduct is not converted to the desired product. Coupling constants and the results of an nOe experiment

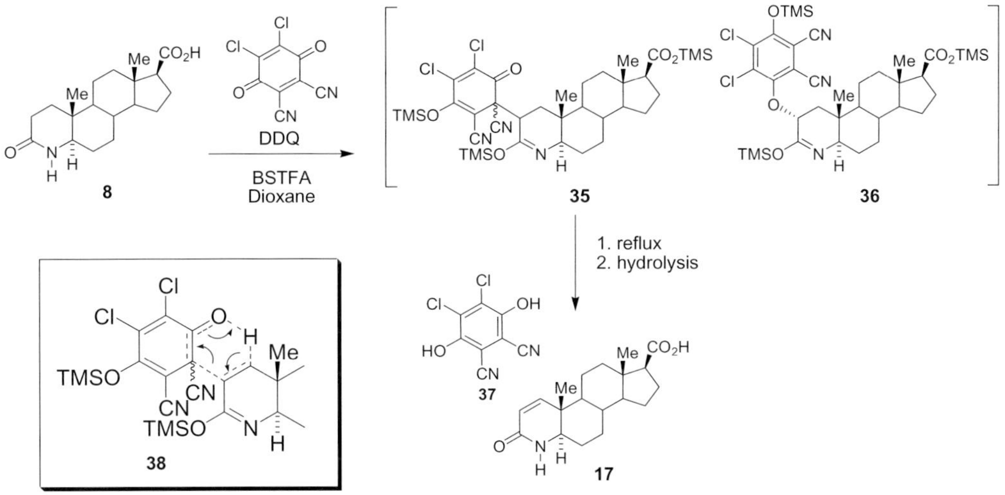

Scheme 3.14 Adducts in the silylation-mediated reaction of DDQ with the azasteroid.

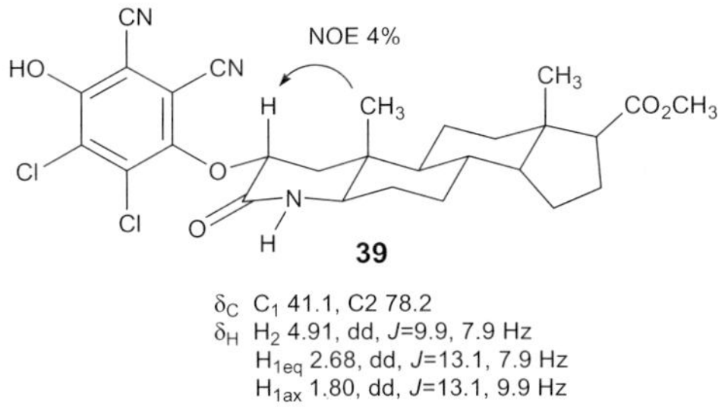

Figure 3.3 NMR characterization of the *O*-adduct.

were consistent with adduct formation on the face opposite the angular methyl group, as shown in Figure 3.3.

Development of the dehydrogenation reaction There remained obstacles in making the dehydrogenation reaction practical for manufacturing. The reaction worked best in dioxane, not a suitable solvent for production. BSTFA and DDQ were both expensive and difficult to obtain in the purity required. Considerable effort was invested at this stage in search of a reagent/solvent system that was practical for scale-up.

We hoped to define conditions that would allow us to use toluene as solvent for the reaction. The toluene solution of **8** could be taken directly into the dehydrogenation following distillation to remove water and alcohol solvents. The boiling point of toluene was suitable for the thermolysis but silylation in toluene was very slow.

A breakthrough came when trifluoromethanesulfonic acid (triflic acid, TfOH) was found to catalyze the silylation reaction in toluene. Using TfOH in toluene, silylation is complete within minutes at 25 °C. Other strong acids, FSO_3H and MsOH gave lower overall yields in the dehydrogenation.

During the dehydrogenation, DDQ reacts with toluene to give **40** consuming up to 6% of the total DDQ charge during the 25 °C stage even though toluene solutions of DDQ were stable at 25 °C for several days. The reaction is observed only on addition of BSTFA and TfOH. Hydrolysis of **40** gives **41** which could be removed by basic extraction (Scheme 3.15).

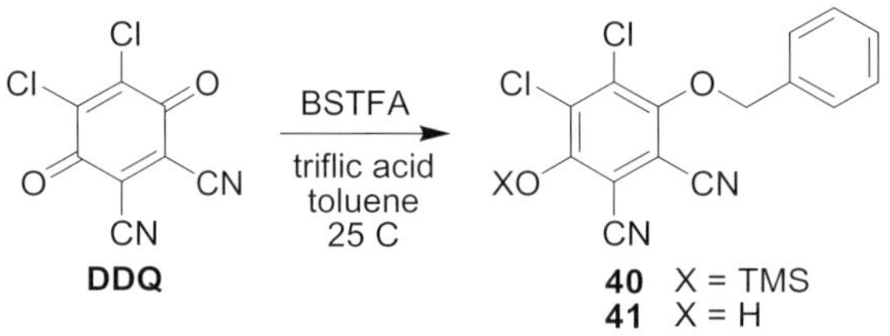

Scheme 3.15 The reaction of DDQ with toluene.

BSTFA provided the highest yield of product in comparison to other silylating agents. BSA [bis(trimethylsilyl)acetamide] reacted with DDQ. TMSOTf with lutidine or collidine in toluene offered the best alternative but the yield was lower than the yield achieved with BSTFA.

Alternatives for DDQ were also considered. Of the other quinones investigated, *o*-chloranil emerged as the most promising. The yield was dependent on solvent, concentration, and TfOH charge. The best results were obtained at high concentration in 1,2-dichlorobenzene or 1,2-dichloroethane using 20 mol% TfOH. The yield with *o*-chloranil following extensive optimization, however, did not match the best yield observed with DDQ (85–90% assay yield cf. 94% for DDQ).

It was clearly important to drive adduct formation to completion; any unreacted starting material **32** was not rejected in isolation of the product **33** and the result was an impurity in the finasteride product. Excess DDQ was required to ensure complete conversion of **32**, but any DDQ remaining after adduct formation produced over-oxidation products in the thermolysis. The three over-oxidation products shown in Figure 3.4 were identified. Sequential dehydrogenation in the B-ring

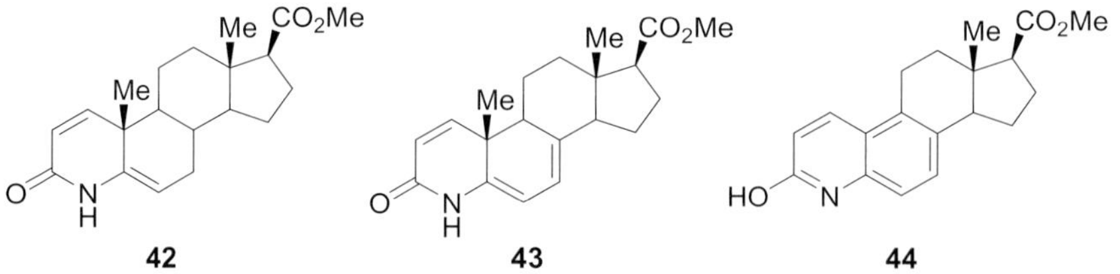

Figure 3.4 Three over-oxidization products.

gives **42** and **43**. Remarkably, further oxidation leads to loss of the angular methyl giving the aromatic compound **44**. Reference samples of each were prepared to allow quantitation of these impurities [11].

If unreacted DDQ was quenched prior to thermolysis, over-oxidation products could be limited to less than 1%. 1,3-Cyclohexanedione was used to quench unreacted reagent prior to thermolysis. Purity following thermolysis was significantly improved and no problems were found on the laboratory scale. The fully optimized conditions are provided in Scheme 3.16. The reaction was run on pilot scale and performed as expected.

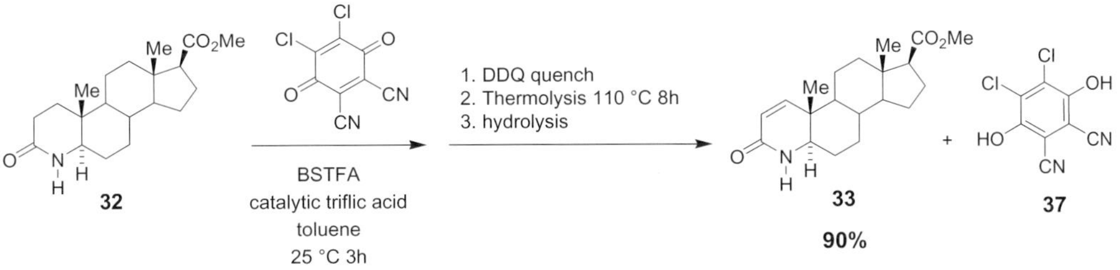

Scheme 3.16 Fully optimized dehydrogenation using DDQ.

In the second campaign in our pilot plant, however, there was a surprise. Adduct formation was completed with less than 0.1% starting material **32** remaining as expected, but after thermolysis the chemical engineers came to us with a problem. An assay of the reaction showed about 1% (and not 0.1% as we expected) of **32**! Considering that **32** would not be rejected in isolation of the product, this was a big problem.

The first thing to do was to sample the reaction again to confirm the result. The result was confirmed. At that point we started looking for an explanation and our first thought was to blame the equipment. Maybe a charge line held starting material so that it was introduced to the reactor during thermolysis. The engineers assured us that this was not possible. Could there be a problem with the chemistry?

The next step was to reproduce the problem in the laboratory. Using all the materials from the pilot plant, including a solution of the starting material from the esterification, we saw complete conversion in the DDQ adduct formation but 1% starting material reappeared following thermolysis, just as in the pilot plant. So what was different from our prior laboratory experience where we had never seen this problem? The carboxylic acid **8** from a previous pilot plant campaign had been esterified in the laboratory and carried through the dehydrogenation with no problem. Analytically, there appeared to be no difference in the starting materials used, and then we discovered 10 ppm Pd from the hydrogenation in the most recent batch of ester **32** prepared in the pilot plant. Adding palladium to a Pd-free batch prior to the thermolysis also resulted in reduction of the product. Apparently, silylation of 1,3-cyclohexanedione **46** under the reaction conditions was giving the silyl enol ether **47** which was serving as a hydrogen source for the Pd-catalyzed reduction of **49** to **50**, as shown in Scheme 3.17.

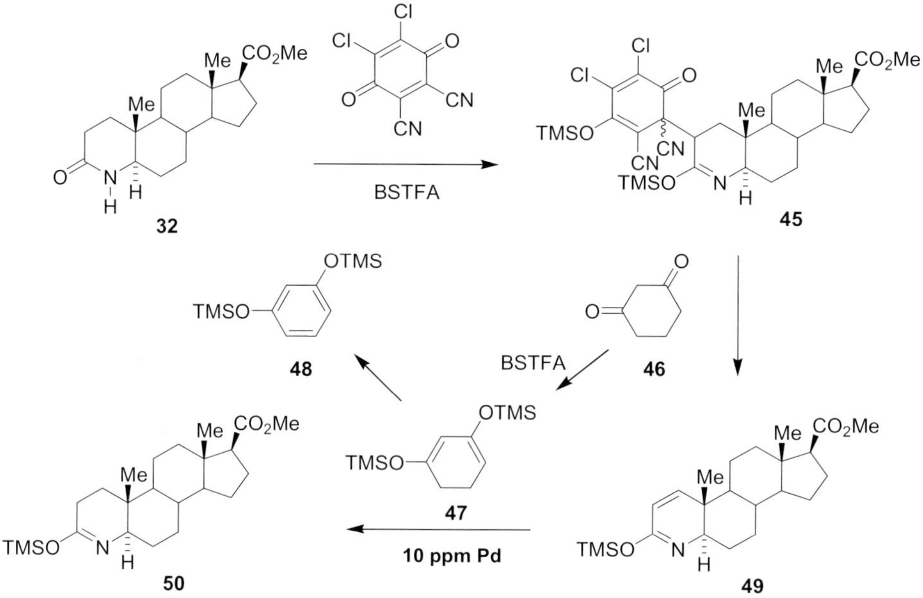

Scheme 3.17 Cyclohexanedione as a transfer hydrogenation agent.

The batch containing 1% starting material was resubjected to the dehydrogenation reaction using a reduced charge of DDQ and the batch was saved. This story has been told many times and has become legend in Merck Process Research. There are clearly some important lessons here that every process chemist should learn – and best not the hard way. We later showed that methyl acetoacetate could be used as a quench, eliminating the potential for reduction.

Committed to reducing the cost of the process and minimizing the waste stream, we developed an efficient procedure for recovering the hydroquinone by-product of the dehydrogenation reaction and oxidizing it to DDQ that could be reused. Acidification of the aqueous waste from extractions following the dehydrogenation gave a 96% recovery of the hydroquinone **37** (Scheme 3.18). Oxidation

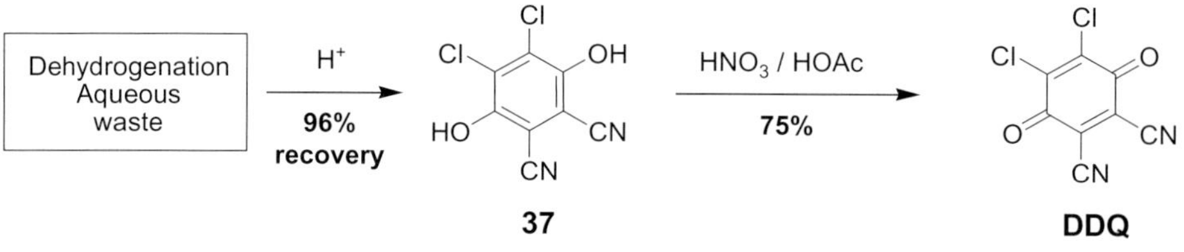

Scheme 3.18 DDQ recycle.

using nitric acid in AcOH afforded high quality DDQ (99% pure) in 75% yield. Performance in the dehydrogenation reaction using recovered DDQ was comparable to commercial DDQ.

Amidation–the Bodroux reaction Finally, amidation of methyl ester **33** was required to complete the process. To accomplish this transformation, we turned to a reaction first reported in 1904 by Bodroux [12]. The reaction involves activation of an amine by conversion to the magnesium amide using a Grignard reagent. Subsequent reaction of the magnesium amide with the ester produces the carboxamide.

We knew that it would be critical to achieve high conversion; any unreacted starting material would be isolated as an impurity in the product. We found that deprotonation of the lactam in some cases led to precipitation of the resulting salt, limiting conversion of ester to amide. Also, deprotonation of *t*-BuNH$_2$ gives salts that are not freely soluble in the solvents that could be used for this reaction making transfer difficult. To overcome these problems, we found that the Grignard reagent could be charged directly into a slurry of ester **33** containing *t*-BuNH$_2$ in THF (Scheme 3.19). EtMgBr gave better results than the chloride. The lithium amide could be pre-formed but did not give good conversion in the reaction, possibly due to solubility problems in the deprotonation of **33**. The magnesium salt of **33** is soluble in THF and the lithium salt is not.

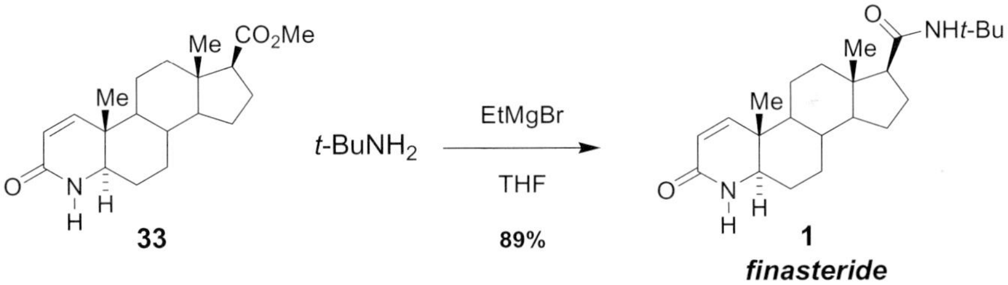

Scheme 3.19 The Bodroux reaction.

The reaction is carried out by consecutively adding *t*-BuNH$_2$ and EtMgBr to a slurry of **33** in THF at 5 °C. The temperature is maintained below 10 °C to avoid formation of ketone **51** and alcohol **52** (Figure 3.5). The reaction will not go to completion at lower temperatures and reflux is required to reduce the level of **33** to less than 0.1%. In theory, the reaction requires 3 equiv of EtMgBr. In practice, at least 4 equiv were required to drive the reaction to completion. The amount of *t*-BuNH$_2$ charged is critical. Excess *t*-BuNH$_2$ slows the reaction. When the reaction was conducted using 3.6 equiv of EtMgBr and 6.0 equiv of *t*-BuNH$_2$, only 10% conversion had been achieved after 6 h at reflux. With 3.6 and 8.2 equiv of the reagents, respectively, there was less than 5% product formed after 6 h at reflux.

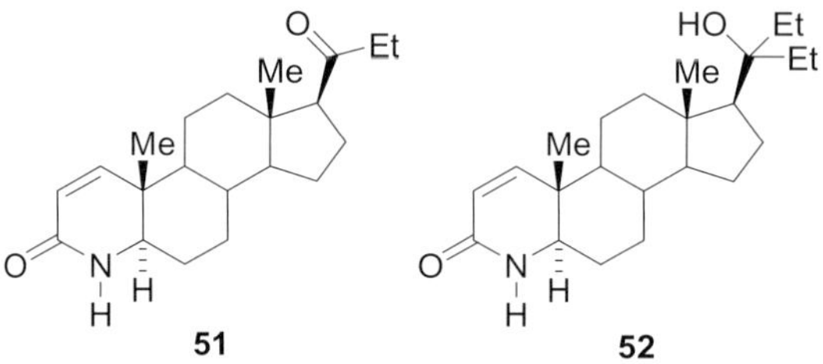

Figure 3.5 Impurities formed in the amidation reaction.

The over-oxidation products produced in the dehydrogenation reaction, **42**, **43**, and **44**, are converted to the corresponding amides in the amidation. Due to the high optical rotation of these impurities, trace levels in finasteride can result in the optical rotation being out of specification. Crystallization does not remove these impurities at any stage. We found that the level of these impurities could be reduced using activated carbon or by taking advantage of the propensity of the Δ^5 double bond to hydrate under mildly acidic conditions.

A carbon treatment also removes residual color. THF proved to be the best solvent for this operation. One antioxidant often used to stabilize commercial THF is BHT **53** (butylated hydroxy toluene). The carbon promotes oxidation of the BHT to give a yellow dimer **54** (Scheme 3.20). After sufficient washing of the carbon with THF, the dimer is no longer formed. Although the dimer is removed in crystallization of the product, it is best to wash the carbon prior to use.

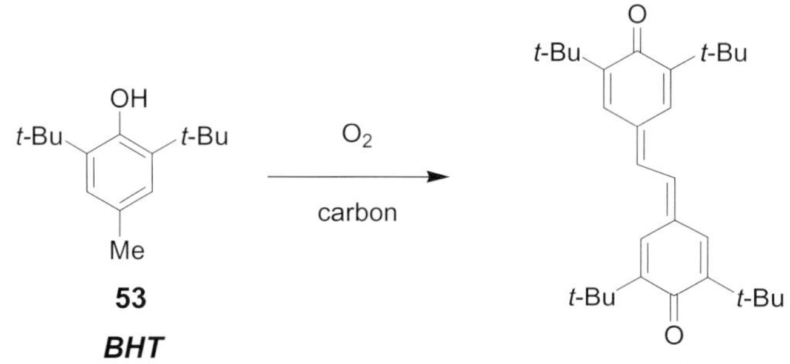

Scheme 3.20 Oxidation of BHT.

Our Chemical Engineering colleagues had developed an elegant impinging jet crystallization which provided excellent particle size control for the finasteride process [13]. In the final pilot plant campaign just before the factory start-up, the crystallization suddenly started producing a different particle size distribution and lower recovery. The problem was traced to a new finasteride solvate which reduced the solubility in the crystallization solvent system. Fortunately, only relatively minor adjustments were required to bring the process back under control and delay was once again averted.

3.1.1.2.4 **Factory Start-up**

A factory was designed and built specifically for the finasteride process. Performance through the dehydrogenation was comparable to our best pilot plant experience. Laboratory runs of the final step, however, revealed a problem; the product was pink. There was obviously some difference in performance that we had not recognized.

After investigating numerous possible explanations, we concluded that the problem resulted from less efficient rejection of impurities in the crystallization of enone ester **33**. The design of the vessel used for crystallization of **33** in the factory appeared to be the cause of the problem. The particle size was smaller compared with material produced in the pilot plant. Filtration was slower and washing the solids was not efficient in removing the mother liquors. The color of the finasteride from different lots of **33** roughly correlated with the level of the *O*-adduct impurity. Subjecting isolated *O*-adduct to the amidation conditions gave predominantly two intensely colored isomeric products (76 : 24). The major product **55** was identified using ^{15}N NMR (Scheme 3.21). The presence of only trace amounts of these impurities was sufficient to give the finasteride bulk a pink color. The problem was resolved by making relatively minor adjustments in the process and the start-up came to a successful conclusion.

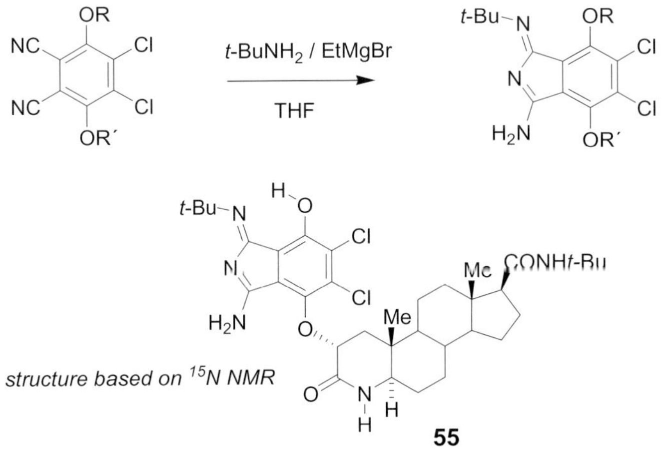

55

Scheme 3.21 Intensely colored impurities.

The final manufacturing process consists of four chemical steps with two isolated intermediates, as illustrated in Scheme 3.22. The overall yield is 74%. A commercial source for the ene lactam **7** was developed.

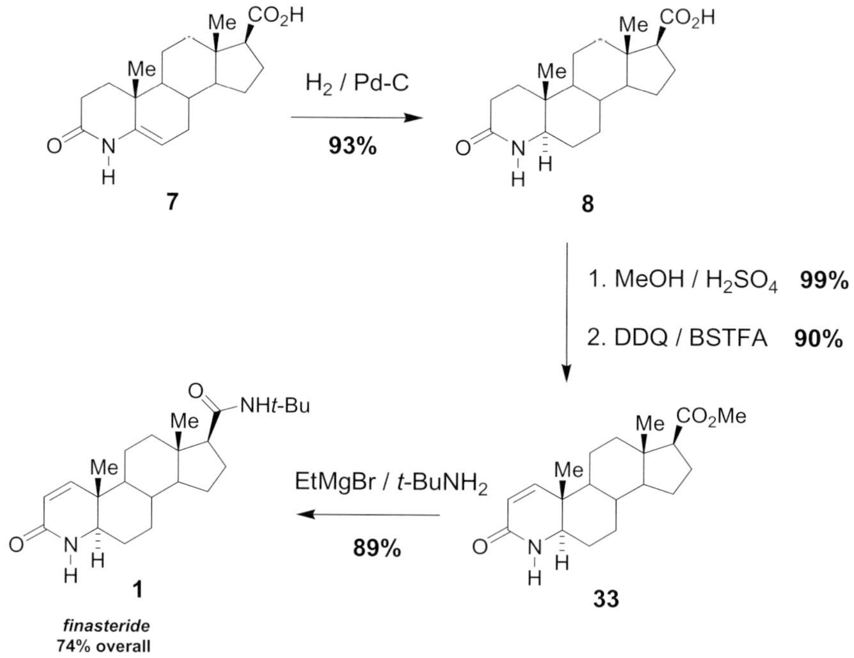

Scheme 3.22 Finasteride manufacturing process.

3.1.2
The Second Generation Candidates

3.1.2.1 The Medicinal Chemistry Route

During the development of finasteride, three back-up candidates differing from finasteride only in the C17 position of the azasteroid core were identified. Each of these ketones presented a different synthetic challenge. All had been prepared in Medicinal Chemistry from the thiopyridyl ester **9**, followed by dehydrogenation using $(PhSeO)_2O$, in good yields (the yield in preparation of the *s*-Bu ketone **2** was 71%) as illustrated in Schemes 3.1 and 3.23 [14]. This method, however, required

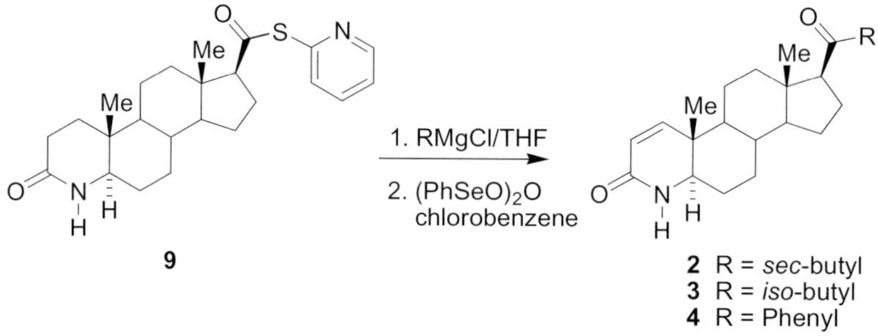

Scheme 3.23 The Medicinal Chemistry synthesis of the C17 ketones.

chromatographic purification and was not practical for scale-up. These compounds were developed in parallel with finasteride and the most efficient strategy was to make use of an intermediate from the finasteride process as starting material. With changes in the finasteride process came new challenges and opportunities for synthesis of the next-generation candidates.

3.1.2.2 Process Development

3.1.2.2.1 Saturated Acyl Imidazolide Route

With the discovery of a new method for introducing the required double bond, two routes to the *s*-Bu ketone **2**, the lead second generation candidate at the time, were evaluated (Scheme 3.24). Both routes relied on conversion of an acyl imidazolide to the ketone [15]. The conversion of acid **8** to acyl imidazolide **56** required heating in DMF and gave product in 90% yield. At least 4 equiv of the reagent was required for complete reaction and the secondary alcohol accounted for 40 to 60% of the product, depending on how the reaction was quenched (adding acid into the reaction mixture gave more alcohol).

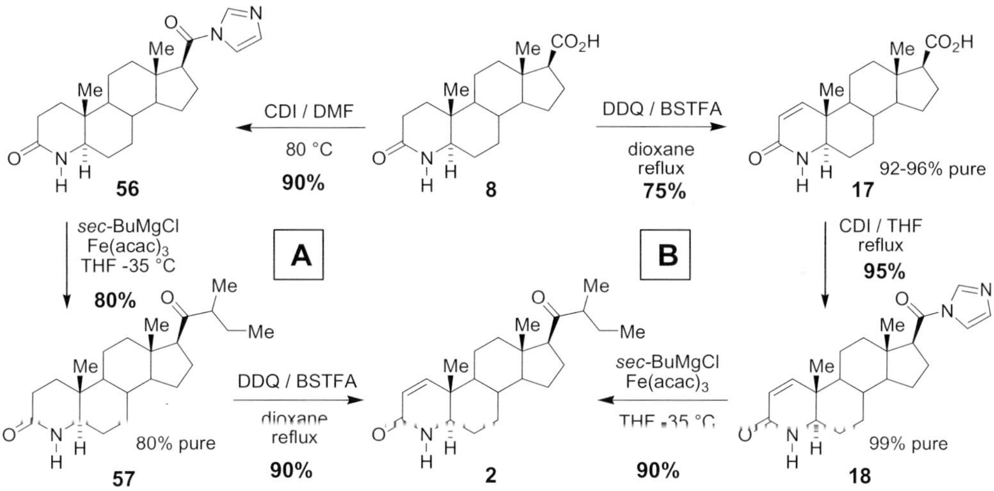

Scheme 3.24 A comparison of routes to the s-Bu ketone.

Marchese had recently reported that 3 mol% Fe(acac)$_3$ suppressed the formation of secondary alcohols in the reaction of a Grignard reagent with an acid chloride [16]. Using 30 mol% Fe(acac)$_3$ at 0 °C reduced the amount of alcohol formed to 1% and improved the yield to 80%. The Fe(acac)$_3$ charge could be reduced to 10 mol% by running the reaction at −35 °C. The reaction, however, required 5 equiv of Grignard and produced about 2% of an impurity that resulted from reaction of acetylacetonate with acyl imidazolide **56**. Finally, dehydrogenation using DDQ gave the drug candidate **2** (path **A**, Scheme 3.24).

3.1.2.2.2 Unsaturated Acyl Imidazolide Route

Reaction of **17** with CDI to give the acyl imidazolide **18** could be carried out in THF rather than DMF as required for **8**. The reaction did not become homogeneous as a result of the acyl imidazolide crystallizing and we found conversion to be variable. This problem was solved by using dichloromethane as solvent, which afforded a homogeneous reaction endpoint. Following completion of the reaction, the solvent was switched to THF to crystallize **18** in 95% yield. Reaction of **18** with *s*-BuMgCl was also improved by using $Fe(acac)_3$ (path **B**, Scheme 3.24).

The *s*-Bu ketone **2** was prepared as a mixture of diastereomers. Ketones lacking this liability were showing comparable activity and pharmaceutical properties, and the decision was made to drop **2**. The *i*-Bu and *i*-Pr ketones (**3** and **58**) were the clear favorites among a number of possibilities and we began work on these compounds (Figure 3.6).

There were several problems to address in developing the conversion of the unsaturated acyl imidazolide **18** to these ketones. The acetylacetonate ligand was found to add to **18** leading to more than 2% of a by-product, believed to be **59** (Figure 3.7), which proved difficult to remove. The reaction also consumes far more organomagnesium reagent than should be necessary; 5 equiv are required for complete conversion (the theoretical is 2.0). Also, the reaction provided best results when carried out at low temperature (−35 °C).

In an effort to overcome these problems, a series of catalysts was compared with the results provided in Table 3.1. Ferric chloride gave a dramatic improvement

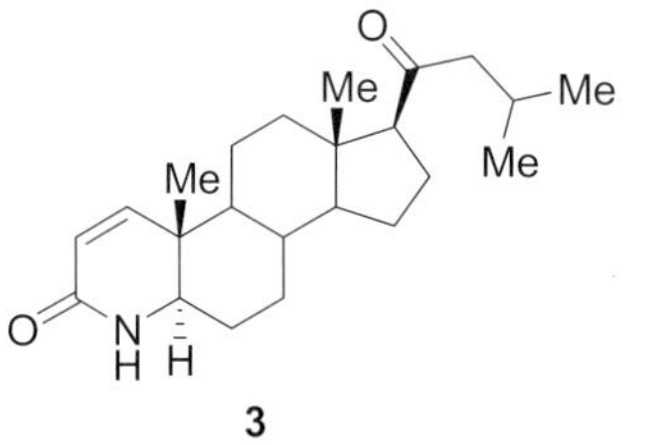

3

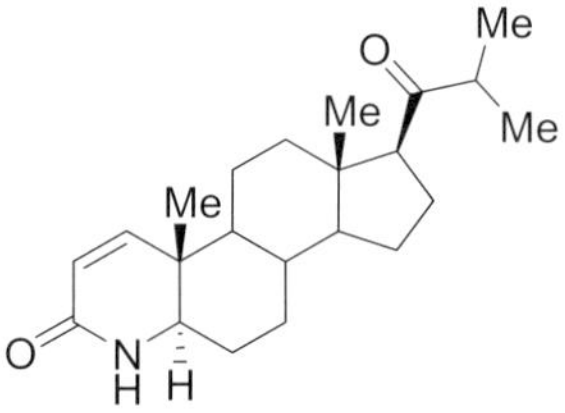

58

Figure 3.6 New targets.

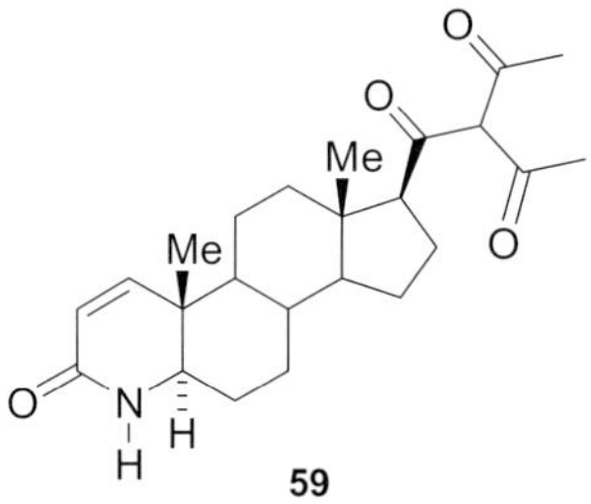

59

Figure 3.7 Proposed by-product formed in the $Fe(acac)_3$ reaction.

Table 3.1 Catalysts for the ketone synthesis.

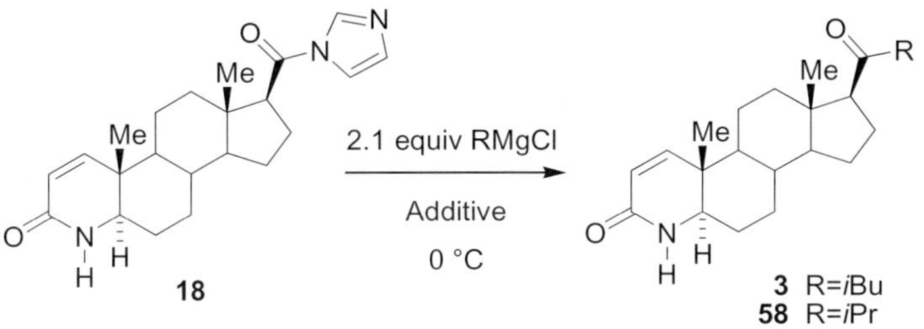

	$3^{a)}$	$58^{a)}$
No catalyst	38 (58)	79 (84)
$Fe(acac)_3$	27 (61)	58 (77)
$FeCl_3$	77 (88)	80 (83)
CuCl	28 (48)	70 (86)

a) % yield (area % purity excluding **18**).

over $Fe(acac)_3$ for the isobutyl ketone **3**. The control experiment with no catalyst, showed that there was no advantage in using a catalyst for preparation of the isopropyl ketone **58**.

At this point, the isobutyl ketone **3** was chosen as the new second generation development candidate. After quickly defining reaction conditions and a work-up, the acyl imidazolide reaction was scaled up. Complete conversion could be achieved using 3.0 equiv of *i*-BuMgCl and 10 mol% $FeCl_3$ in THF at 0 °C. The major side-product was the aldehyde **60** (Figure 3.8) [17]. The amount of **60** observed ranged from 2 to 6% and depended on the quality of **18** and the temperature of the reaction. Unexpectedly, more aldehyde was observed at lower temperature compared with the reaction at 0 °C. The product was crystallized directly by quenching the reaction into HCl and isolated by filtration in 86% yield on the laboratory scale. On scale-up, the yield was somewhat lower at 79%. A series of crystallizations was required to get the purity up to 99.0% with painful losses reducing the overall yield to 48%.

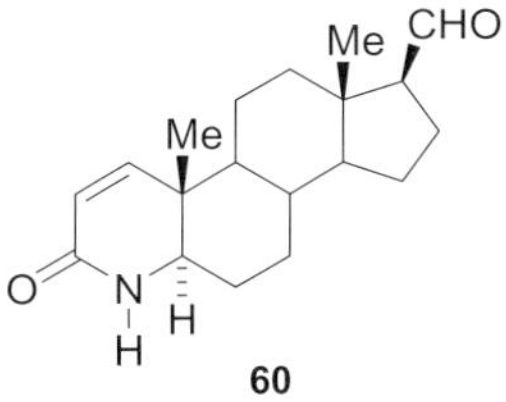

Figure 3.8 Aldehyde side-product.

3.1.2.2.3 The Methyl Ester as a Key Intermediate for Divergent Synthesis

The alkyl ketone candidates With definition of the finasteride manufacturing process came a new challenge. The most efficient synthesis of the second generation candidate would make use of the penultimate in the finasteride process **33** as starting material (Scheme 3.25). With a pilot plant campaign to make **3** scheduled, realizing this objective became a priority. None of the methods reported in the literature for ester to ketone conversion had been applied to a hindered steroidal C17 ester.

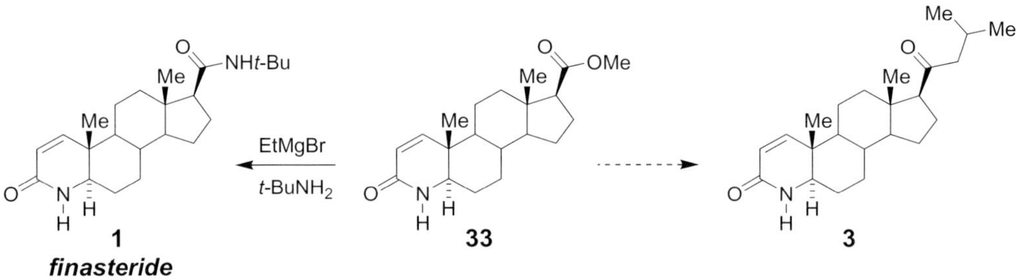

Scheme 3.25 Improved efficiency through divergent synthesis.

In preliminary experiments, the ester **33** was reacted with organolithium and magnesium reagents. The reaction of *i*-BuLi with **33** at low temperature gave predominately tertiary alcohol **61** with only traces of ketone **3** (Figure 3.9). The ester **33** was found to react slowly with *i*-BuMgCl in THF requiring 3 days at 20–25 °C to consume **33**. The principle side reaction was formation of the tertiary alcohol **61** (~15%), which was easily rejected in crystallization of the ketone product. Only traces of the diastereomeric secondary alcohols **62**, the main side product in reaction of the acyl imidazolide **17** with *i*-BuMgCl, were observed. The product **3** could be isolated in good purity and only the yield suffered as a result of the amount of **61** that was formed. Given that **33** was quite valuable at this stage of the process, we hoped to improve the yield by minimizing formation of **61**.

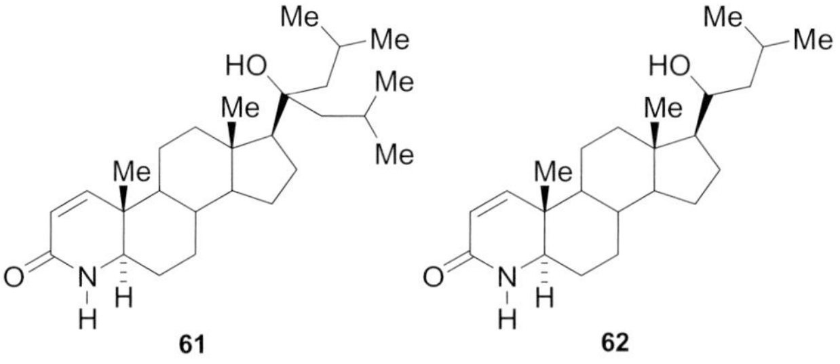

Figure 3.9 Alcohols from reaction of **33** with *i*-BuMgCl.

Continuing to age the reaction after the ester **33** had been consumed or adding more *i*-BuMgCl did not convert the ketone **3** to alcohol **61**. We believed that the ketone was protected as the magnesium enolate **63**, as shown in Figure 3.10. If this were true, then there must be a competition between deprotonation to give the enolate and addition to give the tertiary alcohol [18].

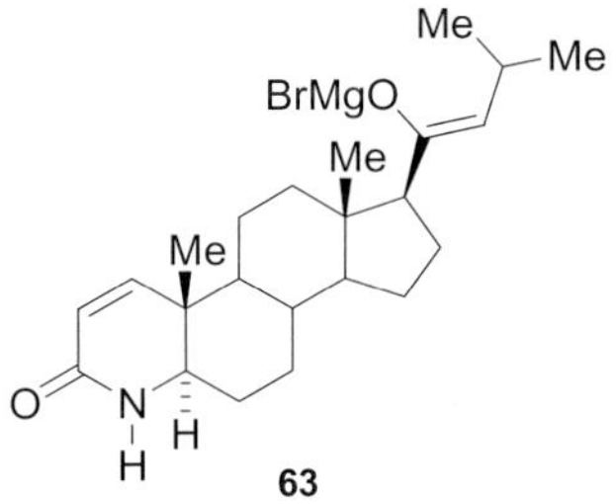

63

Figure 3.10 The magnesium enolate.

Chelating ligands for magnesium were expected to influence the relative rate of deprotonation and addition [19]. TMEDA gave a modest improvement in the reaction affording a 92:8 ratio of **3**:**61** under the best conditions. *N,N′*-Dimethylimidazolidinone was not as effective in suppressing the formation of **61**. No reaction was observed in DME, even on heating to 50 °C.

Screening amide and alkoxide bases produced a lead [20].[2)] With *t*-BuOK or potassium hexamethyldisilazide (KHMDS), the level of **61** was reduced from 15% to less than 2%. Higher conversion was achieved using KHMDS, but the availability and lead times for the quantity that would be required for scale-up were an issue. Substituting hexamethyldisilazane (HMDS) for KHMDS gave a ratio of **3** to **61** that was nearly as good (94:6) with comparable yield and impurity profiles. Commercial suppliers of *i*-BuMgCl at that time provided the reagent only in ether solution, which would have created environmental and safety isssues. *i*-BuCl was nearly three times the cost of *i*-BuBr on a molar basis. These considerations led to the decision to make *i*-BuMgBr in THF in house.

Optimization of the reaction using *i*-BuMgBr and HMDS focused on charges and temperature. Increasing the temperature increased the rate without increasing the amount of **61**. The reaction time was reduced from days at 20 °C to 4 h at reflux. Although the theoretical amount of *i*-BuMgBr required should be 5 equiv, at least 8 equiv of *i*-BuMgBr was required to drive the reaction to completion, even at reflux. Three equivalents of HMDS was required to minimize formation of **61** and maximize yield.

To verify that the ketone is enolized under the reaction conditions, reactions were carried out with and without HMDS quenching both with AcOH-d_4. ^{13}C NMR and MS analysis of the product mixtures showed complete mono-deuteration at

2) House and Traficante suggested that magnesium alkoxides promote deprotonation of ketones in the reaction with Grignard reagents leading to by-products. See Reference 18(a).

C21 in ketone **3**. No deuterium incorporation at C17 was observed and only one of the two possible diastereomeric monodeutero ketones was observed.

With the reaction performance improved, attention turned to isolation of **3**. The work-up of this reaction was complicated by the solubility of THF in water and the low solubility of **3** in most organic solvents. Using extraction to remove residual magnesium salts would have severely limited volumetric productivity. We found that **3** could be isolated by quenching the reaction into aqueous HCl. Crude **3** was isolated after concentrating the organic layer. Residual THF and magnesium salts were then removed by recrystallization from AcOH/water with less than 1% loss of **3**.

On pilot plant scale, the reaction, shown in Scheme 3.26, gave a yield of 83% (cf. 86% laboratory yield). The overall yield for conversion from the acid **8** was improved from 40 to 75% by this route compared with the route through the acyl imidazolide.

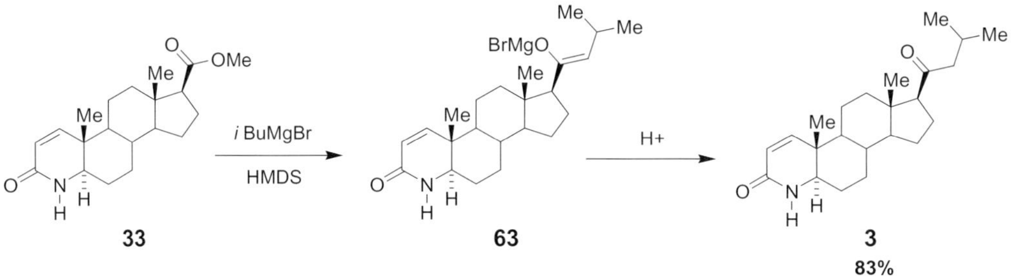

Scheme 3.26 Conversion of the ester to isobutyl ketone–enolate formation.

This method was later adapted for the large-scale preparation of the LTD_4 antagonist **64** by another Merck Process Research group (Figure 3.11) [21]. Conversion of a methyl benzoate to the corresponding acetophenone was required. Formation of the tertiary alcohol was again minimized with the addition of HMDS and excellent reaction performance was achieved.

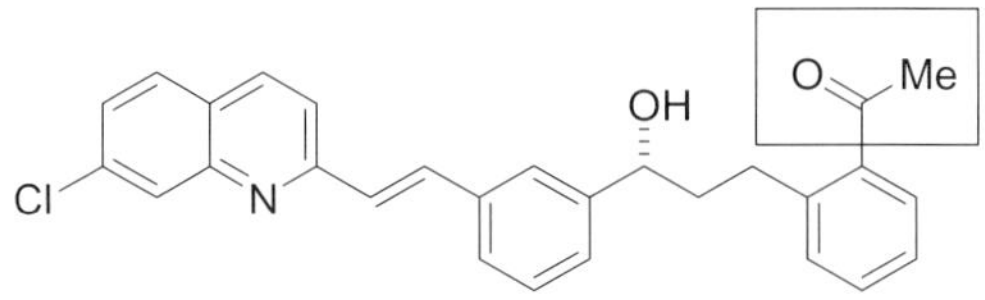

64

Figure 3.11 An LTD_4 Antagonist prepared through conversion of an ester to the methyl ketone.

The phenyl ketone – a new second generation candidate The problem in making the isobutyl ketone **3** had been solved by designing a reaction in which the product was rapidly deprotonated to prevent addition to the ketone carbonyl. When development of **3** was discontinued, the phenyl ketone **4** was approved for development. As shown in Scheme 3.27, we again hoped to make use of methyl ester **33** as starting material. Of course, a new strategy would be required.

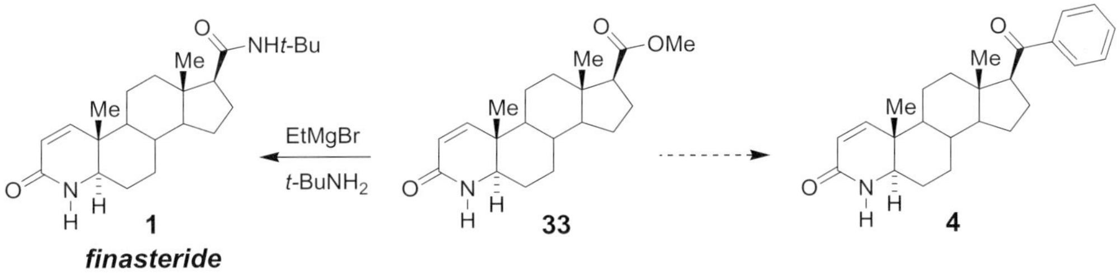

Scheme 3.27 Divergent synthesis of the phenyl ketone.

Preliminary results using only 1 equiv of PhMgBr in reaction with **33** gave the tertiary alcohol and not ketone **4** as the major product. Considering options for suppressing formation of the tertiary alcohol from among the methods reported in the literature, it seemed that the best approach would involve generating a stable tetrahedral intermediate. A method for the conversion of esters to *N*-methoxy-*N*-methyl amides had been reported by Weinreb and Nahm [22]. These amides, now commonly known as Weinreb amides, could be reacted with organomagnesium reagents to give the ketone **4**. A stable tetrahedral intermediate was proposed to account for the low level of addition products observed. The method for preparation of the Weinreb amides involved the reaction of Me_3Al with *N*-methoxy-*N*-methylamine HCl (Weinreb amine) to give the aluminum amide which was then reacted with the ester. The Weinreb amide was isolated following an aqueous work-up. Our attempts to apply this reaction to the azasteroid gave none of the desired Weinreb amide **65** (Scheme 3.28). We thought that **33** might be too hindered and the aluminum amide not sufficiently reactive.

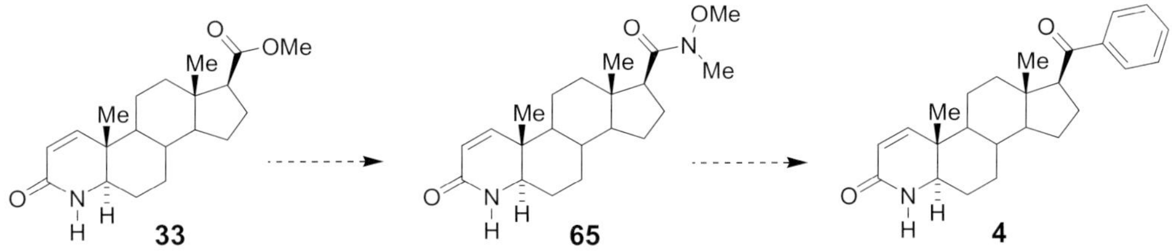

Scheme 3.28 The Weinreb amide strategy.

From our experience with the Bodroux reaction, we knew that the magnesium amide of *t*-$BuNH_2$ reacts with **33**. Recall that, in order to avoid solubility problems in preparation of the magnesium amide reagent, EtMgBr was added to a mixture of **33** and t-$BuNH_2$ in THF. Unsure whether solubility would be a problem, two reactions were conducted at the same time. In one experiment, we attempted to pre-form the magnesium amide reagent by adding EtMgBr to Weinreb amine in THF. The ester **33** was subsequently added but none of the desired product was observed. A second experiment, following the finasteride amidation protocol in which EtMgBr was added to a suspension of Weinreb amine and ester **33** in THF, produced a remarkable 91% yield of the Weinreb amide **65** [3a].

It appears that the magnesium amide reagent **66** formed from Weinreb amine is not stable and undergoes unimolecular decomposition or reacts with the organomagnesium through displacement of methoxide (Scheme 3.29) [23]. Had we known this at the time, the key reactions might never have been run. We concluded that the magnesium amide **66** is formed but must react with the ester faster than it is consumed through other reactions.

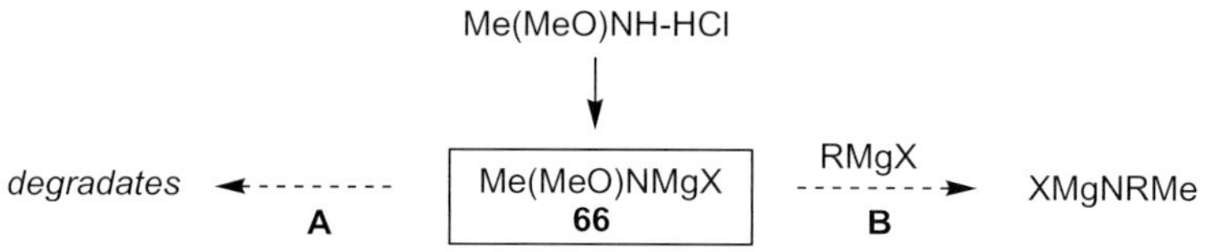

Scheme 3.29 Reactions potentially explaining poor conversion with preformed reagent[3].

We showed that the isolated Weinreb amide **65** could be efficiently converted to the phenyl ketone **4** with less than 2% of the tertiary alcohol formed. We had originally hoped to develop a direct conversion from **33** to **4**, and although the two-step procedure solved the problem, this solution did not meet our original goal. Then it occurred to us that the stability of the two tetrahedral intermediates **67** and **68** might be different enough that we could avoid isolating **65** (Scheme 3.30) [24]. To test this idea, a reaction was performed in which PhMgCl was added to a mixture of **33** and Weinreb amine at 5 °C. Within minutes, ester **33** was consumed and amide **65** was observed by HPLC analysis. On warming to 20 °C, **65** was converted to **4**. Complete conversion required 8 h, but less than 2% of the tertiary alcohol was formed. The major steroidal side-product was identified as the aldehyde **60**. It was interesting that the aldehyde was not observed when isolated amide **65** was reacted with PhMgCl, suggesting that it is ester **33** that was being reduced [25]. The aldehyde formed was proportional to the excess of amine-HCl used and was minimized (<2%) by determining the minimum amount of amine-HCl required (1.25 equiv) With the optimal charges of amine-HCl and PhMgCl, the direct conversion afforded 87% yield of **4** after isolation. The process was demonstrated on pilot plant scale.

3) PhNHMe was observed in the reactions involving PhMgCl.

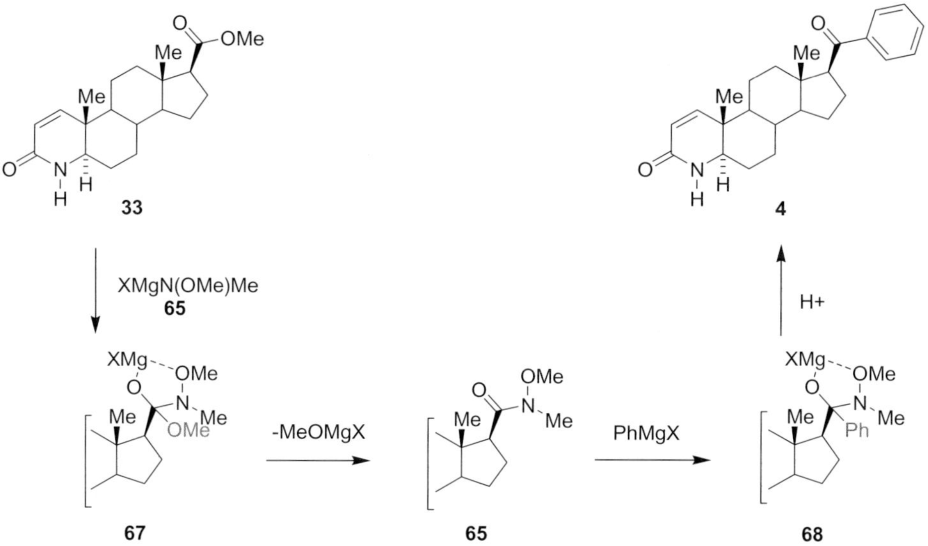

Scheme 3.30 Direct conversion of the ester to phenyl ketone.

3.2 Chemistry Development

3.2.1 Mechanistic Studies – the DDQ Oxidation

Oxidation of the A-ring lactam is clearly the key step in the finasteride process. Although this reaction could have been optimized through trial and error, experiments designed to provide insight into the mechanism gave us the understanding needed to identify the best conditions most efficiently.

In discussion of the dehydrogenation of silyl enol ethers using DDQ, Jung and Murai had proposed that DDQ abstracts a hydride producing a stabilized cation **70** which loses TMS^+ to give the enone **23** (Scheme 3.31) [8].

To account for adduct formation in the dehydrogenation of the azasteroid, we proposed a mechanism involving single electron transfer (SET) through a DDQ-silyl imidate charge transfer complex **71** (Scheme 3.32) [10a]. The resulting radical cation **72** transfers a proton to the radical anion of DDQ affording a radical pair **73**. Bond formation can then lead to formation of the observed *C–C* or *C–O* adducts. This mechanism is consistent with ^{1}H, ^{13}C, and ^{29}Si NMR studies that showed rapid formation of the silyl imidate. The *C–C* adduct is first observed as the lactam **74** which is then silylated to give the corresponding silyl imidate **45**.

Reinvestigating the reaction of cyclohexanone silyl enol ether **22** with DDQ, adducts were also observed [26]. At 25 °C, both *C–C* and *C–O* adducts (**75** and **76**) were formed along with the dehydrogenation products **23** and **77**. The ratio of

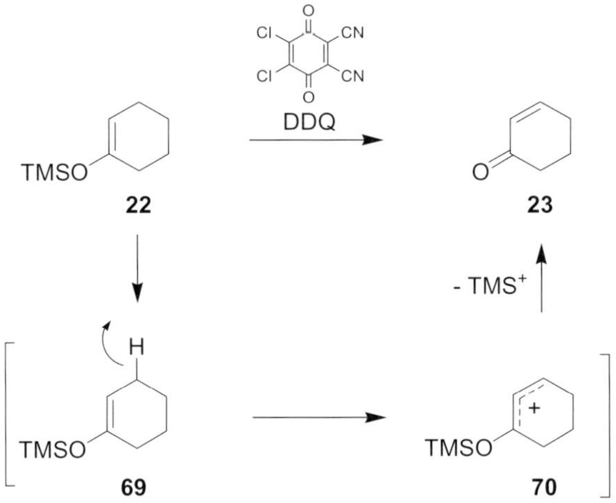

Scheme 3.31 Mechanism proposed by Jung and Murai for the reaction of DDQ with a silyl enol ether.

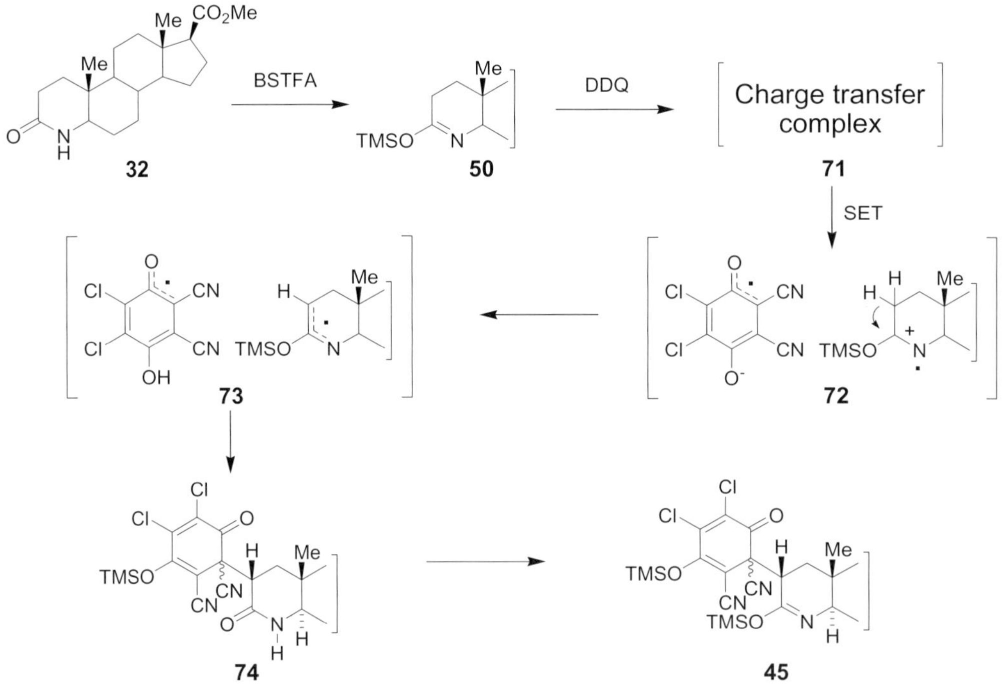

Scheme 3.32 The proposed mechanism for DDQ adduct formation.

Table 3.2 Solvent and temperature dependence in the reaction of DDQ with a silylenol ether.

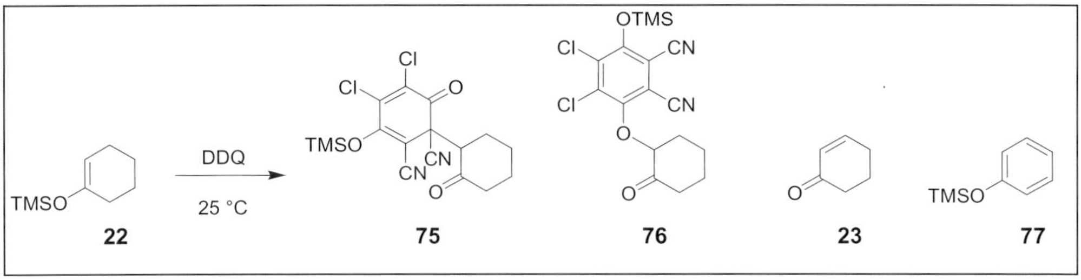

Solvent	75	76	23	77
Benzene	34	38	23	5
Dioxane	55	30	9	6
CH_2Cl_2	51	43	6	–
THF	69	23	8	–
THF (–40 °C)	95	5	–	–
$MeNO_2$	81	17	2	–
CH_3CN	91	7	?	–
CH_3CN (–20 °C)	98	2	–	–

adducts and the amount of dehydrogenation were sensitive to solvent and temperature (Table 3.2). Jung and Murai had reported GC yields for the enones in their work and the *C–C* adducts probably underwent thermolysis in the GC injector, thus they never observed the intermediate adducts.

The reaction with valerolactam **24** was also investigated, with surprising results. The reaction with BSTFA gave the silyl lactam **78** rather than the silyl imidate **25**, as shown in Scheme 3.33. Subsequent reaction with DDQ gave a *C–N* adduct **79**

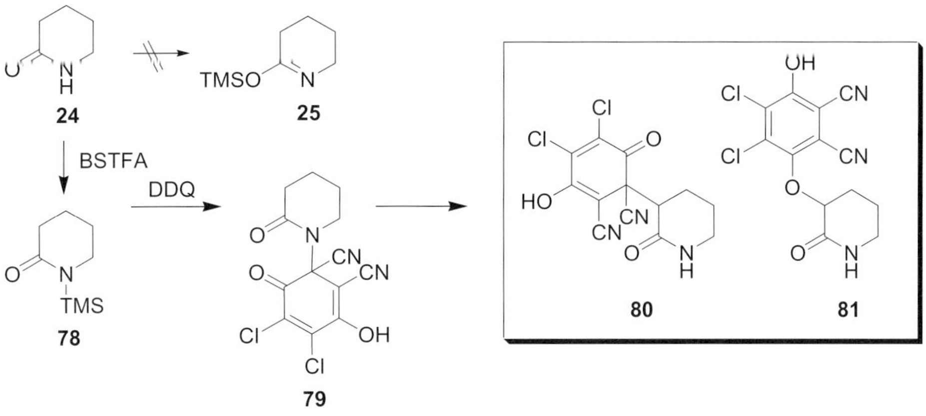

Scheme 3.33 The reaction of valerolactam with DDQ.

which rearranged to give *C–C* **80** and *C–O* **81** adducts. *C–C* adduct **80** gave the unsaturated amide upon thermolysis, but the thermolysis was not as clean as with the azasteroids. It appears that the steric environment about the lactam nitrogen plays an important role in determining the course of the reaction.

Reactions using other quinones were also studied. The results are illustrated in Scheme 3.34. With *o*-chloranil **82**, there appeared to be two mechanisms operating to give the dehydrogenation product, one of which does not involve adduct formation. *O*-Adduct formation was very much dependent on solvent, concentration, and TfOH charge. The best results (2.5–3.0 mol% *O*-adduct) were obtained at high concentration in 1,2-dichlorobenzene or 1,2-dichloroethane using 20 mol% TfOH. Both *p*-chloranil **85** and *p*-benzoquinone **88** formed *C–C* adducts, **86** and **89** respectively, which aromatize without pericyclic elimination. No dehydrogenation product was observed with these quinones.

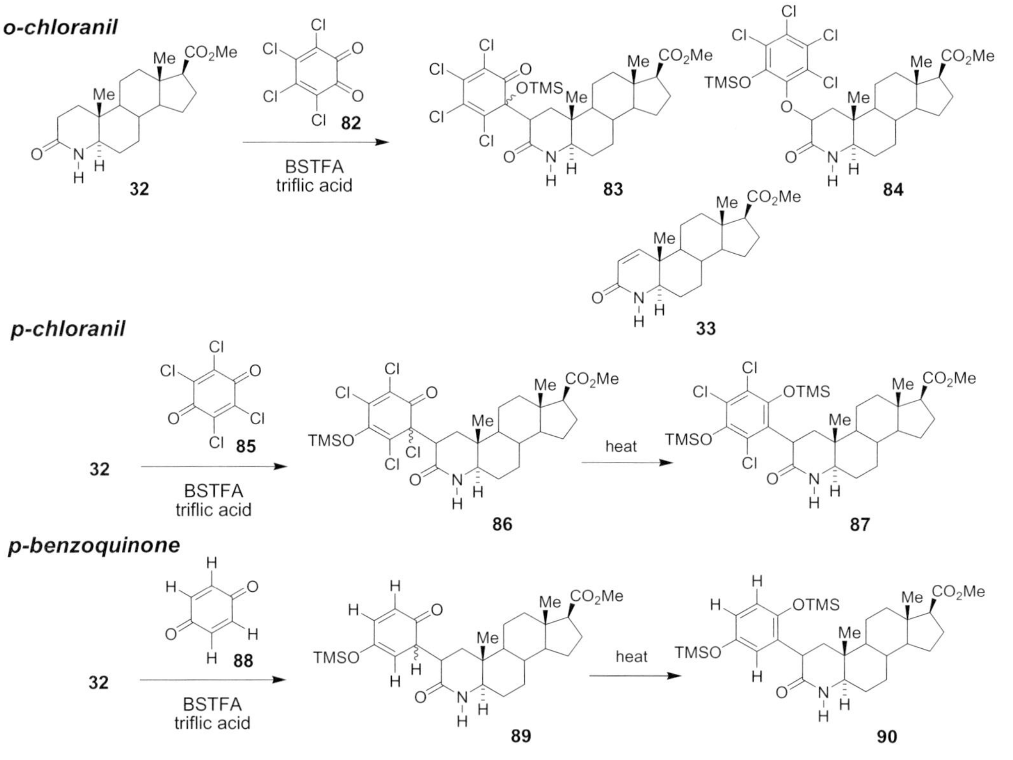

Scheme 3.34 Adduct formation with other quinones.

The structure of the *o*-chloranil *O*-adduct **91** was determined by single crystal X-ray (Figure 3.12).

Physical chemical studies provided insight into the mechanism of the reactions and aided in optimization of key parameters. The order of the reaction in BSTFA,

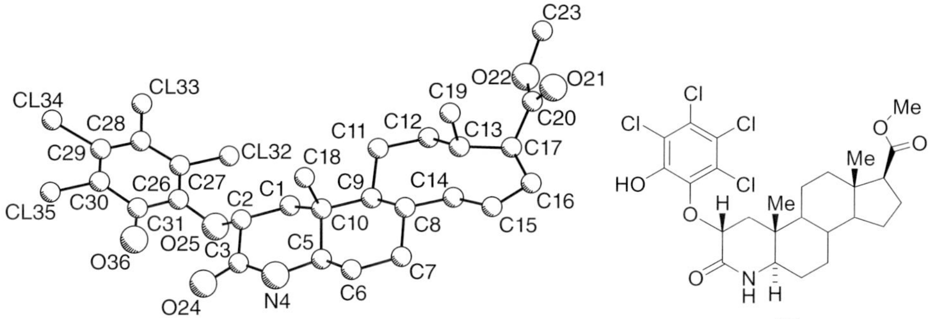

91

Figure 3.12 Single crystal X-ray structure of O-adduct with *o*-chloranil.

TfOH, and DDQ was determined by independently varying the concentration and measuring the rate. The order for each was found to be in the range 0.5 to 0.7. Since silyl imidate formation is rapid, we did not expect the rate of adduct formation to depend on BSTFA and TfOH concentration. The role of TfOH in the reaction must be complex. We found that not only does the acid catalyze silyl imidate formation but, at low concentration, it also catalyzes adduct formation. NMR experiments showed that TfOH can protonate the silyl imidate at higher concentration and under these conditions adduct formation was inhibited. The optimal amount of TfOH was found to be 15–20 μl (g of substrate)$^{-1}$ or 5–8 mol%. Other strong acids, FSO_3H and MsOH, gave higher levels of the unproductive *O*-adduct (15 and 14%, respectively, compared with less than 3% with TfOH). The kinetic deuterium isotope effect was determined by comparing rates for the 2,2-dideuterio and 2,2-diprotio azasteroid under identical conditions. A kinetic isotope effect of 12 was measured consistent with breaking the C2–H bond being rate determining [27]. (Scheme 3.35)

The dehydrogenation reaction was generally monitored by taking samples for reversed phase HPLC analysis. Diode array detectors for HPLC were relatively new at that time and proved valuable for quickly getting structural information on products of the reaction produced under different conditions. Key reaction parameters for adduct formation, overall concentration, BSTFA, TfOH, and DDQ charges, were optimized using a thermostated HPLC autosampler to sample reactions directly for analysis. Comparison of reaction profiles provided rate and reaction time information that was used to select a more limited number of reaction conditions that were scaled up to compare yields.

In the early 1990s, FTIR was being evaluated at Merck for the *in situ* monitoring of reactions. This new technology was expected to provide a powerful means to study a reaction as well as a method for analytical control in production [28]. Both silyl imidate formation and the reaction with DDQ could be conveniently monitored by FTIR, as shown in Figure 3.13. Silyl imidate formation was indicated by the appearance of an absorbance at 1667.5 cm^{-1} with concomitant disappearance of the absorbance corresponding to BSTFA at 1324.0 cm^{-1}. A new absorbance

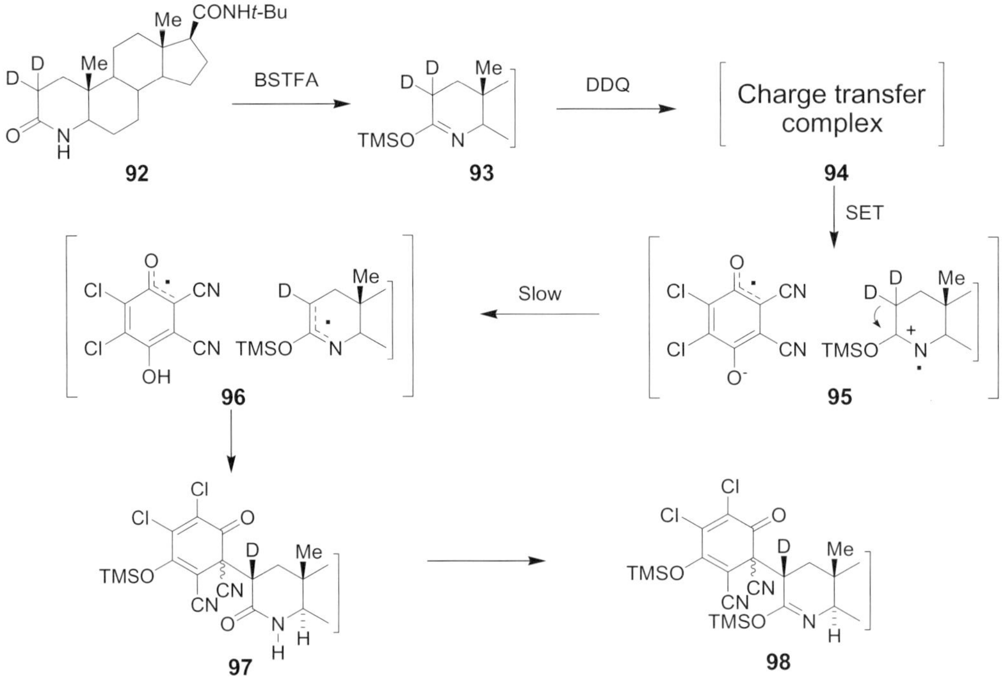

Scheme 3.35 Deuterium isotope effect.

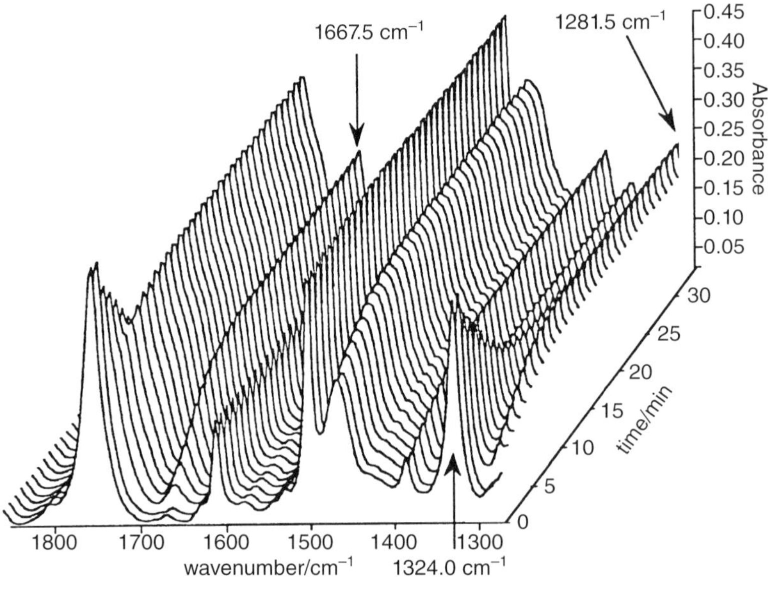

Figure 3.13 FTIR monitoring of the silylation reaction. Reprinted with permission from [28], copyright Merck & Co., Inc., Whitehouse Station, NJ.

at 1281.5 cm^{-1} grows in with time as a result of the formation of monosilyltrifluoroacetamide (MSTFA). The concentrations of silyl imidate, BSTFA, and MSTFA over the course of the reaction can be extracted from the FTIR data, as shown in Figure 3.14.

With the addition of DDQ, adduct formation could be monitored by following the appearance of an absorbance at 1544.0 cm^{-1} and disappearance of the absorbance corresponding to DDQ at 1563.3 cm^{-1} (Figure 3.15).

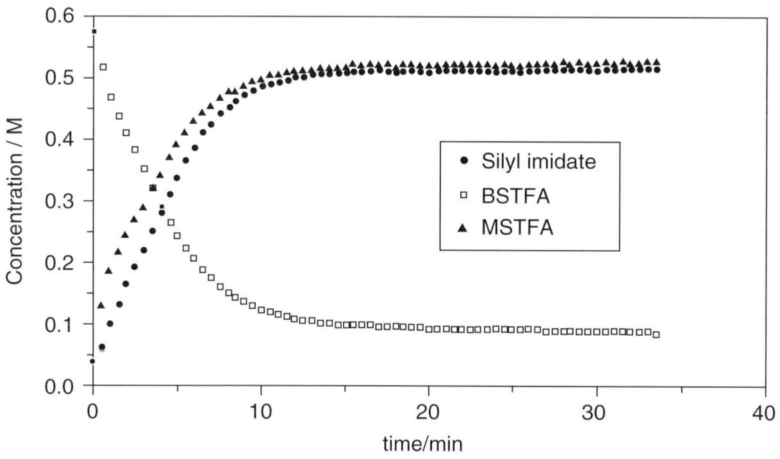

Figure 3.14 Silylation reaction profile extracted from the FTIR data. Reprinted with permission from [28], copyright Merck & Co., Inc., Whitehouse Station, NJ.

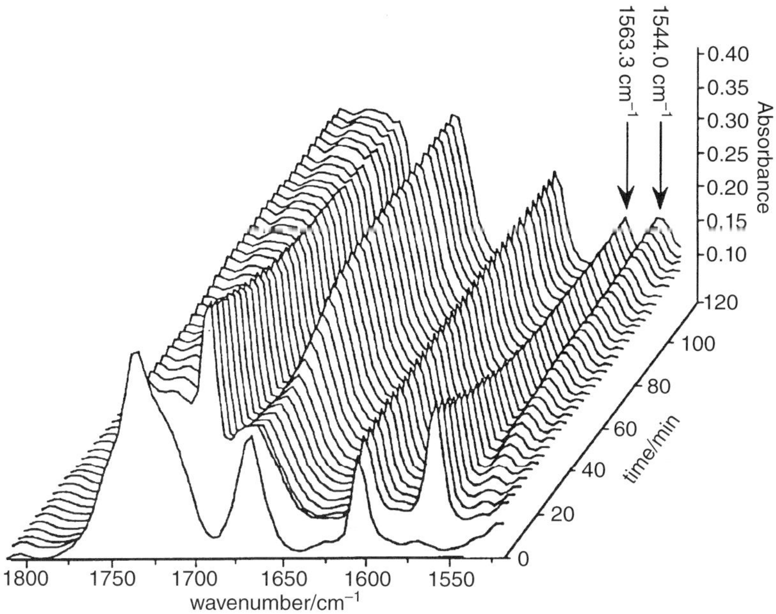

Figure 3.15 FTIR monitoring of the reaction with DDQ. Reprinted with permission from [28], copyright Merck & Co., Inc., Whitehouse Station, NJ.

Clearly, our effort toward understanding the course of the dehydrogenation reaction was key in achieving optimal performance for production. The understanding of this reaction, however, goes beyond the finasteride process changing the way that we think about quinone oxidations.

3.2.2 A New General Method for the Preparation of Weinreb Amides from Esters

The success achieved in the one-pot process for making the phenyl ketone **4** from the ester **33** was remarkable. We believe that the success of this method depends primarily on the relative stability of tetrahedral intermediates. Unfortunately, the direct conversion of esters to ketones did not prove to be general [29].[4] As shown in Table 3.3, a general method for the conversion of esters to Weinreb amides, however, was realized. Using MeMgCl and EtMgBr, 2 to 5% of the corresponding ketones was observed in some cases and this problem was completely eliminated by using a more hindered base. Although the non-nucleophilic bases mesitylmagnesium bromide and LHMDS gave comparable yields, *i*-PrMgCl provided the cleanest crude product simplifying purification. The chemistry worked well with both enolizable and hindered esters (Table 3.3) [3].

The reaction of an ester with Weinreb amine and *i*-PrMgCl provides a general method for the preparation of Weinreb amides that has been widely used in the

Table 3.3 Preparation of Weinreb amides from esters.

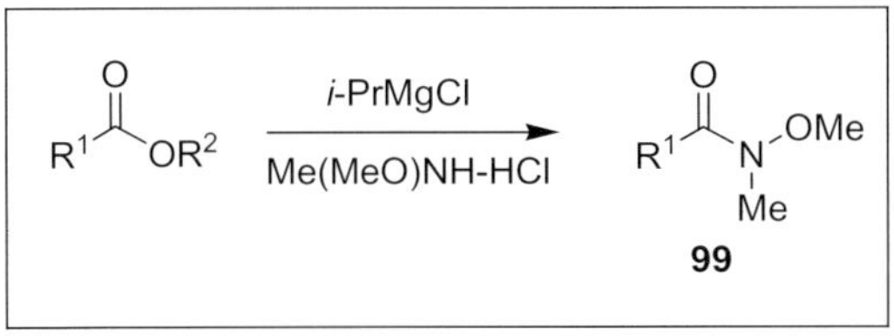

Entry	R^1	R^2	Amide 99	Yield[a] (%)
1	Ph	Me	a	100[b]
2	$PhCH_2$	Me	b	92
3	$PhCH_2CH_2$	Et	c	97
4	$PhCH_2CMe_2$	Et	d	85
5	$PhCH = CH$	Me	e	88
6	C_6H_{11}	Me	f	94
7	$3,5\text{-}(MeO)_2Ph$	Me	g	98

a) Yield after chromatographic purification.
b) HPLC assay yield.

4) Broader utility has been realized since the publication of our work.

synthesis of natural products and pharmaceutical agents [3]. Even though development of the phenyl ketone was discontinued, a valuable new synthetic method had been discovered as a result of the challenging problem and our demanding objectives.

3.3 Conclusion

The challenge in developing a practical manufacturing process for these 5α-reductase inhibitors gave us new methods for the dehydrogenation of lactams and the conversion of esters to ketones. The mechanistic understanding that came from the study of these new methods was critical in achieving optimal performance. Strategic thinking about the problems of solubility and purification played an important part in the decisions that led to an efficient process. We focused on solving problems to ensure that the immediate needs of the project were met, but never stopped thinking about the ultimate goal. We were always willing to seize opportunities and make substantial changes in direction rather than settling for optimization of a process with limited potential.

Acknowledgments

The list of Process Chemists, Chemical Engineers, and Analytical Chemists who contributed to the success of this effort is too long to include here. I would, however, like to acknowledge the contribution of Apu Bhattacharya in discovering the silylation-mediated dehydrogenation of the azasteroids and Ulf Dolling whose guidance made our success possible.

References

1 (a) Farnsworth, W.E., and Brown, R.J. (1963) *JAMA*, **183**, 436–439. (b) Price, V.H. (1975) *Arch. Dermatol.*, **111**, 1496.

2 (a) Rasmusson, G.H., Reynolds, G.F., Utne, T., Jobson, R.B., Primka, R.L., Berman, C., and Brooks, J.R. (1984) *J. Med. Chem*, **27**, 1690–1701. (b) Rasmusson, G.H., Reynolds, G.F., Steinberg, N.G., Walton, E., Patel, G.F., Liang, T., Cascieri, M.A., Cheung, A.H., Brooks, J.R., and Berman, C. (1986) *J. Med. Chem.*, **29**, 2298–2315.

3 (a) Williams, J.M., Jobson, R.B., Yasuda, N., Marchesini, G., Dolling, U.H., and Grabowski, E.J.J. (1995) *Tetrahedron Lett.*, **36**, 5461–5464. Selected applications appearing in more recent publications: (b) Ley, S.V., Tackett, M.N., Maddess, M.L., Anderson, J.C., Brennan, P.E., Cappi, M.W., Heer, J.P., Helgen, C., Kori, M., Kouklovsky, C., Marsden, S.P., Norman, J., Osborn, D.P., Palomero, M.Á., Pavey, J.B.J., Pinel, C., Robinson, L.A., Schnaubelt, J., Scott, J.S., Spilling, C.D., Watanabe, H., Wesson, K.E., and Willis, M.C. (2009) *Chem. Eur. J.*, **15**, 2874–2914. (c) Hagiwara, H., Suka, Y., Nojima, T., Hoshi, T., and Suzuki, T. (2009) *Tetrahedron*, **65**, 4820–4825. (d) Barber, C.G., Blakemore, D.C., Chiva,

J.-Y., Eastwood, R.L., Middleton, D.S., and Paradowski, K.A. (2009) *Bioorg. Med. Chem. Lett.*, **19**, 1075–1079. (e) Wang, W., Dai, M., Zhu, C., Zhang, J., Lin, L., Ding, J., and Duan, W. (2009) *Bioorg. Med. Chem. Lett.*, **19**, 735–737. (f) Sattely, E.S., Meek, S.J., Malcolmson, S.J., Schrock, R.R., and Hoveyda, A.H. (2008) *J. Am. Chem. Soc.*, **131**, 943–953. (g) Balasubramaniam, S., and Aidhen, I.S. (2008) *Synthesis*, 3707–3738. (h) Evans, D.A., Kværnø, L., Dunn, T.B., Beauchemin, A., Raymer, B., Mulder, J.A., Olhava, E.J., Juhl, M., Kagechika, K., and Favor, D.A. (2008) *J.Am. Chem. Soc.*, **130**, 16295–16309. (i) Barbazanges, M., Meyer, C., and Cossy, J. (2008) *Org. Lett.*, **10**, 4489–4492. (j) Ribes, C., Falomir, E., Carda, M., and Marco, J.A. (2008) *J. Org. Chem.*, **73**, 7779–7782. (k) Clark, R.C., Yeul Lee, S., and Boger, D.L. (2008) *J. Am. Chem. Soc.*, **130**, 12355–12369. (l) Salit, A.-F., Meyer, C., Cossy, J., Delouvrié, B., and Hennequin, L. (2008) *Tetrahedron*, **64**, 6684–6697. (m) Scribner, A., Dennis, R., Lee, S., Ouvry, G., Perrey, D., Fisher, M., Wyvratt, M., Leavitt, P., Liberator, P., Gurnett, A., Brown, C., Mathew, J., Thompson, D., Schmatz, D., and Biftu, T. (2008) *Eur. J. Med. Chem.*, **43**, 1123–1151. (n) Alimardanov, A., Schmid, J., Afragola, J., and Khafizova, G. (2008) *Org. Proc. Res. Dev.*, **12**, 424–428. (o) Timmons, A., Seierstad, M., Apodaca, R., Epperson, M., Pippel, D., Brown, S., Chang, L., Scott, B., Webb, M., Chaplan, S.R., and Breitenbucher, J.G. (2008) *Bioorg. Med. Chem. Lett.*, **18**, 2109–2113. (p) Wang, Y., Gang, S., Bierstedt, A., Gruner, M., Fröhlich, R., and Metz, P. (2007) *Adv. Synth. Catal.*, **349**, 2361–2367. (q) He, W., Huang, J., Sun, X., and Frontier, A.J. (2007) *J. Am. Chem. Soc.*, **130**, 300–308. (r) Denmark, S.E., and Fujimori, S. (2005) *J. Am. Chem. Soc.*, **127**, 8971–8973. (s) Arai, N., Chikaraishi, N., Ikawa, M., Omura, S., and Kuwajima, I. (2004) *Tetrahedron Asym.*, **15**, 733–741. (t) Ley, S.V., Diez, E., Dixon, D.J., Guy, R.T., Michel, P., Nattrass, G.L., and Sheppard, T.D. (2004) *Org. Biomol. Chem.*, **2**, 3608–3617. (u) Evans, D.A., Rajapakse, H.A., and Stenkamp, D. (2002) *Angew. Chem. Int. Ed.*, **41**, 4569–4573. (v) Colletti, S.L., Li, C., Fisher, M.H., Wyvratt, M.J., and Meinke, P.T. (2000) *Tetrahedron Lett.*, **41**, 7825–7829. (w) Andrés, J.M., Pedrosa, R., and Pérez-Encabo, A. (2000) *Tetrahedron*, **56**, 1217–1223. (x) Evans, D.A., Trotter, B.W., Coleman, P.J., Côté, B., Dias, L.C., Rajapakse, H.A., and Tyler, A.N. (1999) *Tetrahedron*, **55**, 8671–8726.

4 Carlsen, P.H.J., Katsuki, T., Martin, V.S., and Sharpless, K.B. (1981) *J. Org. Chem.*, **46**, 3936–3938.

5 Piatak, D.M., Bhat, H.B., and Caspi, E. (1969) *J. Org. Chem.*, **34**, 112–116.

6 (a) Trost, B.M., Salzmann, T.N., and Hiroi, K. (1976) *J. Am. Chem. Soc.*, **98**, 4887–4902. (b) Magnus, P., and Pappalardo, P.A. (1986) *J. Am. Chem. Soc.*, **108**, 212–217.

7 Minami, I., Takahashi, K., Shimizu, I., Kimura, T., and Tsuji, J. (1986) *Tetrahedron*, **42**, 2971–2977.

8 (a) Fleming, I., and Paterson, I. (1979) *Synthesis*, 736–738. (b) Jung, M.E., Pan, Y.-G., Rathke, M.W., Sullivan, D.F., and Woodbury, R.P. (1977) *J. Org. Chem.*, **42**, 3961–3963. (c) Ryu, I., Murai, S., Hatayama, Y., and Sonoda, N. (1978) *Tetrahedron Lett.*, 3455–3458. (d) Turner, A.B., and Ringold, H.J. (1967) *J. Chem. Soc. (C)*, 1720–1730. (e) Walker, D., and Hiebert, J.D. (1967) *Chem. Rev.*, **67**, 153–195.

9 Bhattacharya, A., Williams, J.M., Amato, J.S., Dolling, U.H., and Grabowski, E.J.J. (1990) *Synth. Commun.*, **20**, 2683–2690.

10 (a) Bhattacharya, A., Dimichele, L.M., Dolling, U.H., Douglas, A.W., and Grabowski, E.J.J. (1988) *J. Am. Chem. Soc.*, **110**, 3318–3319. For reactions of quinones where adducts were also observed see. (b) Becker, H.-D. (1965) *J. Org. Chem.*, **30**, 989–994. (c) Becker, H.-D. (1969) *J. Org. Chem*, **34**, 1203–1210. (d) Becker, H.-D., and Turner, A.B. (1988) *The Chemistry of Quinoid Compounds*, vol. II (eds S. Patai and Z. Rappoport), John Wiley & Sons Inc., New York, NY, pp. 1351–1384. (e) Fu, P.P., and Harvey, R.G. (1978) *Chem. Rev.*, **78**, 317–361.

11 Williams, J.M., Marchesini, G., Reamer, R.A., Dolling, U.-H., and Grabowski, E.J.J. (1995) *J. Org. Chem.*, **60**, 5337–5340.

12 Bassett, H.L., and Thomas, C.R. (1954) *J. Chem. Soc.*, 1188–1190 and references cited therein.

13 (a) Midler, M., Jr., Paul, E.L., Whittington, E.F., Futran, M., Liu, P.D., Hsu, J., and Pan, S.-H. (1994) U.S. Patent 5 314 506. (b) Dauer, R., Mokrauer, J.E., and McKeel, W.J. (1996)U.S. Patent 5 578 279.

14 Araki, M., Sakata, S., Takei, H., and Mukaiyama, T. (1974) *Bull. Chem. Soc. Jpn.*, **47**, 1777.

15 Staab, H.A., and Jost, E. (1962) *Ann. Chem.*, **655**, 90–94.

16 Fiandanese, V., Marchese, G., Martina, V., and Ronzini, L. (1984) *Tetrahedron Lett.*, **25**, 4805–4808.

17 Reduction is also observed in the reaction of acid chlorides with Grignard reagents: Whitmore, F.C., Whitaker, J.S., Mosher, W.A., Brevick, O.N., Wheeler, W.R., Miner, C.S., Jr., Sutherland, L.H., Wagner, R.B., Clapper, T.W., Lewis, C.E., Lux, A.R., and Popkin, A.H. (1941) *J. Am. Chem. Soc.*, **63**, 643–654.

18 (a) House, H.O., and Traficante, D.D. (1963) *J. Org. Chem.*, **28**, 355–360. (b) Percival, W.C., Wagner, R.B., and Cook, N.C. (1953) *J. Am. Chem. Soc.*, **75**, 3731–3734.

19 (a) Chastrette, M., and Amouroux, R. (1970) *J. Chem. Soc. D, Chem. Commun.*, 470–471. (b) Chastrette, M., and Amouroux, R. (1970) *Bull. Soc. Chim. Fr.*, 4348. (c) Georgoulis, C., Gross, B., and Ziegler, J.C. (1971) *C. R. Seances Acad. Sci., Ser. C*, **273**, 378–381.

20 (a) Georgoulis, C., Gross, B., and Ziegler, J.C. (1971) *C. R. Seances Acad. Sci., Ser. C*, **273**, 378–381. (b) Eaton, P.E., Lee, C.-H., and Xiong, Y. (1989) *J. Am. Chem. Soc.*, **111**, 8016–8018.

21 King, A.O., Corley, E.G., Anderson, R.K., Larsen, R.D., Verhoeven, T.R., Reider, P.J., Xiang, Y.B., Belley, M., and Leblanc, Y. (1993) *J. Org. Chem.*, **58**, 3731–3735.

22 Nahm, S., and Weinreb, S.M. (1981) *Tetrahedron Lett.*, **22**, 3815–3818.

23 (a) Beak, P., Basha, A., Kokko, B., and Loo, D. (1986) *J. Am. Chem. Soc.*, **108**, 6016–6023. (b) Beak, P., and Selling, G.W. (1989) *J. Org. Chem.*, **54**, 5574–5580.

24 The difference in stability of tetrahedral intermediates was used in developing ureas that could be used to prepare ketones: (a) Hlasta, D.J., and Court, J.J. (1989) *Tetrahedron Lett.*, **30**, 1773–1776. (b) Whipple, W.L., and Reich, H.J. (1991) *J. Org. Chem.*, **56**, 2911–2912.

25 (a) Sanchez, R., and Scott, W. (1988) *Tetrahedron Lett.*, **29**, 139. (b) Majewski, M. (1988) *Tetrahedron Lett.*, **29**, 4057.

26 Bhattacharya, A., Dimichele, L.M., Dolling, U.H., Grabowski, E.J.J., and Grenda, V.J. (1989) *J. Org. Chem.*, **54**, 6118–6120.

27 A deuterium isotope effect is also observed in the reaction of quinones with acenaphthenes. Trost, B.M. (1967) *J. Am. Chem. Soc.*, **89**, 1847–1851.

28 (a) Landau, R.N., Penix, S.M., Donahue, S.M., and Rein, A.J. (1992) *Optically Based Methods for Process Analysis*, vol. 1681, 1st edn, SPIE, Somerset, NJ, USA, pp. 356–373. (b) Landau, R.N. (1995) *Automated Laboratory Reactors & Calorimeters*, 2nd edn, Ralph N. Landau, Merck & Co., Inc., Whitehouse Station, NJ.

29 (a) Yang, S.-B., Gan, F.-F., Chen, G.-J., and Xu, P.-F. (2008) *Synlett*, 2532–2534. (b) Tartaglia, S., Padula, D., Scafato, P., Chiummiento, L., and Rosini, C. (2008) *J. Org. Chem.*, **73**, 4865–4873. (c) Kuboki, A., Yamamoto, T., Taira, M., Arishige, T., Konishi, R., Hamabata, M., Shirahama, M., Hiramatsu, T., Kuyama, K., and Ohira, S. (2008) *Tetrahedron Lett.*, **49**, 2558–2561. (d) Ma, Z., and Zhai, H. (2007) *Synlett*, 161–163. (e) Kim, J., and Thomson, R.J. (2007) *Angew. Chem. Int. Ed.*, **46**, 3104–3106. (f) Canales, E., and Corey, E.J. (2007) *J. Am. Chem. Soc.*, **129**, 12686–12687. (g) Doroh, B., and Sulikowski, G.A. (2006) *Org. Lett.*, **8**, 903–906. (h) Blanc, A., and Toste, F.D. (2006) *Angew. Chem. Int. Ed.*, **45**, 2096–2099. (i) Pedrosa, R., Andrés, C., Gutiérrez-Loriente, A., and Nieto, J. (2005) *Eur. J. Org. Chem.*, 2449–2458. (j) Davis, J.M., Truong, A., and Hamilton, A.D. (2005) *Org. Lett.*, **7**, 5405–5408. (k) Davis, F.A., Lee, S.H., and Xu, H. (2004) *J. Org. Chem.*, **69**, 3774–3781. (l) Wallace, O.B., Smith, D.W., Deshpande, M.S., Polson, C., and Felsenstein, K.M. (2003) *Bioorg. Med. Chem. Lett*, **13**, 1203–1206. (m) Francavilla, C., Chen, W., and Kinder, F.R. (2003) *Org. Lett.*, **5**, 1233–1236. (n) Bourghida, A., Wiatz, V., and Wills, M. (2001) *Tetrahedron Lett.*, **42**, 8689–8692. (o) Wallace, O.B. (1997) *Tetrahedron Lett.*, **38**, 4939–4942.

4
Rizatriptan (Maxalt®): A 5-HT$_{1D}$ Receptor Agonist

Cheng-yi Chen

Sumatriptan, a selective 5HT$_{1B/1D}$ receptor agonist from Glaxo Smith Kline, was the first triptan drug approved for the treatment of migraine headaches [1]. The introduction of sumatriptan represented a major advance both in the understanding and treatment of migraine and, more importantly, it intensified research efforts towards development of more efficacious triptan drugs. As a result, rizatriptan benzoate (MK-0462, **1**) was developed by Merck Research Laboratories [2] and obtained approval by the United States Food and Drug Administration on June 29, 1998 (Figure 4.1). Today, it is widely prescribed as a second generation triptan for the treatment of migraine headaches. Over the past decade, rizatriptan has emerged as one of the most efficacious among the 5HT$_{1B/1D}$ receptor agonists or "triptans" [3]. In addition, our understanding of the mechanism of action of these triptans has increased significantly [4]. This chapter will detail the chemical development of a manufacturing process for rizatriptan. We have previously described our initial efforts for the synthesis of this compound [5]. In Section 4.1 a full account of the chemical development for the synthesis is presented here with emphasis on the

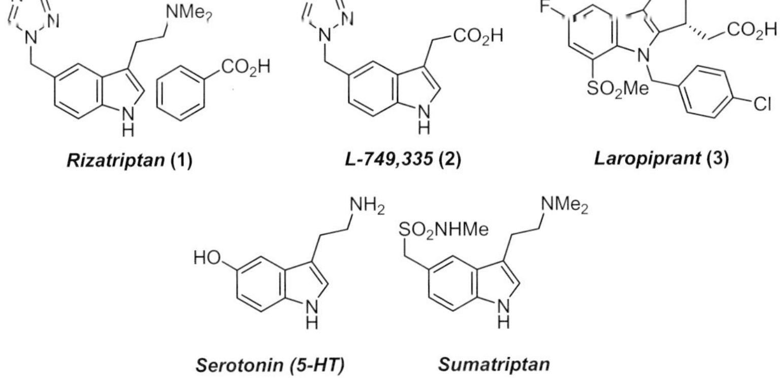

Figure 4.1 5-HT$_{1D}$ agonists and PGD_2 receptor antagonist.

The Art of Process Chemistry. Edited by Nobuyoshi Yasuda

ISBN: 978-3-527-32470-5

process development for the palladium-catalyzed indole synthesis. In Section 4.2 an application of the methodology to the synthesis of a metabolite of rizatriptan, indole acetic acid (L-749,335, **2**), will be described. In addition, during the optimization of the key coupling step for the synthesis of rizatriptan, a novel indole synthesis via the palladium-catalyzed coupling of *ortho*-haloanilines and ketones was discovered. Its scope, limitations and further application to the synthesis of a PGD_2 receptor antagonist, laropirant (MK-0524A, **3**) [6], will also be described (Figure 4.1).

4.1 Project Development

4.1.1 Medicinal Chemistry Route

The drug discovery route to rizatriptan (**1**) began with the preparation of 1-(4′-nitrobenzyl)-1,2,4-triazole **5** using 4-nitrobenzyl bromide (**4**) and 1,2,4-triazole. (Scheme 4.1). Benzylation of the sodium salt of 1,2,4-triazole prepared with NaH was not regioselective and afforded a 1.5 : 1 mixture of 1-(4′-nitrobenzyl)-1,2,4-triazole (**5**) and its regioisomer, 4-(4′-nitrobenzyl)-1,2,4-triazole. The desired isomer **5** was isolated in 52% yield after silica gel chromatography. Hydrogenation

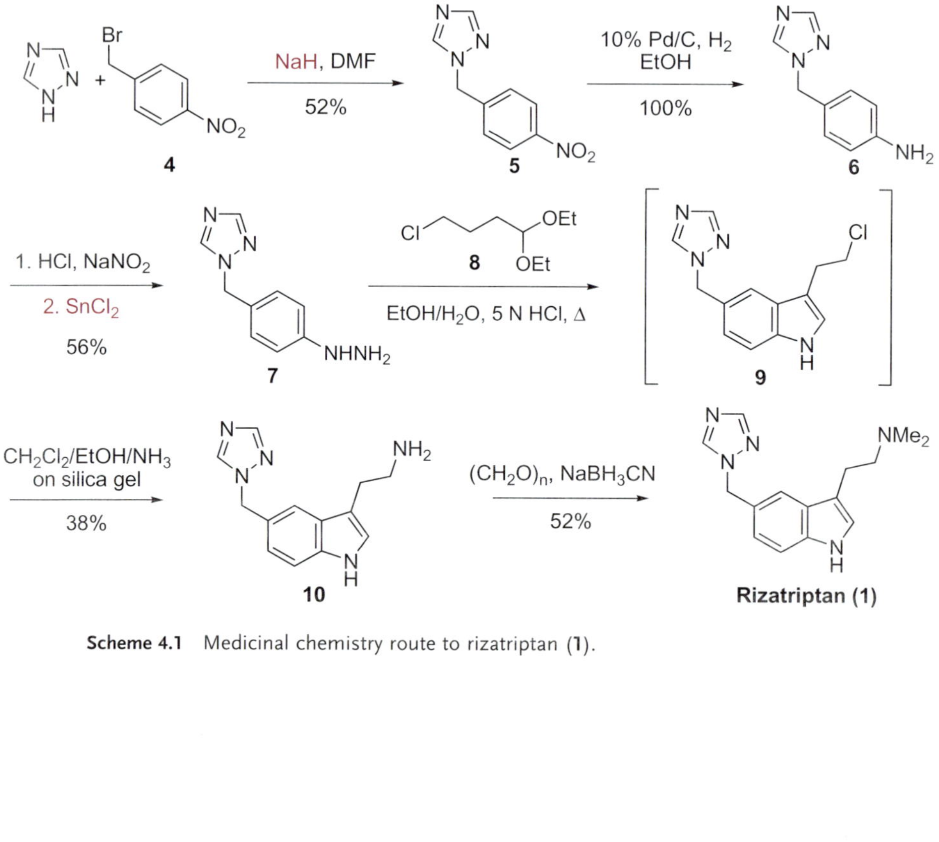

Scheme 4.1 Medicinal chemistry route to rizatriptan (**1**).

of **5** over 10% Pd/C in EtOH gave aniline **6** quantitatively. Treatment of aniline **6** with $NaNO_2$ followed by reduction of the intermediate diazonium salt with 3.75 equiv of $SnCl_2$ monohydrate afforded hydrazine free base **7** in 56% yield after saponification and silica gel chromatography. Removal of large amounts of insoluble tin oxide from the reaction mixture made this isolation challenging. Next, the Fisher reaction of hydrazine **7** using diethyl 4-chlorobutanal (**8**) in refluxing EtOH/H_2O and 5 M HCl provided 3-chloroethyl indole **9**. Chloroethyl indole **9**, presumably unstable, was not isolated. Rather it was directly converted to the corresponding 2-aminoethyl indole **10** during chromatography in silica gel by eluting with a mixture of solvents: CH_2Cl_2/EtOH/NH_3 (30 : 8 : 1) in 38% isolated yield. Finally, reductive amination of tryptamine **10** using formaldehyde and $NaBH_3CN$ in AcOH led to the desired rizatriptan (**1**) free base in 52% yield after silica gel chromatography. Overall, the synthesis of rizatriptan (**1**) free base was completed in five linear steps but only afforded the product in 5.7% overall yield [7].

4.1.1.1 Problems of the Original Route

Though the medicinal chemistry route allowed for the preparation of rizatriptan (**1**) in a straightforward manner, it suffered from several shortcomings that prohibited implementation of the synthesis on a kilogram scale:

1) Benzylation of sodium 1,2,4-triazole was not regioselective and the step utilized NaH, which is not suitable for large scale preparation, for generating the sodium triazole species.
2) The use of excess $SnCl_2$ monohydrate as the reducing agent for the diazonium salt was not desirable because of environmental concerns and tedious separation issues.
3) The key Fisher indole synthesis using diethyl 4-chlorobutanal (**8**) suffered from poor yield to make the primary tryptamine **10**, which then calls for reductive amination to complete the synthesis.
4) Multiple chromatographies coupled with several low yielding steps in the synthesis deem this route not suitable for scale-up.

4.1.1.2 Advantages of the Original Route

Perhaps the advantage of the medicinal chemistry route lies in the flexibility of introducing different alkyl groups on the primary amine through reductive amination on 2-aminoethyl indole **10** and hence allows access to various *N, N*-dialkyl tryptamine derivatives for structure–activity relationship (SAR) studies.

4.1.2 Process Development

4.1.2.1 Convergent Fisher Indole Synthesis

We envisioned that improvement of the key Fisher indole reaction using diethyl 4-(*N,N*-dimethylamino)butanal (**11**) instead of diethyl 4-chlorobutanal (**8**) would lead to formation of the desired product directly. The approach would circumvent the necessity of amine displacement and reductive amination steps (Scheme 4.1). The diethyl 4-(*N,N*-dimethylamino)butanal (**11**) was readily prepared from chloroacetal **8** via S_N2 displacement using 40% aqueous dimethyl amine; later **11** became commercially available. This more convergent Fisher indole synthesis using the functionalized acetal **11** was previously developed for another 5HT$_{1D}$ receptor agonist L-695,894 (**13**) and was extended to a wide variety of tryptamines, as shown in Scheme 4.2 [5]. For example, condensation of cyanobenzohydrazine **12** with acetal **11** in 4% H_2SO_4 afforded the desired tryptamine **14** in 81% yield. The cyano group in **14** subsequently served as a functional handle for the introduction of the heterocycle via ethyl ester **15** in an efficient manner to produce L-695,894 (**13**). This improved Fisher indole synthesis was readily applied to the preparation of other tryptamines (**16 a–g**, **17** and **18**) in multi-gram quantities in 82–100% yields.

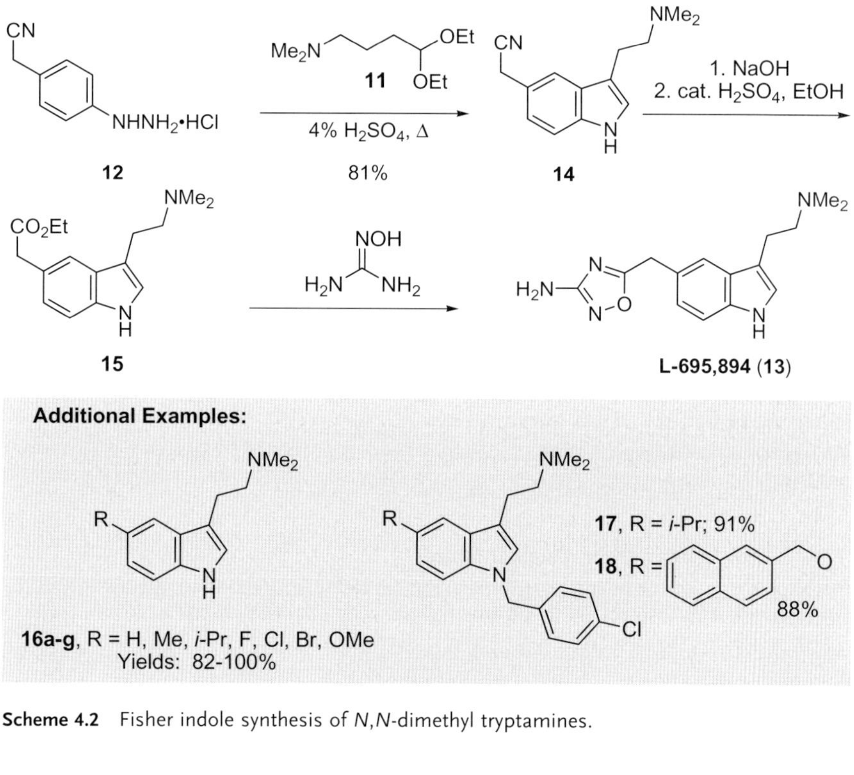

Scheme 4.2 Fisher indole synthesis of *N,N*-dimethyl tryptamines.

As shown in Scheme 4.3, the Fisher indolization was thought to involve (i) hydrolysis of diethyl dimethylamino acetal **11**, (ii) formation of hydrazone **22**, (iii) isomerization of hydrazone to *ene*-hydrazine **23**, and (iv) [3.3] sigmatropic rearrangement followed by ring closure to give indole **16b**. Acetal **11** is stable in AcOH at room temperature, but can be readily hydrolyzed to aldehyde **19** at 100 °C, with subsequent cyclization to hemiaminal **20**. Hemiaminal **20** was also formed readily

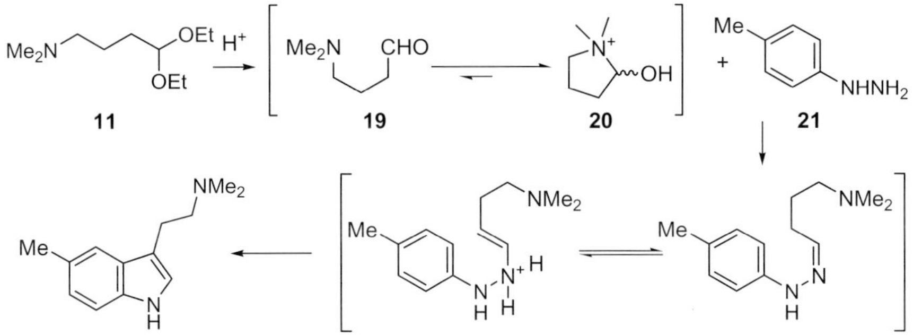

Scheme 4.3 Mechanism of the improved Fisher indole synthesis using *N, N*-dimethylaminobutanal.

under acidic conditions such as 8% HCl, 4% H_2SO_4, or 8% TFA at room temperature. A mixture of **19** and **20** in a ratio of 5 : 95 (by NMR in DMSO-d_6) was observed. We chose *para*-methylphenylhydrazine (**21**) as a model compound to study the catalytic efficiency of acids in the Fisher indole synthesis. Since the formation of hydrazone **22** occurred readily for all of these acids, the successful indolization had to rely on step (iii): the isomerization of hydrazone **22** to *ene*-hydrazine **23**. Indolization of hydrazine free base **22** proceeded slowly in AcOH to give indole product **16b** and was still incomplete after heating for 24 h as the hydrazone intermediate **22** was observed by NMR. The reaction when carried out in 8% HCl led to the desired indole **16b** but was plagued by the formation of *para*-toluidine, presumably due to the N–N bond cleavage. Finally the reaction was found to proceed cleanly in either 4% H_2SO_4 or 8% TFA in 2 h to give indole **16b** in 89% and 80%, respectively. These results indicated that H_2SO_4 is superior to other protic acids because it efficiently catalyzed the isomerization of hydrazone **22** to *ene*-hydrazine **23**. Although TFA worked equally well for the *para*-methylphenylhydrazine, the generality of the TFA-catalyzed indolization has yet to be investigated.

Application of this methodology to the synthesis of rizatriptan (**1**), however, was ineffective because of the low yield of the key indolization step, as shown in Scheme 4.4. In this particular case the triazole group acted as a good leaving group under the acidic conditions generating a large amount of oligomers. As a result, a new indole synthesis was required for rizatriptan (**1**).

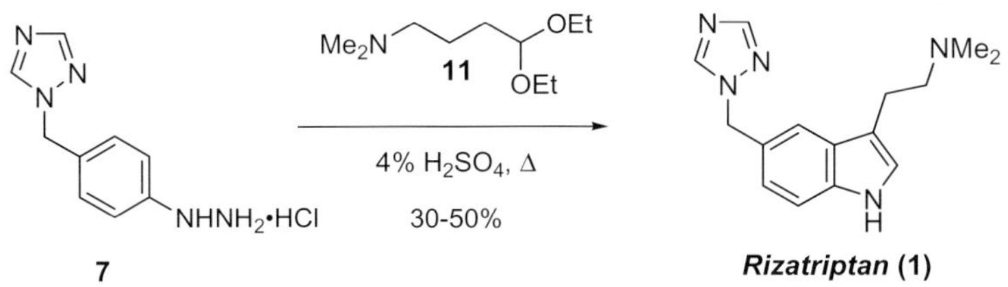

Scheme 4.4 Application of the improved Fisher indole synthesis for rizatriptan.

4.1.2.2 Palladium-Catalyzed Indole Synthesis

Larock *et al.* have shown that the coupling of iodoaniline **24** with internal acetylenes **25** using palladium catalysis gives 2,3-disubstituted indoles **26** in good to excellent yields (Scheme 4.5) [8]. This annulation has proven to be highly regioselective for unsymmetrical alkynes with the more sterically bulky group ending up in the 2-position of the indole ring. We therefore envisioned that rizatriptan (**1**) could be derived from the installation of the dimethyl amine moiety to 5-(1,2,4-triazol-1-ylmethyl)tryptol **27** which in turn could be prepared by the coupling of 4-substituted-2-iodoaniline **28** with a suitably protected 3-butyn-1-ol derivative **29** (Scheme 4.6). The application of a palladium-catalyzed coupling methodology to the synthesis of tryptophol **27** had not been reported prior to our work.

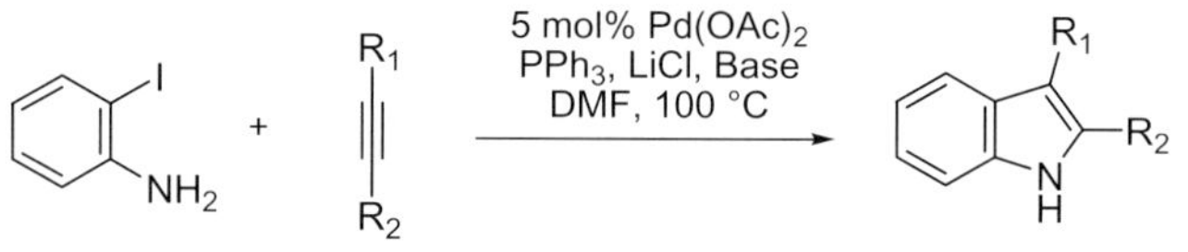

Scheme 4.5 Lorock's synthesis of indoles.

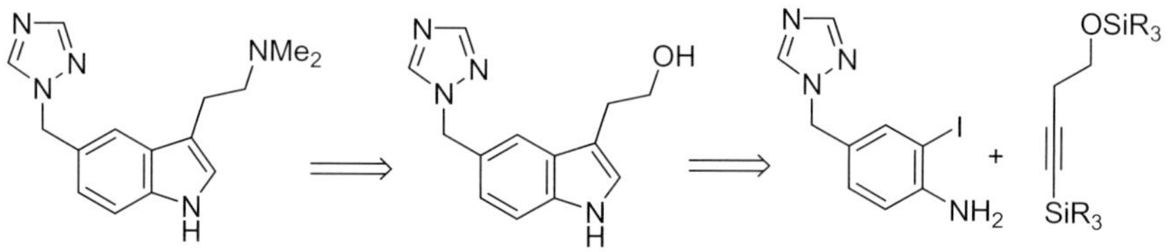

Scheme 4.6 Retrosynthetic analysis of **1**.

The discussion of process development will focus on the following:

1) Efficient preparation of iodoanile **28**.
2) Optimization of the Pd-catalyzed coupling reaction between iodoaniline **28** and bis-trialkylsilyl butynol ether **29**.

4.1.2.2.1 Preparation of Iodoaniline 28

Process development of the synthesis of iodoaniline **28** began with an improved synthesis of 1-(4′-aminobenzyl)-1,2,4-triazole (**6**) (Scheme 4.7), which was prepared in the medicinal chemistry synthesis, albeit with poor regioselectivity (Scheme 4.1). We found that this aniline intermediate **6** could be readily prepared in three steps in >90% overall yield from 4-amino-1,2,4-triazole (**30**) and 4-nitrobenzyl bromide (**4**) based on a modified literature procedure [9]. The condensation of **30** and **4** in isopropyl alcohol followed by deamination gave the nitro

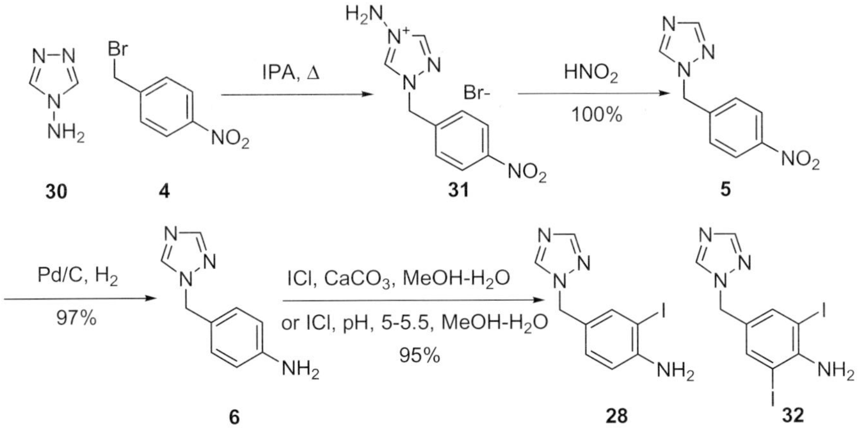

Scheme 4.7 Synthesis of iodoaniline **28**.

species **5** quantitatively. The use of 4-amino-1,2,4-triazole led to complete regioselectivity for the alkylation step. Apparently, the amino group in 4-amino-1,2,4-triazole (**30**) prevents alkylation at the 4-position nitrogen, leading exclusively to the desired product **31**. Hydrogenation of **5** afforded aniline **6** in 97% overall yield from **4**. This sequence required very little further optimization and the synthesis was readily outsourced.

Aniline **6** was converted to iodoaniline **28** through a selective *ortho*-iodination since the *para*-position is blocked. Reaction of **6** with neat iodine monochloride (ICl) in the presence of powdered $CaCO_3$ in aqueous methanol at 0 °C for 6 h furnished iodoaniline **28** in 91% yield; some over-iodination occurred to provide 3% of diiodoaniline **32**. The over-iodination was not difficult to control since it occurred much more slowly than the first iodination. For example, treatment of iodoaniline **28** with 1 equiv of ICl at room temperature for 12 h only generated 30% of **32**. Even with 5 equiv of ICl under prolonged aging at room temperature, **32** could be produced in only 75% yield from **6**. Alternatively, the iodination could be carried out using an aqueous 5 M ICl solution in the presence of $CaCO_3$ at ambient temperature and this avoided the handling of corrosive neat ICl. It was critical to use powdered $CaCO_3$ for the reaction as granular $CaCO_3$ failed to give effective iodination and other inorganic bases such as Na_2CO_3 or K_2CO_3 also failed (Scheme 4.7).

The ICl–$CaCO_3$ procedure required a filtration to remove insoluble, inorganic by-products prior to biphasic extraction. In an effort to develop a homogeneous process for the iodination step, a pH control protocol was later implemented in the manufacturing process. The pH-controlled iodination was run in a single phase in a MeOH–water system by simultaneous addition of the aqueous ICl solution and 1 M NaOH. Citric acid was added to increase the buffer capacity to the optimal pH (5–5.5) for robust operation. Under these conditions, the iodoaniline **28** was typically obtained in >99 A% with <1% of diiodoaniline **32**. Residual

ICl was quenched with sodium thiosulfate. A solvent switch from MeOH to EtOAc followed by water wash allowcd isolation of thc crudc iodoaniline **28** in solution in EtOAc. Subsequent crystallization from EtOAc–heptane afforded iodoaniline **28** in 95% yield in >99A% purity. The procedure was quite robust and produced high quality **28** that could readily be used in the coupling step.

4.1.2.2.2 Optimization of the Pd-Catalyzed Coupling Reaction Between Iodoaniline 28 and Bis-TES Butynol Ether

Larock indole synthesis of tryptophols The coupling of an alkyne with an *ortho*-iodoaniline for the formation of tryptophol requires protection of the terminal position on the alkyne. A protecting group that was robust enough to survive the coupling reaction, but labile enough to be removed from the resulting indole was required. In addition, a sterically bulky protecting group would provide strong preference for the 3-alkyl regioisomer relative to the 2-alkyl regioisomer. We reasoned that a trialkylsilyl protecting group would work best because of the ease of formation of the carbon–silicon bond, steric size, stability during the coupling reaction, and facile deprotection from the 2-postion of the indole under mild conditions. For example, the trimethylsilyl (TMS) protected alkyne **36** reacted with 4-substituted iodoaniline **28** to afford a 94:6 mixture of **37** and **38.** In contrast, the reaction of iodoaniline **24** and 2-pentyne (**33**) only provided a 2:1 mixture of regioisomers (**34** : **35**) (Scheme 4.8). In this reaction the more sterically bulky silyl

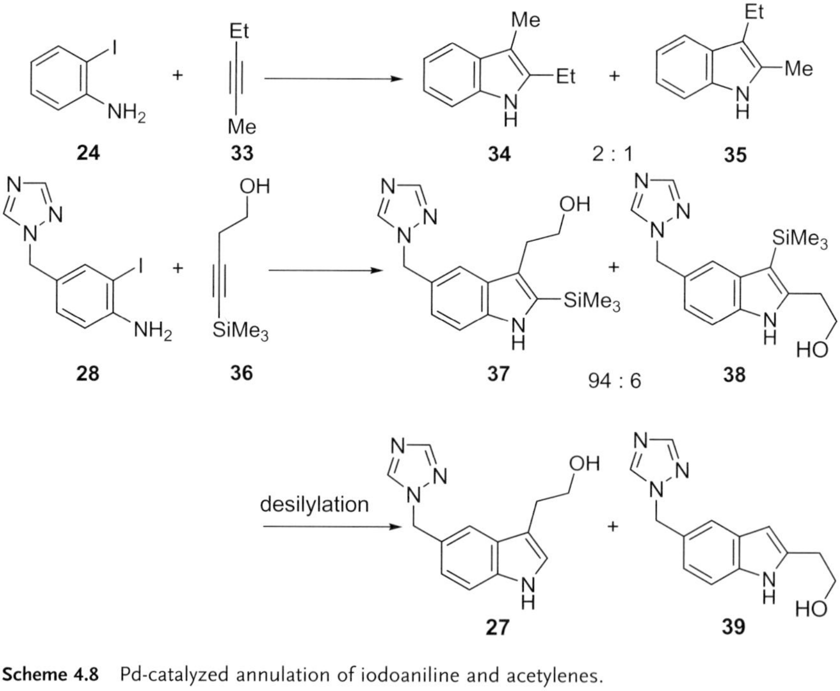

Scheme 4.8 Pd-catalyzed annulation of iodoaniline and acetylenes.

group ended up adjacent to the nitrogen as expected. This very first attempt using TMS-butynol **36** proved to be a valid method for the preparation of tryptophol **27** (56% assayed yield after desilylation).

Although the least expensive and most readily available silyl protecting group was TMS, the low yield of indole product **37** and the generation of many impurities was a disadvantage. The liability of the TMS group was responsible for the low yield and several of the impurities were identified from this reaction (Scheme 4.9): Siloxane **44** was presumably formed from hypervalent silicon species **43**, derived from the reaction of the free alcohol with the silyl group followed by a methyl migration; this by-product was not formed when the alcohol was protected.

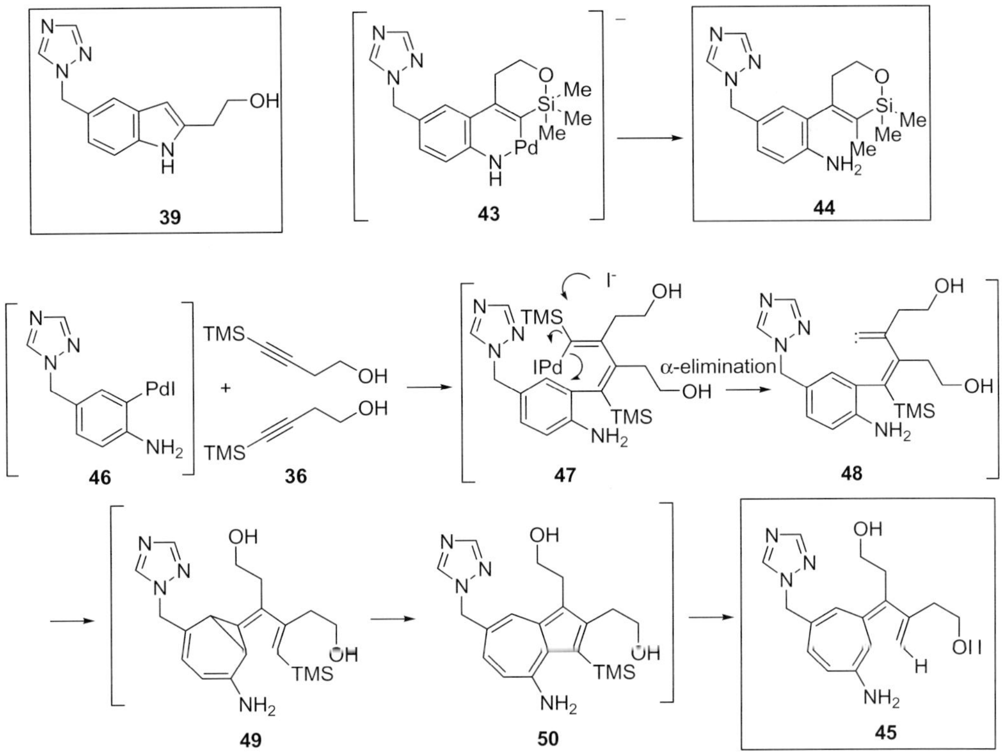

Scheme 4.9 Mechanism for the formation of impurities derived from coupling of iodoaniline **28** and TMS-butynol **36**.

C-TMS protection of the alkyne provided acceptable yields of 3-substituted indole as long as the hydroxy group was protected with a stable group. Purple colored impurities, one of which has been identified as azulene **45**, were seen in both coupling reactions using C-TMS-alkynes such as **36** and **40d** (Scheme 4.9). The azulene was presumably formed through the dimerization of acetylenes

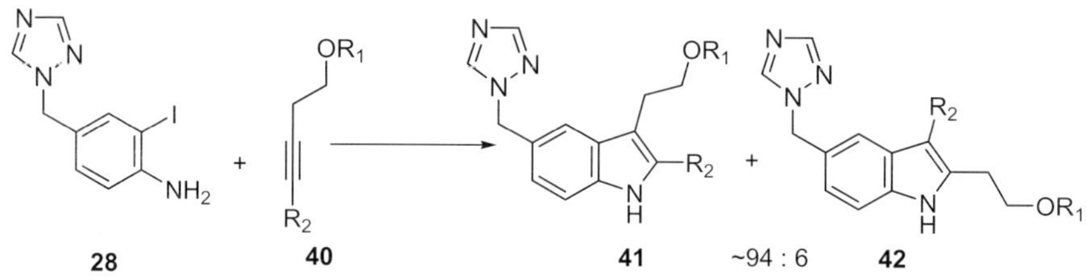

Entry[a]		Acetylenes	Yields of Indoles	
1	40a	R_1, R_2 = TES	41a	(80%)*
2	40b	R_1 = H, R_2 = TES	41b	(74%)
3	36	R_1 = H, R_2 = TMS	37	(56%)
4	40c	R_1, R_2 = TBDMS	41c	(78%)
5	40d	R_1 = TBDMS, R_2 = TMS	41d	(77%)
6	40e	R_1 = H, R_2 = TBDMS	41e	(60%)

[a]Conditions: 2 mol% $Pd(OAc)_2$, Na_2CO_3, DMF, 100 °C; Ratio of **28** : **40** = 1:1.05-1.2
* Mixture of OH and OTES

Figure 4.2 Optimization of protective group.

during the coupling reaction to form intermediate **47**. α-Elimination of **47** and carbene insertion to the benzene ring followed by ring-expansion and desilylation led to the formation of azulene **45**. In contrast, the purple coloration was much less prominent in the reactions with *C*-TES-alkynes (**40a** and **40b**) and *C*-TBDMS-alkynes (**40c** and **40e**). Very likely, the bulkiness of the TES and TBDMS group suppressed the dimerization.

As shown in Figure 4.2, protection of the hydroxy group also played an important role in the yield of the coupling. For instance, coupling of iodoaniline **28** with **40d** (R_1 = TBDMS) gave **41d** (R_1 = TBDMS, 77%) as compared to 56% yield with **36** (R_1 = H). The *C, O*-bis-TBDMS-protected butynol **40c** provided an 18% higher yield than the *C*-mono-TBDMS butynol **40e**, 78% and 60%, respectively.

By-product generation with TMS-alkynes and the sluggish coupling rate with TBDMS-alkynes rendered the triethylsilyl (TES)-alkyne **40a** the best reactant for the coupling reaction. Indeed, *C*-protection with the TES group gave indole **41a** in 80% yield and also provided sufficient hydrolytic stability and satisfactory reaction kinetics for use in large scale synthesis.

The mechanism of the coupling reaction The chemistry of the coupling involves a Pd-catalyzed heterocylization (Scheme 4.10). $Pd(OAc)_2$ is added to the reaction mixture which undergoes *in situ* reduction to Pd(0) upon heating. The Pd(0) is necessary for the oxidative addition to occur with the iodoaniline **28** (Step A) to form the arylpalladium (II) iodide species **46**. The palladium then forms a π-complex with the alkyne (Step B) followed by a carbopalladation reaction (Step C). Reductive elimination then occurs to generate the indole and regenerate Pd(0) (Step D). As part of the ring formation and reductive elimination to Pd(0), hydrogen iodide is generated, and neutralized with a base, such as Na_2CO_3, in order to maintain adequate catalyst turnover.

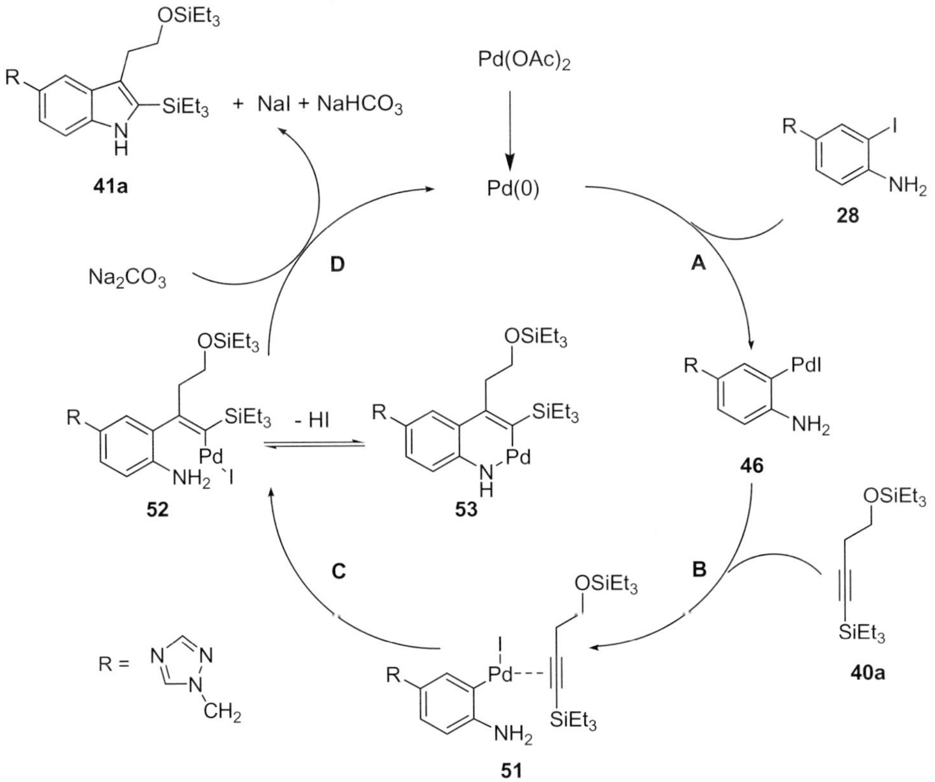

Scheme 4.10 Mechanism for Pd-mediated indole formation.

Preparation of *C,O*-bis-TES butynol ether 40a Experimentally, preparation of the bis-TES-butynol **40a** was carried out by deprotonation of 3-butyn-1-ol (**54**) with *n*-BuLi at −20 °C to generate dianion **55**. Subsequent addition of TESCl yields the desired bis-protected butynol **40a**, as shown in Scheme 4.11. Reaction optimization revealed that 2.0 equiv of *n*-BuLi was optimal for this reaction. With excess *n*-BuLi several unidentified impurities were formed, resulting in a decrease in yield and product purity. Furthermore, an overcharge of *n*-BuLi cannot be rectified by addition of more butynol **54** later. For example, an overcharge (2.6 equiv) of *n*-BuLi

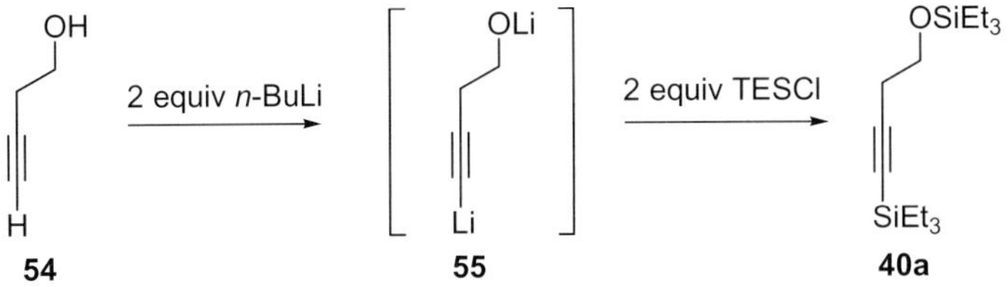

Scheme 4.11 Preparation of *C,N*-bis-silylated 3-butyn-1-ol **40a**.

reduced the isolated yield of **40a** to only 57% due to side reactions. In an alternative procedure the dianion can be generated by reaction of **54** with MeMgCl. The *C*-silylation occurred readily to provide mono-*C*-protected alkynol. However, in this case *O*-silylation required 48 h to complete, presumably due to the low nucleophilicity of the magnesium alkoxide.

Optimization of the coupling reaction The optimized coupling reaction was conducted by heating a degassed mixture of iodoaniline **28**, a slight excess of *C,O*-bis-TES-butynol **40a** (1.05 equiv), powdered Na_2CO_3 (5 equiv), $MgSO_4$ (1.5 equiv) and $Pd(OAc)_2$ (2 mol%) in DMF at 105 °C for 6–7 h. No benefit was realized with the use of various phosphines as ligands. This was presumably due to the use of the reactive aryl iodide substrate and the stabilization of the Pd intermediates via internal chelation of the amine as well as DMF solvation. The reaction mixture contained a 95 : 5 mixture of **41a** and **41b** along with 3–4% of regioisomer **42a** (Scheme 4.12). Other reaction parameters were screened during the optimization and each is briefly discussed below.

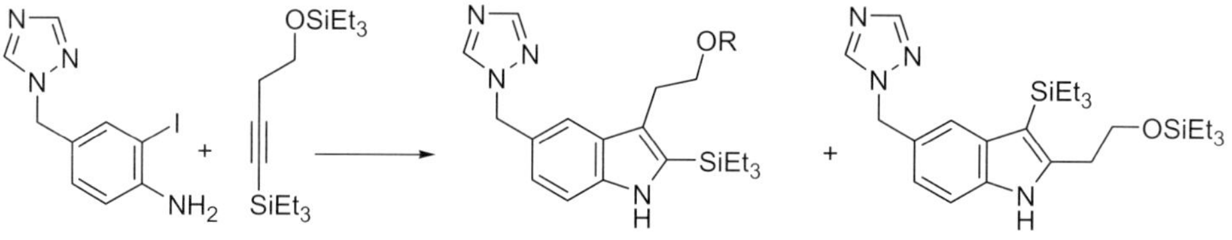

Scheme 4.12 Optimized Pd-catalyzed indole formation.

The use of Na_2CO_3 in the process followed the literature method for the palladium-catalyzed indolization [8]. A variety of bases and solvents were tested during optimization studies [6]. Interestingly, both Li_2CO_3 and K_2CO_3 gave low conversions (30–45%) and Cs_2CO_3 and $CaCO_3$ were completely ineffective. Although some amines were suitable bases for the coupling reaction, the generation of amine-derived impurities rendered this class of bases undesirable. These side reactions with amines, though not useful for the rizatriptan process, were notable since they later led to the discovery of a novel indole ring construction method via palladium catalyzed annulation of iodoaniline with ketones. This surprising discovery and application to the synthesis of indoles will be elaborated on in Section 4.2.

$MgSO_4$ was not needed for the coupling reaction itself, rather, its presence minimized the desilylation of the *C,O*-bis-TES-tryptophol **41a** to *C*-mono-TES-tryptophol **41b**, thereby, improving the yield of product. The $MgSO_4$ charge can be varied from zero, where the desilylation reached 30%, to 3 equiv where the desilylation was minimized at 4%. A charge of 1.5 equiv of $MgSO_4$ was set to minimize both the desilylation and the total amount of solids in the reaction mixture in order to allow adequate mixing. The use of $MgSO_4$ decreased the reaction rate slightly but this could be compensated by increasing the temperature by

5 °C to 105 °C. Above 110 °C, the yield of the reaction began to decrease and the amount of the regioisomer **42a** began to increase.

4.1.2.2.3 **Residual Palladium Removal and Desilylation to Tryptophol 27**

After cooling the reaction mixture, the inorganic salts were removed from the mixture by filtration. The filtrate was then treated with n-Bu_3P (20 mol%) to sequester and solubilize residual palladium in the organic phase. If n-Bu_3P was not used, colloidal Pd would continue to precipitate during the work-up. Furthermore, without this additional work-up step the isolated tryptophol **27** would be contaminated with high levels of Pd (800–1300 ppm). At this level of Pd contamination, the residual Pd specification of <10 ppm in the final drug product could not be attained. The use of n-Bu_3P consistently provided tryptophol **27** contaminated with <100 ppm of Pd. This was readily reduced to <10 ppm in the final product. The mixture was then diluted with isopropyl acetate (IPAc) and water. The bis-silylated product **41a** and mono-silylated product **41b** were partitioned into the IPAc phase and the aqueous phase was discarded. The IPAc phase, containing **41a**, **41b** and some solubilized Pd, was treated with a 2 M aqueous solution of citric acid. This treatment removed the majority of highly colored impurities while also desilylating most of **41a** to **41b**. The majority of the regioisomer **42a** was converted to **39** and was removed with the aqueous layer. Evidently, the TES group at the C-3 position was much more easily deprotected than that at the C-2 position. The aqueous citric acid phase was removed and desilylation at the C-2 position was completed by treating the organic phase with 2.5 M aqueous hydrochloric acid. The now completely desilylated tryptophol **27** remained in the acidic aqueous phase (Scheme 4.13).

Tryptophol **27** was isolated by neutralization of the aqueous phase and extraction into IPAc/MeOH (85 : 15). Methanol was used to increase the solubility of tryptophol **27** in the organic layer and only two extractions were required to obtain >95% recovery. The organic layers were combined and treated with activated carbon (Darco-KB, 10 wt%) to further remove colored impurities. After filtration to remove the carbon, removal of the methanol from the organic phase by distillation led to crystallization of **27**. Addition of heptane to the IPAc slurry completed crystallization of the product. Tryptophol **27** was isolated by filtration in 70–72% isolated yield. The isolated **27** was only contaminated by 80–100 ppm of residual Pd and contained ~3A% of regioisomer **39**, which was reduced to <0.1% during down stream processing. This refined work-up provided high quality tryptol **27** in a reproducible manner.

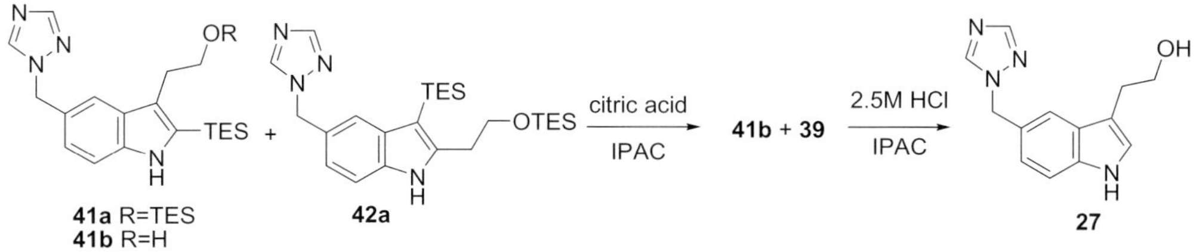

Scheme 4.13 Desilylation.

4.1.2.2.4 End-Game Chemistry for Rizatriptan (1)

Conversion of tryptophol **27** to rizatriptan benzoate (**1**) was fairly straightforward, as shown in Scheme 4.14. The formation of mesylate **56** from tryptophol **27** followed by displacement with dimethylamine afforded rizatriptan as the free base. Mesylate **56** was prone to polymerization from intermolecular alkylation by the triazole, therefore crude mesylate **56** was treated directly with 40% aqueous dimethylamine. Experimentally, the reaction of tryptophol **27** with triethylamine (1.3 equiv) and MsCl (1.2 equiv) over 15 min at −20 °C in THF (10 ml g^{-1}) gave a solution of mesylate **56** together with a precipitate of triethylamine hydrochloride. Direct reaction of the crude reaction mixture with 40% aqueous dimethylamine resulted in formation of rizatriptan in 75–85% yield. Major impurities observed in the crude rizatriptan were the regioisomer **57** (2–3 A%), indole-*N*-mesyl-rizatriptan (3–5 A%) **58** and unreacted tryptophol (**27**, 1 LCAP%) as shown in Figure 4.3. All of these impurities were removed in the benzoate salt formation step. Addition of a solution of benzoic acid in IPAc-isopropyl alcohol (IPA) to the

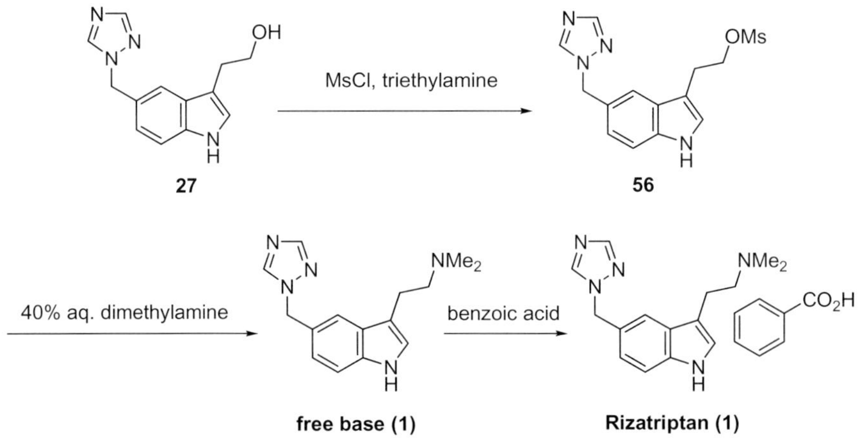

Scheme 4.14 Conversion of tryptophol **21** to rizatriptan (**1**).

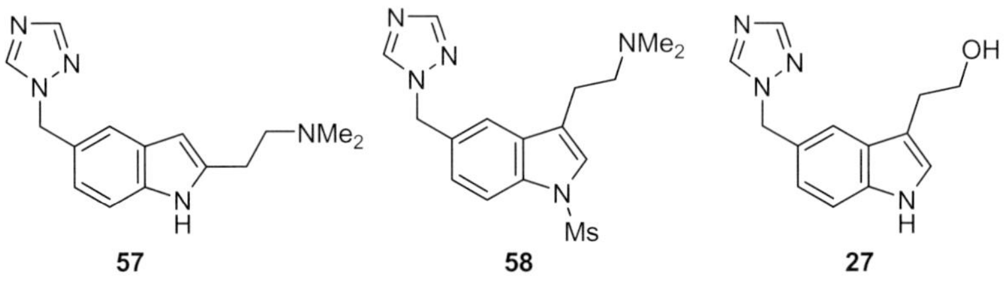

Figure 4.3 Impurities in the end game.

rizatriptan free base in isopropyl alcohol afforded rizatriptan as a benzoate salt in 95% yield and excellent purity.

4.1.2.2.5 Summary of Process Development

1) A much improved, chromatography-free process for aniline **6** was developed with ~90% overall yield. The remarkable efficiency for each transformation rendered this intermediate **6** to be suitable for out-sourcing without further optimization.
2) The pH-controlled iodination for the preparation of iodoaniline **28** not only eliminated the solid waste in the isolation step but also minimized the formation of diiodoaniline **32**.
3) The key Pd-catalyzed coupling step to the indole was optimized by judicious choice of the protection group on 3-butyn-1-ol, screening of different reaction parameters such as bases and additives, and implementing a fine-tuned, yet practical work-up procedure. Tryptophol **27** was isolated in 76–82% yield by crystallization.
4) Overall, rizatriptan was synthesized in excellent purity (>99 A%) and ~60% yield over four steps from iodoaniline **28**.

4.2 Chemistry Development

4.2.1 Application of Pd-Catalyzed Annulation to the Synthesis of the Indole Acetic Acid

The newly developed Pd-catalyzed indole construction method was the cornerstone of the manufacturing process for rizatriptan (**1**). The Pd-catalyzed indole preparation was extended to the synthesis of 3-indole acetic acid (L-749,335, **2**) [10]. 3-Indole acetic acid **2** is a metabolite and a potential oxidative degradate from rizatriptan benzoate. A highly pure sample of **2** was required for a screening test to compare its bioactivity with that of rizatriptan. The efficient synthesis of this seemingly simple target has proven to be challenging using traditional methods, as shown in Scheme 4.15. For example, the Fisher indole synthesis using either aldehyde **59** or acetal **60** with triazole hydrazine **7** in 4% H_2SO_4 failed to give the desired product. Presumably, the products decomposed under the reaction conditions. Oxidation of of tryptophol **27**, a key intermediate for rizatriptan, only led to oligomers.

Eventually, indole acetic acid **2** was prepared from iodoaniline **28** and propargyl alcohol derivative **61** via the newly developed coupling reaction followed by a cyanide displacement-hydrolysis sequence, as shown in Scheme 4.16.

The *C,O*-bis-TES side chain **61** was prepared quantitatively from propargyl alcohol and TESCl using *n*-BuLi as a base. Pd-catalyzed coupling of iodoaniline **28** with bis-TES side chain **61** in the presence of 1.5 equiv of $MgSO_4$ and 5 equiv of

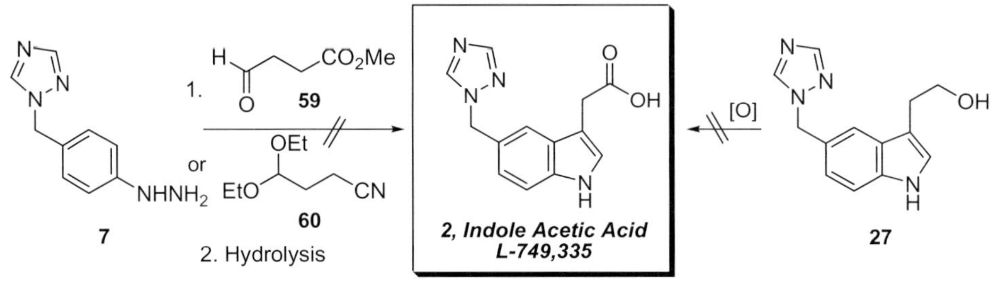

Scheme 4.15 Fisher indole and oxidation approach to indole acetic acid **2**.

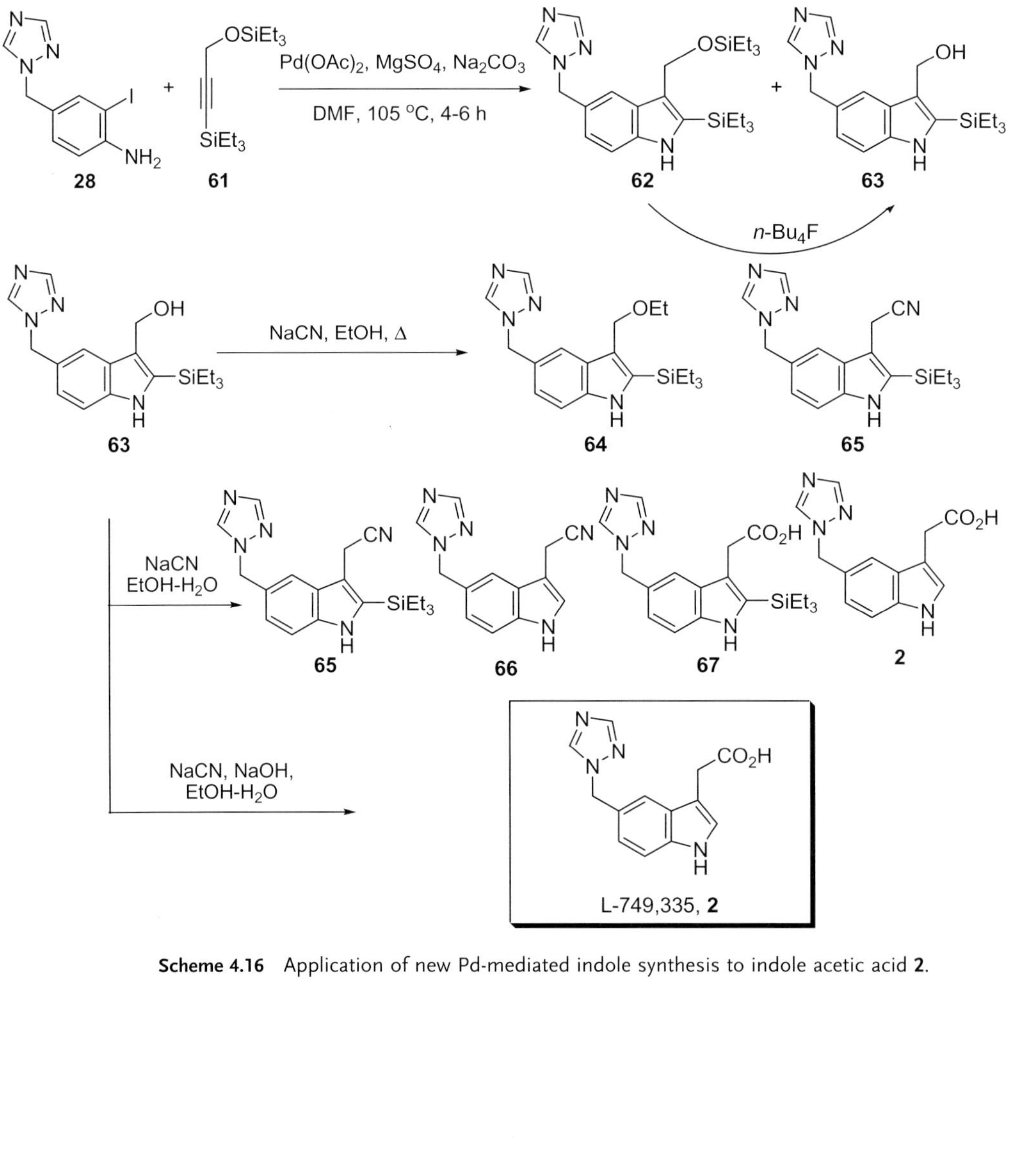

Scheme 4.16 Application of new Pd-mediated indole synthesis to indole acetic acid **2**.

Na_2CO_3 in DMF gave a mixture of **62** and **63** in a ratio of 96 : 4 (by HPLC A%). Selective *O*-desilylation of this mixture with *n*-Bu_4NF afforded 2-triethylsilylindole **63**. 2-TES-indole **63** was next converted to acid **2** through intermediates **65–67**.

Treatment of alcohol **63** with excess NaCN in EtOH upon heating gave a 1 : 1 mixture of indole acetonitrile **65** and indole ethyl ether **64**, presumably formed due to the competitive cyanide and EtOH exchange with the OH group in **63**. The exchange most likely proceeded through an intermediate such as **68**. (Scheme 4.17). In fact, reaction of alcohol **63** with EtOH in the presence of a catalytic amount of NaOH afforded ethyl ether **64** quantitatively. The formation of **64**, however, could be suppressed when 9 : 1 EtOH:H_2O was used as the solvent. The reaction in EtOH–H_2O afforded a mixture of desired acid **2**, and intermediates **65–67** due to incomplete hydrolysis of the nitrile moiety and desilylation. We reasoned that addition of NaOH to the reaction would drive indoles **65–67** to the desired acid **2** and, indeed, when alcohol **63** was treated with a mixture of NaCN and NaOH in EtOH–H_2O under prolonged reflux, the sodium salt of **2** was formed in good yield. The cyanide exchange, hydrolysis of the cyanide group, and desilylation all occurred in this step. Acidification of the reaction mixture with 2 M HCl to pH 3 precipitated the crude acid **2**. This material, however, contained two major impurities: dimer **69** (7A%) and trimer **70** (2A%). Both of which were presumably formed through **68** [5] and were extremely difficult to remove from **2** by silica gel chromatography or traditional crystallization techniques.

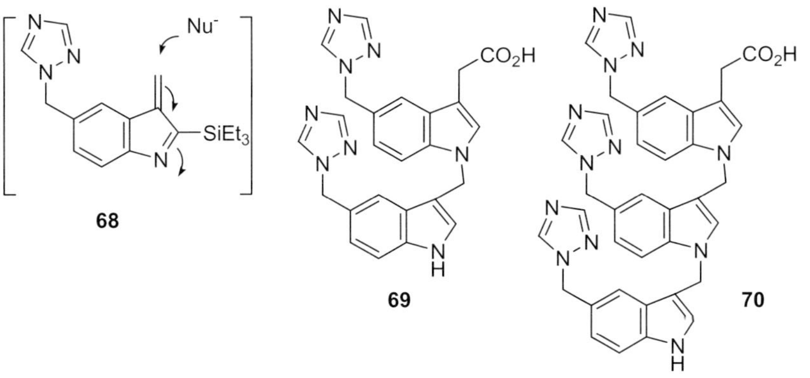

Scheme 4.17 Reaction intermediate and structures of dimer and trimer.

For the large scale synthesis, the sodium salt of **2** formed in the NaCN–NaOH reaction could be purified by brominated polystyrene resin SP-207 chromatography to avoid acidic work-up which generates HCN. The SP-207 resin was first saturated with 1 M NaCl, and the crude reaction mixture was loaded onto the column. The column was then eluted with 1 M NaCl to remove inorganic salts such as excess NaCN and NaOH and other polar impurities. Eluant switching to MeOH–H_2O eluted the sodium salt of **2**. Fractions containing >98.5 A% of sodium salt of **2** and no contamination with dimer **69** and trimer **70** were combined, concentrated, and pH adjusted to 3–4 with 2 M HCl to precipitate out acid **2**. The preparation of analytically pure acid **2** was achieved in 38% overall yield from iodoaniline **28** utilizing an efficient one-pot reaction sequence from alcohol **63** to **2**.

4.2.2
New Indole Chemistry from Development of Pd Chemistry

4.2.2.1 Discovery of New Indole Synthesis from Amines

As mentioned, during the optimization of the Pd-catalyzed coupling reaction between iodoaniline **28** and acetylene **40a**, various amine and inorganic bases were investigated. In one experiment, diisopropylamine (DIPA) proved to be a suitable base, providing tryptophol **27** in 76% yield after desilylation of **41a**. The regioisomer ratio (94 : 6) was similar to that obtained when using Na_2CO_3 as the base. The reaction had many benefits: the reaction was faster, taking place in 2 h versus 4 h, and the work-up was cleaner and simpler. A severe disadvantage, however, was the generation of ~9% of a new impurity **71** (Scheme 4.18).

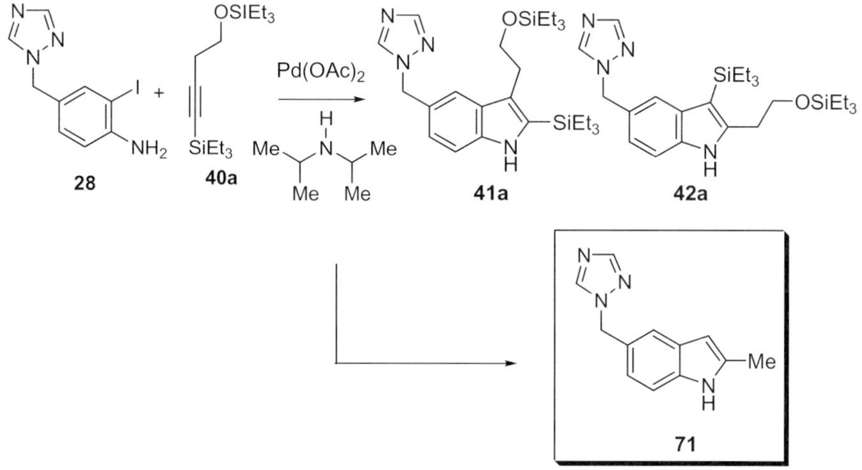

Scheme 4.18 Coupling reaction using amines as base-formation of 2-methyl-indole.

This new impurity proved to be derived from the Pd-catalyzed oxidation of DIPA to the enamine via β-hydride elimination. In fact, mixing $Pd(OAc)_2$ with DIPA in DMF-d_7 readily formed Pd black along with two species, primary amine and acetone, presumably derived from the enamine through hydrolysis. The resulting enamine or acetone then underwent a coupling reaction with iodoaniline **28**. Heterocyclization through the arylpalladium(II) species provided 2-methyl indole **71**, as shown in Scheme 4.19.

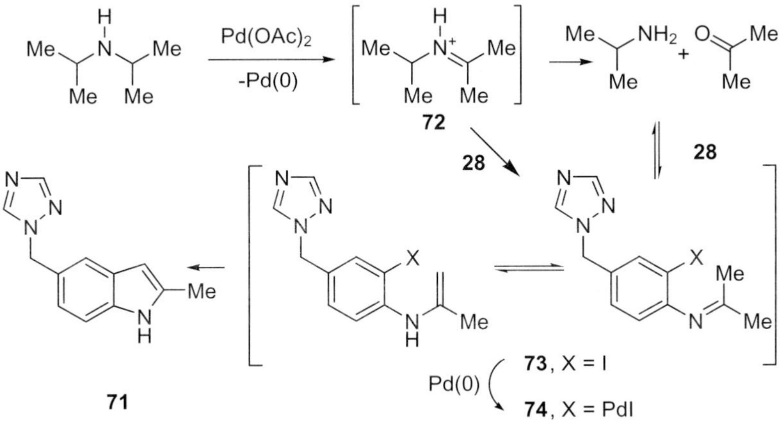

Scheme 4.19 Mechanism for the formation of 2-methyl indole **71**.

Similarly, tryptophol **27** was obtained in 60–70% yield with other amines. Each by-product related to the corresponding amine was also observed (Scheme 4.20). For example, using triethylamine as base forms indole impurity **75** while using dicyclohexyl amine leads to the formation of bicyclic indole impurity **76**.

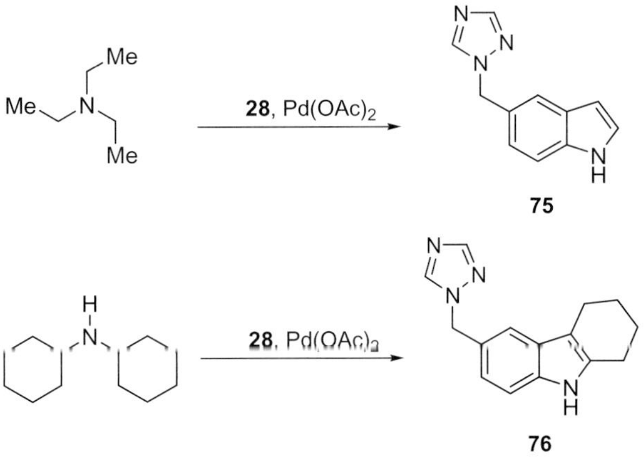

Scheme 4.20 Amine-derived indole byproducts.

In order to eliminate this Pd-catalyzed amine oxidative pathway, reaction conditions were further optimized and the problem could be circumvented by using a base that would not undergo this oxidation. DABCO and quinuclidine were candidates because oxidation will not occur due to Bredt's rule [11]. The coupling reaction (Scheme 4.18) using DABCO as base gave a 60% yield of tryptophol **27** over 2 h with 2% regioisomer formation after acid hydrolysis. Performing the same coupling reaction using quinuclidine as the base gave a 76% yield of tryptophol **27**. This is not competitive with the carbonate process but due to the high cost and lack of availability of quinuclidine this base was not developed further. The coupling reaction employing proton sponge (1,8-bis(dimethylamino)napthalene), an amine base containing methyl groups which would not lead to indole by-products even if they underwent oxidation, only gave 20% conversion to indoles after 3 days.

Interestingly, using amine bases in solvents other than DMF did not generate the enamine-derived indoles. For example, running the reaction in propionitrile gave the desired product using either *n*-Bu_3N or diisopropylethylamine as bases. Indoles **41a** and **42a** were obtained in 65% and 1.8% assay yield, respectively. Using amine bases in solvents other than DMF apparently led to an increase in the regioselectivity. Propionitrile, however, was not suitable for large scale reaction due to potential β-elimination with release of HCN. Reaction in acetonitrile, on the other hand, was very sluggish.

4.2.2.2 Direct Coupling of Iodoaniline with Ketone

Since the amine by-product formation was essentially derived from the reaction of an enamine or a ketone with iodoaniline, the direct use of a ketone as the substrate instead of an amine, would also be expected to yield the indole (Scheme 4.21). Indeed, we were gratified to find that direct condensation of *o*-iodoaniline **24** (77, R_1 = H) with cyclohexanone (in the presence of 5 mol% $Pd(OAc)_2$ and 3 equiv DABCO as a base at 0.3 M and 105 °C afforded the tetrahydrocarbazole **81a** in 77% yield with no other major impurities (Figure 4.4) [5]. The use of DMF as a solvent is crucial to the success of this reaction; other solvents such as acetonitrile and toluene were ineffective.

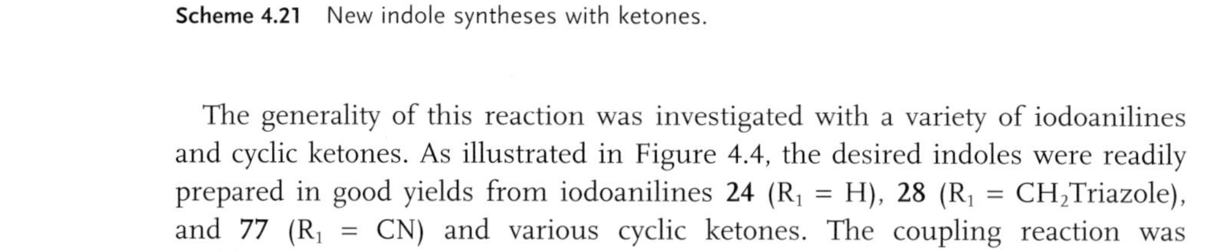

Scheme 4.21 New indole syntheses with ketones.

The generality of this reaction was investigated with a variety of iodoanilines and cyclic ketones. As illustrated in Figure 4.4, the desired indoles were readily prepared in good yields from iodoanilines **24** (R_1 = H), **28** (R_1 = CH_2Triazole), and **77** (R_1 = CN) and various cyclic ketones. The coupling reaction was highly regioselective. For instance, condensation of iodoaniline **24** with 2-methylcyclohexanone gave 2-methyltetrahydrocarbazole **81j** in 68% yield. Reaction of 3-methylcyclohexanone with iodoaniline favored formation of carbazole **81i**, with only 5% regioisomer observed. The coupling of the indole nucleus onto a

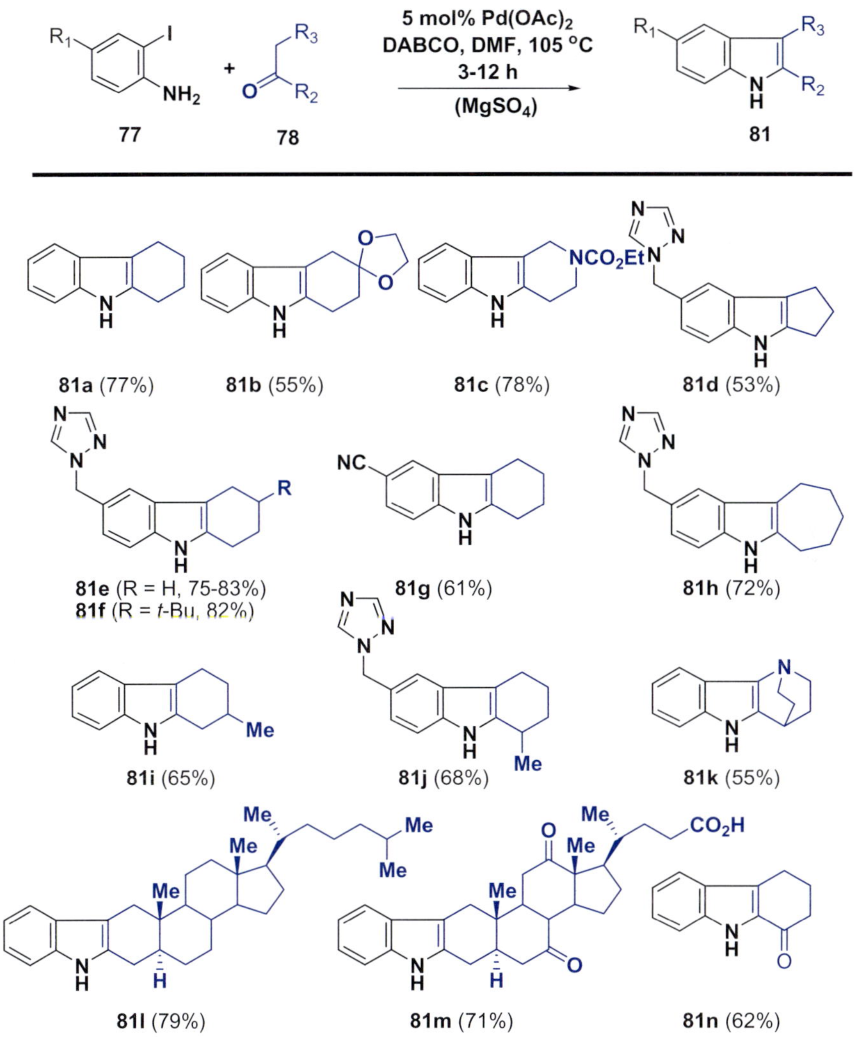

Figure 4.4 Scope and limitation of new indole construction method.

steroid was also achieved with 5α-cholestanone (1.0 equiv) affording **81l** exclusively in 79% yield. Coupling using another steroid bearing three carbonyl groups occurred regioselectively to afford **81m** in 71% yield. The reaction is also compatible with cyclopentanone and cycloheptanone as both **81d** and **81h** were isolated in 53% and 72% yield, respectively.

The reaction tolerates a variety of functional groups, especially the acid-sensitive acetal (**81b**), carbamate (**81c**) and the benzyl triazole (**81d–f**, and **81h, j**). These intermediates, which are unstable under the conditions of the traditional Fischer indole reaction, were conveniently synthesized using this method. The structurally interesting indole **81k** was prepared from 3-quinuclidinone hydrochloride (1.0 equiv) in 55% yield. Though most reactions proceeded efficiently in DMF at 105 °C, the additive $MgSO_4$ (1.5 equiv), presumably acting as a dehydrating agent, was found to promote the annulation in the more sluggish cases (**81d**, and **81j–m**). Finally, coupling of 1,2-cyclohexane-dione with iodoaniline **24**, afforded ketocarbazole **81n** in 62% yield.

Compared to the cyclic ketones, the coupling of aliphatic aldehydes to prepare 3-substituted indoles was less successful, except for phenyl acetaldehyde, which afforded 3-phenyl indole **83** in 76% yield (Scheme 4.22). The lack of imine formation or the instability of the aliphatic aldehyde towards the reaction conditions may be responsible for the inefficiency of these reactions. Therefore, a suitable aldehyde equivalent was considered. With the facile removal of a 2-trialkylsilyl group from an indole, an acyl silane was tested as a means of preparing 3-substituted indoles. Indeed, coupling of acetyl trimethylsilane with the iodoaniline **24** gave a 2 : 1 mixture of 2-TMS-indole **84** and indole (**85**) in a combined 64% yield. Evidently, the reaction conditions did lead to some desilylation. Regardless, the silyl group of **84** was quantitatively removed upon treatment with HCl to afford indole (**85**).

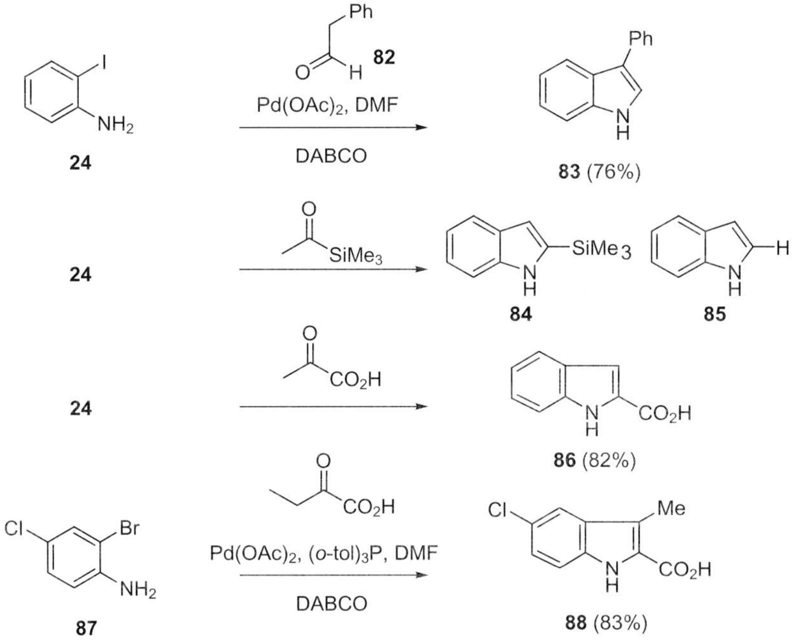

Scheme 4.22 Coupling reaction using aldehyde, acylsilane and oxo-acid.

Indole 2-carboxylic acids can be readily decarboxylated to afford an indole. Hence, using pyruvic acid as an aldehyde equivalent in the coupling with **24** gave

2-indolecarboxylic acid (**86**) in 82% yield. Furthermore, the reaction was extended to bromoaniline **87** which provided indole carboxylic acid **88** in 83% yield. The coupling reaction with other acyl silanes and 2-oxocarboxylic acid or esters is worthy of further investigation for the synthesis of 2,3-substituted indoles.

Thus far, we have discovered and demonstrated a new and efficient method for the synthesis of indoles from various carbonyl compounds. This, in conjunction with the use of alkynes in the palladium-catalyzed indolization, widens the spectrum of indoles that can be prepared by these means. The simple procedure, mild reaction conditions, and ready availability of the starting materials render these methods valuable additions to indole chemistry. We next extended this method to the synthesis of the indole core of a PGD_2 receptor antagonist, laropiprant **3**.

4.2.2.3 **Application to Laropiprant Indole Synthesis**

Prostaglandin D_2 (PGD_2) is the major cyclooxygenase metabolite of arachidonic acid produced by mast cells in response to antigen challenge [12]. It has been proposed that excess production of PGD_2 causes the inflammation commonly observed in allergic diseases such as allergic rhinitis, asthma, and atopic dermatitis [13]. Efforts to develop a PGD_2 receptor antagonist have identified laropiprant (**3**) as a very promising lead in the alleviation of various allergic disorders [14]. This compound is also being tested in combination with niacin to reduce blood cholesterol [15]. Our thought was to apply the indole synthesis described above to the indole core of the molecule by reaction of bromoaniline **93** and cyclopentanone **90**. In this case, the efficiency of the reaction relied on preformation of an imine followed by an intramolecular heterocyclization. Prior to this approach, Fischer indole cyclization of hydrazine **89** with optically pure ketoester **90** provided the advanced intermediate **91** with complete racemization of the stereogenic center (Scheme 4.23, eq 1) as well as formation of the regioisomer **92** [16]. Likewise, and not surprisingly, condensation of aniline **93** with optically pure **90** followed by intramolecular Heck reaction rapidly provided the desired indole **94** in good yield, but again as a racemic mixture (Scheme 4.23, Eq. (2)). However, these procedures described above showed promise for a very efficient racemic synthesis of the drug

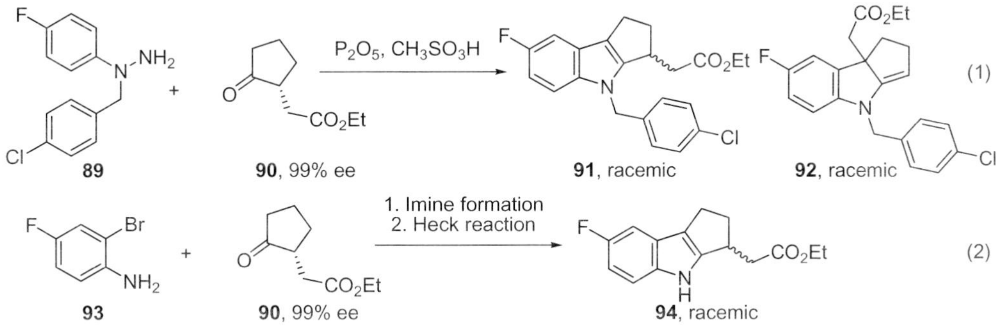

Scheme 4.23 Synthesis of racemic indole ester **94**.

and we recognized that enantiopure **3** could be potentially obtained via a chemical resolution.

Condensation of the commercially available 2-bromo-4-fluoroaniline (**93**) and racemic ethyl 2-oxo-cyclopentylacetate (**90**) in refluxing toluene in the presence of a catalytic amount of *p*-TsOH under standard Dean–Stark azeotropic distillation afforded imine **95** in greater than 95% conversion (Scheme 4.24). The reaction was run uneventfully several times on 100 g scale. However, when we scaled it up to 1 kg, it stalled at 83% conversion and attempts to drive it to completion failed. This prompted us to re-investigate this reaction and we decided to screen different water scavengers including CF_3CO_2*i*-Pr, $(EtO)_3B$, $(EtO)_4Si$, and Ti(O*i*-Pr)$_4$. All of them were found to be ineffective but to our delight $P(OEt)_3$ was tried and proved to be very effective. Removal of the ethanol by-product was not required. After optimization, the reaction was conducted neat with 1.2 equiv of $P(OEt)_3$ and 4 mol% of H_3PO_4 at 70 °C for 4 h to give **95** in 98% conversion. The resulting intermediate imine **95** was subsequently used without purification in the Heck cyclization, but no reaction occurred due to catalyst poisoning. We found that **95** was stable enough to survive extraction into cyclohexane (10% TEA) with water washes to remove the diethyl phosphonite by-product that was poisoning the catalyst.

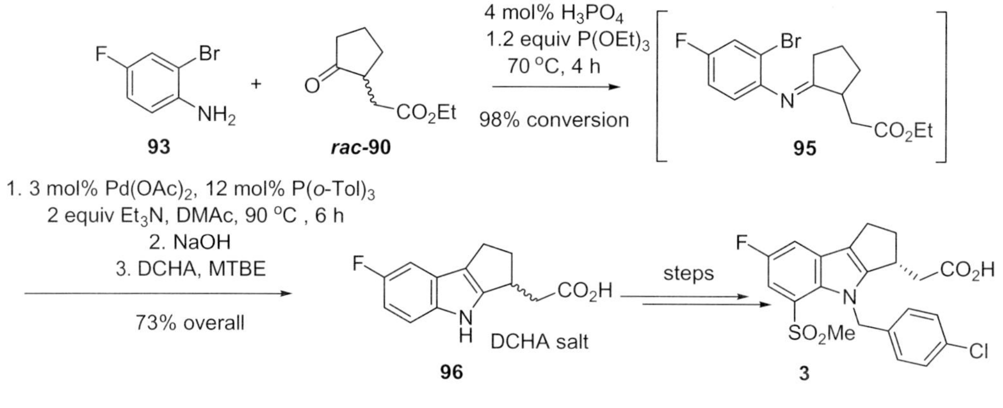

Scheme 4.24 Synthesis of racemic indole acid **96**.

After solvent switching from cyclohexane to dimethylacetamide (DMAc), the Pd-catalyzed heterocyclization was carried out using 3 mol% $Pd(OAc)_2$, 12 mol% tri-*ortho*-tolylphosphine and 2 equiv of TEA at 90 °C for 6 h to give the ethyl ester indole **94**. Finally, the ester was hydrolyzed in the same pot with 5 M aqueous NaOH to give the corresponding acid, which was crystallized as its dicyclohexylamine (DCHA) salt **96** in MTBE in 73% overall isolated yield (Scheme 4.24). The indole **96** serves as the key intermediate in the synthesis of laropirant (**3**) [17].

4.3 Conclusion

A new, efficient and chromatography-free process for the synthesis of rizatriptan (MK-0462, **1**) featuring a palladium-catalyzed coupling of iodoaniline **28** and *C,O*-bis-TES-butynol **40a** to form the indole core has been developed, and is amenable to scale-up. The process has been successfully demonstrated on the multi-kilogram scale to provide product of excellent purity in good yield. This method is currently the manufacturing route for rizatriptan. The key Pd-catalyzed coupling protocol was extended to the efficient synthesis of the indole acetic acid metabolite **2**. During the study of the effect of base on the coupling reaction of iodoaniline and acetylenes, we discovered a new indole synthesis based on the direct coupling between iodoaniline and a variety of carbonyl species such as cyclic ketones, aldehydes, acyl silanes and oxo-carboxylic acids. A wide array of structurally diverse indoles were readily prepared in an efficient manner. Finally, this novel indole synthesis was readily applied to the synthesis of the key indole intermediate for laropiprant (**3**), a potential drug tested in combination with niacin to reduce blood cholesterol.

Acknowledgments

The author thanks David R. Lieberman, Robert, D. Larsen, Thomas R. Verhoeven, Robert A. Reamer, Chris. H. Senanayake, Timothy Bill, Simon Johnson, Peter D. Houghton, Richard G. Osifchin, Michel Journet and Guy Humphrey for their contributions to the projects discussed.

References

1 (a) Ferrari, M.D. (1993) *Neurology*, **43** (Suppl. 3), S43–S47; (b) Plosker, G.L., and McTavish, D. (1994) *Drugs*, **47**, 622.

2 Street, L.J., Baker, R., Davey, W.B., Guiblin, A.R., Jelly, R.A., Reeve, A.J., Routledge, H., Sternfeld, F., Watt, A.P., Beer, M.S., Middlemiss, D.N., Noble, A.J., Stanton, J.A., Scholey, K., Hargreaves, R.J., Sohal, B., Graham, M.I., and Matassa, V.G. (1995) *J. Med. Chem.*, **38**, 1799.

3 For an excellent overview on rizatriptan and references cited therein, see: (a) Hargeaves, L.C.R., Rapoport, A.M., Ho, T.W., and Sheftell, F.D.R.J. (2009) *Headache*, **49**, S3; for selected expert review articles: (b) Mannix, L.K. (2008) *Expert Opin. Pharmacother.*, **9**, 1001; (c) Pascual, J. (2004) *Expert Opin. Pharmacother.*, **5**, 669.

4 (a) Humphrey, P.P.A., and Goadsby, P.J. (1994) *Cephalagia*, **14**, 401; (b) Williamson, D.J., Hill, R.G., Stepheard, S.L., and Hargreaves, R.J. (2001) *Br. J. Pharmacol.*, **133**, 1029; (c) Williamson, D.J., Stepheard, S.L., Hill, R.G., and Hargreaves, R.J. (1997) *Eur. J. Pharmacol.*, **328**, 37; (d) Clumberbatch, M.J., Hill, R.G., Stepheard, S.L., and Hargreaves, R.J. (1997) *Eur. J. Pharmacol.*, **328**, 49.

5 Chen, C.-y., Lieberman, D.R., Larsen, R.D., Reamer, R.A., Verhoeven, T.R., and Reider, P.J. (1994) *Tetrahedron Lett.*, **35**, 6981.

6 Chen, C.-y., Senanayake, C.H., Bill, T.J., Larsen, R.D., Verhoeven, T.R., and Reider, P.J. (1994) *J. Org. Chem.*, **59**, 3738.

7 Chen, C.-y., Lieberman, D.R., Larsen, R.D., Verhoeven, T.R., and Reider, P.J. (1997) *J. Org. Chem.*, **62**, 2676.

8 Larock, E.C., and Yum, E.K. (1991) *J. Am. Chem. Soc.*, **113**, 6689.

9 Astleford, B.A., Goe, G.L., Keay, J.G., and Scriven, E.F.V. (1989) *J. Org. Chem.*, **54**, 731.

10 Chen, C.-y., Lieberman, D.R., Larsen, R.D., and Verhoeven, T.R. (1996) *Synth. Commun.*, **26**, 1977.

11 Bredt, J., Thouet, H., and Schnitz, J. (1924) *Liebigs Ann.*, **437**, 1.

12 Lewis, R.A., Soter, N.A., Diamond, P.T., Austen, K.F., Oates, J.A., and Roberts, L.J. (1982) *J. Immunol.*, **129**, 1627.

13 (a) Matsuoka, T., Hirata, M., Tanaka, H., Takahashi, Y., Murata, T., Kabashima, K., Sugimoto, Y., Kobayashi, T., Ushikubi, F., Aze, Y., Eguchi, N., Urade, Y., Yoshida, N., Kimura, K., Mizoguchi, A., Honda, Y., Nagai, H., Narumiya, S., Kato, M., Watanabe, M., Vogler, B., Awen, B., Masuda, Y., Tooyama, Y., and Yoshikoshi, A. (2000) *Science*, **287**, 2013; (b) Charlesworth, E.N., Kagey-Sobotka, A., Schliemer, R.P., Norman, P.S., and Lichtenstein, L.M. (1991) *J. Immunol.*, **149**, 671; (c) Proud, D., Sweet, J., Stein, P., Settipane, R.A., Kagey-Sobotka, A., Freidlander, M., and Lichtenstein, L.M. (1990) *J. Allergy Clin. Immunol.*, **85**, 896; (d) Murray, J.J., Tonnel, A.B., Brash, A.R., Roberts, L.J., Gosset, P., Workman, R., Capron, A., and Oates, J. (1986) *N. Eng. J. Med.*, **315**, 800.

14 Berthelette, C., Lachance, N., Li, L., Sturino, C., and Wang, Z. (2003) WO 2003062200 A2 20030731 CAN 139.

15 Lai, E., De Lepeleire, I., Crumley, T.M., Liu, F., Wenning, L.A., Michiels, N., Vets, E., O'Neill, G., Wagner, J.A., and Gottesdiener, K. (2007) *Clin. Pharmacol. Ther.*, **81**, 849.

16 Campos, K.R., Journet, M., Lee, S., Grabowski, E.J.J., and Tillyer, R.D. (2005) *J. Org. Chem.*, **70**, 268.

17 Internal communication to Journet, M., Humphrey, G.R. *et al.* on the long term factory process for the production of a prostaglandin D_2 receptor antagonist – unprecedented asymmetric hydrogenation of an indole *Exo*-Cyclic Trisubstituted α,β-Unsaturated Acid.

5
SERM: Selective Estrogen Receptor Modulator

Zhiguo Jake Song

Estrogen plays an important role in human health and disease. Estrogen receptors (ERs) belong to a super family of steroid hormone nuclear receptors. They are ligand-dependent transcription factors that bind to specific DNA sequences and regulate gene expression. There are two known members of the estrogen receptor family, ERα and ERβ, encoded by distinct genes. Estrogen receptor modulators are ER ligands that act like estrogens in some tissues, but block estrogen action in others. Thus, estrogen receptor modulators may exhibit an agonistic or antagonistic activity, depending on the context in which their activity is examined. Selective estrogen receptor modulators (SERMs) are potentially useful agents for treatment or prevention of a variety of conditions related to estrogen functions including bone loss, cartilage degeneration, endometriosis, uterine fibroid disease, increased levels of LDL cholesterol, and cancer [1–3]. Some SERMs are shown in Figure 5.1. Tamoxifen is a SERM drug approved to treat estrogen positive breast cancer and for breast cancer prevention among high risk women. However, its use leads to menopause-like side effects and acquired resistance. Raloxifene is another SERM approved to treat osteoporosis and has been found to be effective for breast cancer prevention with fewer side effects [1]. Research at Merck in the SERM area

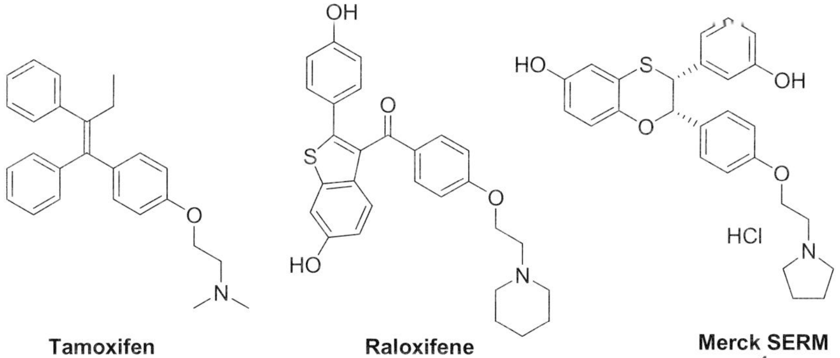

Tamoxifen　　Raloxifene　　Merck SERM 1

Figure 5.1 Structure of a SERM candidate and two SERM drugs.

The Art of Process Chemistry. Edited by Nobuyoshi Yasuda

ISBN: 978-3-527-32470-5

has been focused on finding better therapeutic agents with fewer side effects [4]. Compound **1** is an ERα selective SERM which may offer such advantages [4a].

This chapter will cover the process chemistry development work to support the multiple kilogram delivery of compound **1** and related chemistry development work [5].

5.1 Project Development

5.1.1 Medicinal Route

The key structural features of compound **1** are the chiral *cis*-diaryl benzoxathiin fused ring system, two phenols, and one phenol ether linkage with the pyrrolidinylethanol. Originally, SERM **1** was prepared by medicinal chemists from a key ketone intermediate **5** shown in Scheme 5.1. Compound **5** was prepared in four steps with rather low yield [4a]. Among these steps, the high temperature de-methylation step and the use of extremely toxic MOM-Cl were not particularly suitable for scale-up. The ketone **5** was then brominated with $PhMe_3NBr_3$ (PTAB) and coupled with thiophenol **7** to produce adduct **8**. The key step of the synthesis was the conversion of adduct **8** to *cis*-diaryl benzoxathiin **9** under the Kursanov–Parne reaction conditions (TFA/Et_3SiH). This novel reaction allowed the formation

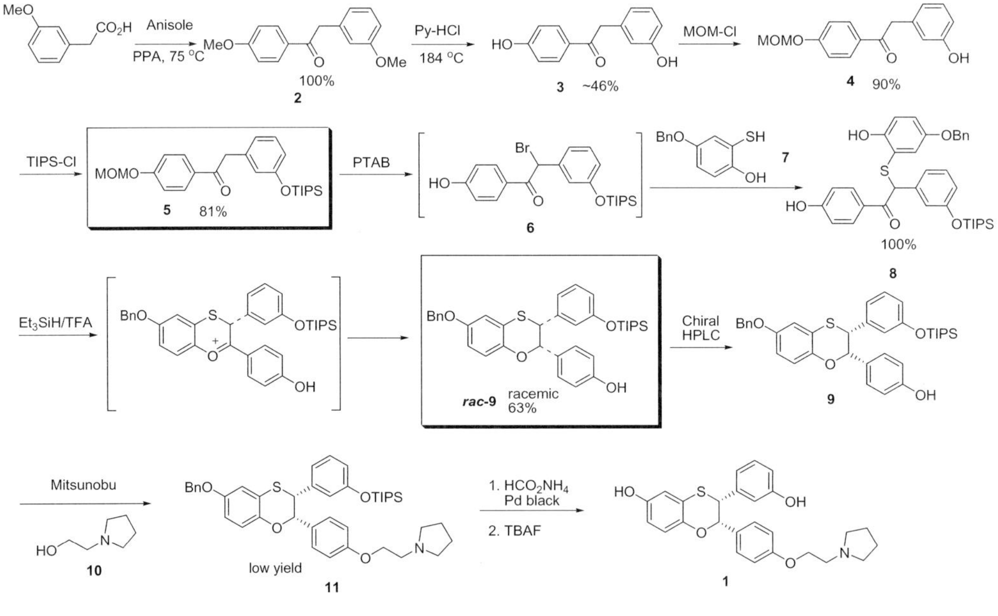

Scheme 5.1 Preparation of **1** by medicinal chemists.

of the *cis*-diaryl benzoxathiin scaffold effectively [6a]. This racemic intermediate **9** was then resolved using a preparative chiral HPLC with very modest separation efficiency. The pyrrolidinylethanol moiety was introduced to the phenol in **9** via a low yielding Mitsunobu reaction with alcohol **10**. Then, the benzyl protecting group in **11** was removed through a sluggish hydrogenolysis which required 100% weight loading of palladium black. Finally, the TIPS protecting group was removed with TBAF and compound **1** isolated through column chromatography. During the synthesis, a number of chromatographic purifications were also required to purify the intermediates.

5.1.1.1 Problems of the Original Route

While the Medicinal Chemistry route was adequate for the initial discovery stage of drug development, viewed against the need to make multiple kilograms or much larger quantities of compound **1** efficiently, the original route suffered from a few obvious shortcomings.

1) Racemic late stage intermediate **9** was resolved using preparative chiral HPLC with low efficiency.
2) Preparation of intermediate **5** was low yielding, required harsh reaction conditions and the carcinogenic reagent MOM-Cl.
3) The Mitsunobu reaction late in the synthesis was low yielding and generated excessive waste.
4) There were excessive protecting group manipulations and the catalyst loading at the final de-benzylation step was very high.
5) Multiple chromatographic purifications were required.

5.1.2 Process Development

To support preparation of multiple kilograms of compound **1** and develop a route that is potentially suitable for long term needs to supply much larger amounts, a few goals were set for the process development of **1** after analyzing the synthetic challenges.

1) Develop an asymmetric synthesis to avoid the late stage resolution.
2) Develop alternatives to the Mitsunobu reaction to improve yield and to reduce reaction waste.
3) Eliminate chromatographic separations.
4) Minimize protecting group manipulations and avoid the use of MOM-Cl or other carcinogenic reagents.

In our retro-synthetic analysis, we envisioned the pyrrolidinylethanol side chain could be installed via the Ullmann ether formation or the analogous reactions from the aryl-iodide functional group. The key intermediate **9** (*cis*) in the Medicinal Chemistry route was not stable under strongly acidic or basic conditions since it was easily isomerized to the thermodynamically more stable *trans*-isomer **9a** via ring opening–closing, as shown in Scheme 5.2. Thus, we focused on the Ullmann-type reaction.

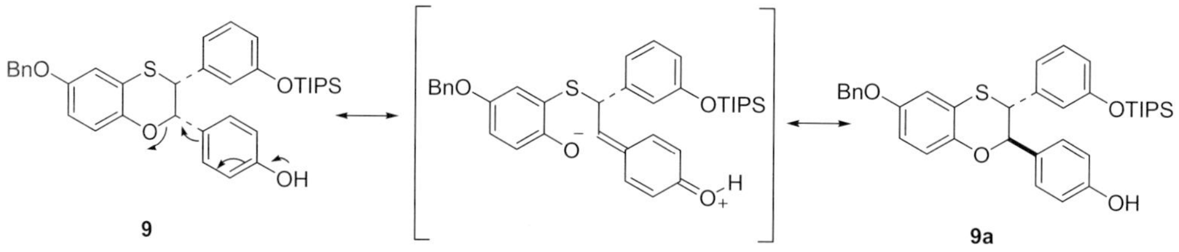

Scheme 5.2 *Cis–trans* isomerization.

The *cis*-2,3-diaryl-2,3-dihydro-1,4-benzoxathiin is a very unique structural motif. Other than scattered reports in the literature on the formation of this scaffold, there was no effective asymmetric synthesis for it [6]. We explored two major synthetic approaches to realize the key chiral *cis*-diaryl dihydrobenzoxathiin scaffold, as shown in Scheme 5.3. One was the quinone ketal route in which the quinone ketal **13** and the chiral mercaptol alcohol **14** were the key intermediates. The other approach was the stereo- and enantioselective reduction of the diaryl benzoxathiin **16**. The key mercaptol alcohol **14** and the diaryl benzoxathiin **16** were both envisioned to be prepared from the key, common iodoketone intermediate **15**.

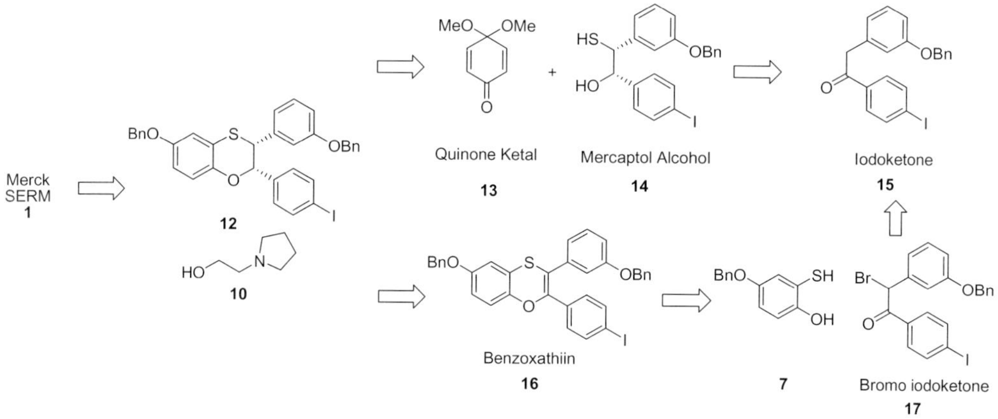

Scheme 5.3 Retro-synthetic analysis of SERM candidate **1**.

This process development section will be subdivided into the following areas.

1) Preparation of the iodoketone intermediate **15**.
2) The quinone ketal route to *cis*-diaryl dihydrobenzoxathiin.
3) Benzoxathiin reduction route to *cis*-diaryl dihydrobenzoxathiin **12**.

4) Installation of the pyrrolidinylethanol.
5) Final benzyl deprotection and isolation of target **1**.

The section will cover various aspects of the synthetic route development and the scale up to make several kilograms of final compound **1**. As will be seen, a novel sulfoxide-directed stereospecific borane reduction was discovered in this effort and the scope and application of this reaction will be discussed in Section 5.2 [5].

5.1.2.1 Preparation of Intermediate 15

As seen in the retro-synthetic Scheme 5.3, intermediate **15** is useful for both routes. The choice of benzyl protection group was made based on the robust stability of benzyl phenol ethers toward most reactions and several possible avenues to remove it, although it was reported from Medicinal Chemistry that benzyl group removal via hydrogenolysis posed challenges in this compound. The choice of iodide substitution was born out of the known high reactivity of iodides in the Ullmann-type coupling reaction with alcohols and the robust stability of aryl iodides in many other common reactions.

The most straightforward route toward **15** is likely to be the Friedel–Crafts reaction of 3-benzyloxyphenylacetic acid (**19**) with iodobenzene (**18**), as shown in Scheme 5.4. However, preliminary attempts at this type of reaction were met with the very low reactivity of iodobenzene (**18**) toward Friedel–Crafts substitutions.

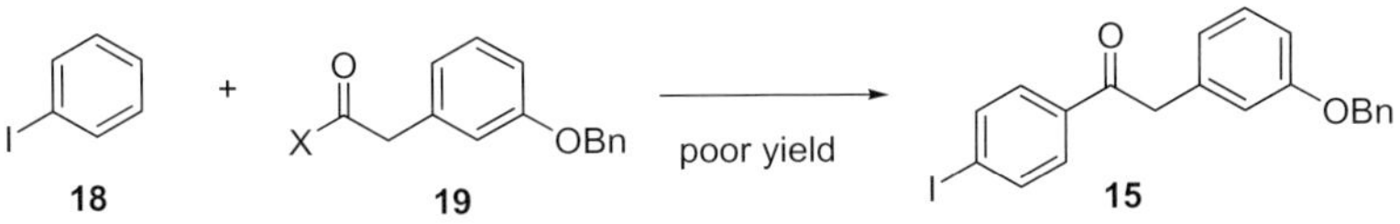

Scheme 5.4 Friedel–Crafts approach to ketone **15**.

The ketone **15** was eventually prepared by Grignard addition to Weinreb amide **21**, as shown in Scheme 5.5. The Weinreb amide **21** was prepared from *p*-iodobenzoic acid (**20**). The phenol of readily available 3-hydroxybenzaldehyde (**22**) was first protected with a benzyl group, then the aldehyde was converted to chloride **24** via alcohol **23** under standard conditions. Preparation of the Grignard reagent **25** from chloride **24** was initially problematic. A large proportion of the homo-coupling side product **26** was observed in THF. The use of a 3 : 1 mixture of toluene:THF as the reaction solvent suppressed this side reaction [7]. The iodoketone **15** was isolated as a crystalline solid and this sequence was scaled up to pilot plant scale to make around 50 kg of **15**.

5.1.2.2 Quinone Ketal Route to *cis*-Diaryl Dihydrobenzoxathiin 30

Several key observations make the quinine ketal route potentially viable (Scheme 5.6):

1) The reduction of α-sulfur substituted ketone **27** and its analogues usually gives the *anti*-mercaptol alcohol like *rac*-**14** which can be elaborated into *cis*-dihydrobenzoxathiin **30**.

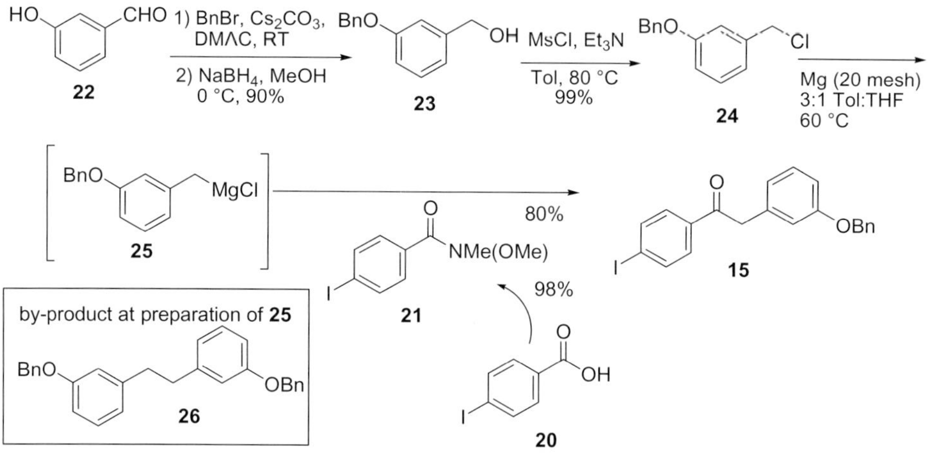

Scheme 5.5 Preparation of iodoketone intermediate **15**.

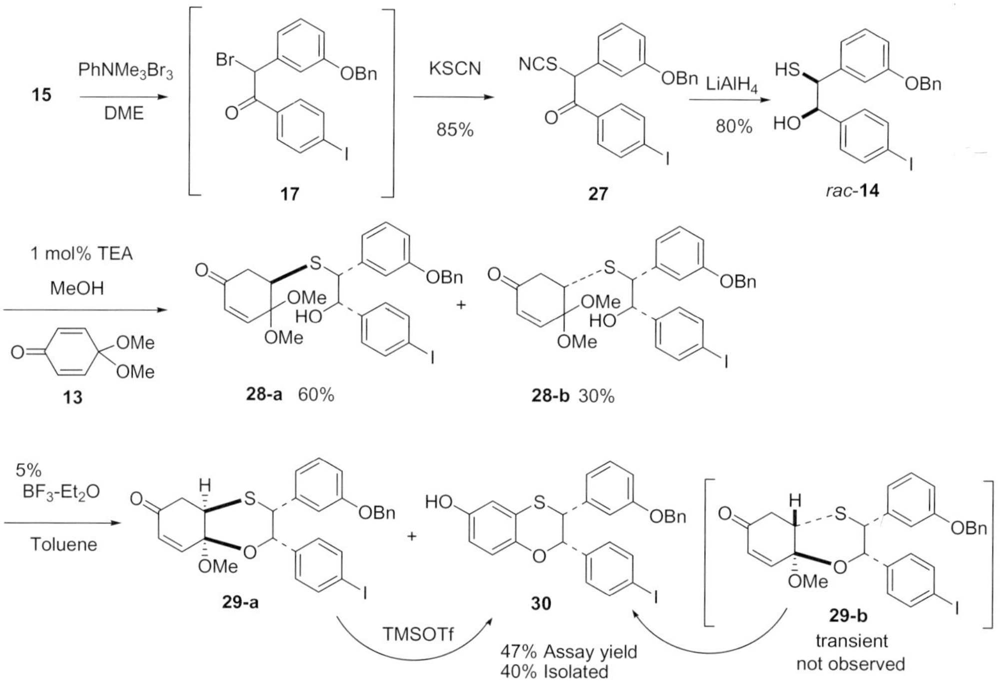

Scheme 5.6 Quinone ketal route to intermediate **30**.

2) The mercaptol alcohol *rac*-**14** undergoes facile Michael addition reaction with quinone ketal **13** which is commercially available or can be readily prepared.
3) The Michael addition products **28-a** and **28-b** can be reketalized under Lewis acidic conditions to form **29-a** and **29-b** which presumably further eliminate methanol to aromatize to form the target **30**.

Modest diastereoselectivity was observed for the Michael addition reaction of *rac*-**14** to **13** and these diasteromers **28-a**/**28-b** could be separated and individually identified. The minor isomer **28-b** was found to readily undergo conversion to benzoxathiin **30** when treated with BF_3 etherate, presumably through the transient intermediate **29-b**. The major isomer **28-a** was converted by BF_3 etherate to intermediate **29-a**. Conversion to **30** required the use of the stronger Lewis acid TMSOTf, presumably due to the *cis*-stereochemistry between the methoxy and the neighboring hydrogen, making it more difficult to eliminate/aromatize.

The above sequence was demonstrated on racemic *anti*-mercaptol alcohol **14** as well on a small amount of optically pure **14** (separated by chiral HPLC separation) and the chiral centers of **14** were completely retained, as expected (Scheme 5.6) [8]. With proof of concept for the ring formation strategy, some efforts were put into developing a chiral synthesis of **14**, as shown in Scheme 5.7.

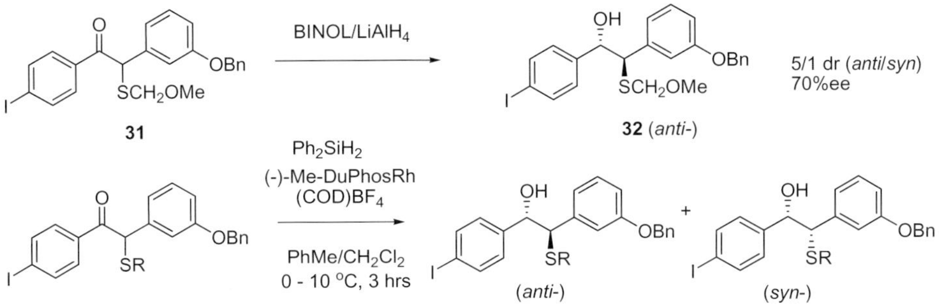

31, R = MOM
33, R = H

32, R = MOM 32% yield, 70% ee
34, R = H 40% yield, 89% ee

not detetced
20% yield, 00%ee

Scheme 5.7 Synthetic approach to chiral mercaptol alcohol **14**.

The dynamic kinetic resolution (DKR) of α-sulfur-substituted ketones such as **31** and **33** was investigated. When the MOM protected mercaptol ketone **31** was treated with the BINOL–$LiAlH_4$ complex, a moderate diastereoselectivity of 5 : 1 favoring the desired *anti* isomer was observed. The major diastereomer had 70% ee.

DKR via hydrosilylation was also investigated in the presence of the methyl-DuPhos rhodium complex. When MOM protected **31** was tested, a low yield of desired diasteromer was observed with modest 70% ee. On the other hand, the unprotected thiol-ketone **33** gave 89% ee of the desired *anti*-diastereomer in 40%

yield along with the undesired diastereomer in 20% yield (93% ee). A few other analogous α-sulfur-substituted ketones were tested in a number of other reduction conditions without success. Cleavage of the sulfur carbon bond was often observed as the main pathway in the hydrosilylation reactions.

Based on the limited success of this approach and the more promising results from the benzoxathiin reduction approach, development of this route was put into lower priority.

5.1.2.3 Benzoxathiin Reduction Route to the *cis*-Diaryl Dihydrobenzoxathiin Intermediate 12

5.1.2.3.1 Preparation of Benzoxathiin Precursor 16

The asymmetric reduction of the benzoxathiin is very appealing because of its simplicity (Scheme 5.3). It was envisioned that intermediate **16** could be prepared from thiol-phenol **7** and bromoketone **17**. Scheme 5.8 summarized the synthesis for **16**. The 1,3-benzoxathiol-2-one **35** was prepared from 1,4-benzoquinone and thiourea following a literature procedure with minor modifications. Benzylation of **35** with benzyl bromide in the presence of KI gave benzyl ether **36** as a crystalline solid. It was observed that the benzylation gave better results when the reaction was run under anaerobic conditions. Hydrolysis of thiocarbonate **36** gave free thiophenol **7** which was used directly in the next reaction.

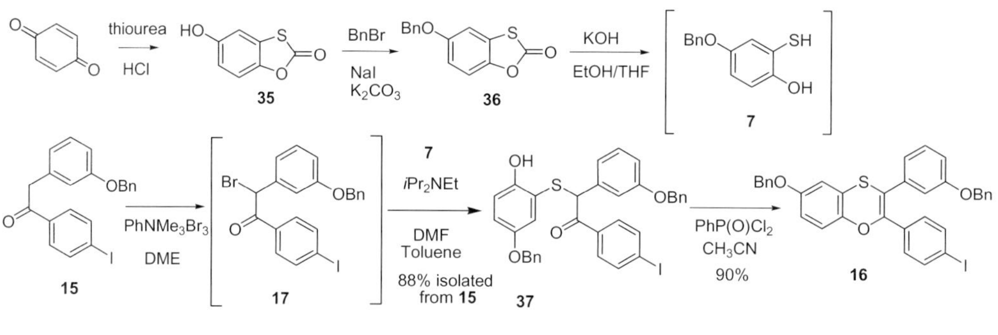

Scheme 5.8 Preparation of benzoxathiin **16**.

The bromoketone **17** was prepared via bromination of **15** with PTAB in DME. Hydrogen bromide, formed during the reaction, reacted with DME to generate methyl bromide and 2-methoxyethanol, both of which could be easily removed from the reaction medium under vacuum. This method was more convenient than the bromination reaction in THF because the resulting 4-bromobutanol by-product formed from THF was not volatile. The bromide **17** was used directly in the next reaction, partly because **17** is rather reactive with limited stability.

The reaction of bromide **17** with thiophenol **7** was instantaneous in the presence of Hünig's base. The adduct **37** was isolated by crystallization in 88% overall yield from **15**. The dehydrative cyclization of **37** to **16** was screened against solvents and acids. Phenylphosphonic dichloride [PhP(O)Cl_2] was found to offer the most convenient process. The reaction was run in acetonitrile under nitrogen sweep removing the HCl gas that was formed. After completion of the reaction, ethanol was added to quench the excess PhP(O)Cl_2 and the product **16** directly crystallized out without further work-up. This was a major advantage with PhP(O)Cl_2 because other dehydrating agents such as P_2O_5 affected clean conversion to **16** but product **16** decomposed during the aqueous work-up.

5.1.2.3.2 Direct Reduction of Benzoxathiin 16

With benzoxathiin **16** in hand, the asymmetric hydrogenation was screened against a set of common reduction conditions, as shown in Scheme 5.9. Not surprisingly, no reaction (**16** → **12**) was observed. The reason is presumably the electron-rich nature of the targeted tetra-substituted olefin double bond. Prior to evaluating the reduction of **16**, a model compound **38** was made and found to be partially hydrogenated with $RuCl_3$ as a catalyst under forcing conditions of 1000 psi hydrogen. The encouraging fact was that product **39** had the desired *cis*-diaryl relative stereochemistry. Nevertheless, the asymmetric reduction of such an electron-rich tetra-substituted olefin was considered to have rather low probability of success at the time, so alternative strategies were pursued [9].

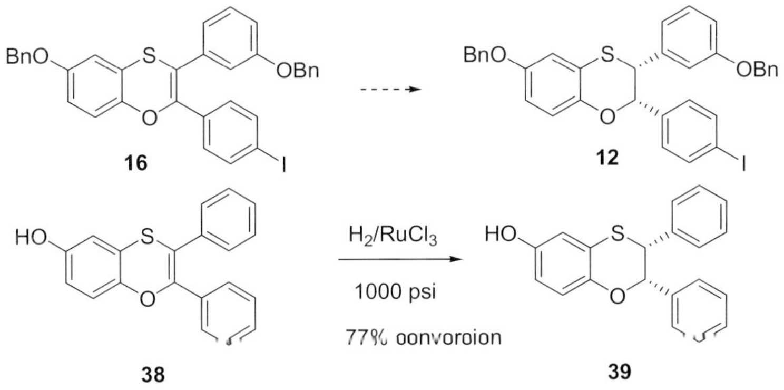

Catalyst screen at 90 psi hydrogen at r.t. and 60 °C for 20 h – No product detected
Pfaltz-Ir-BARF-cat, (Et-Duphos)Rh(COD)BF_4,(BINAP)Ru(II)Cl_2,
Phanephos/$(COD)_2RhBF_4$, Josiphos SL-J009-1/(COD)RhCl,
(BINAP)$RuCl_2$-*p*-cymene

Scheme 5.9 Early attempts for benzoxathiin reduction.

5.1.2.3.3 Sulfoxide Directed Reduction – Proof of Concept

It was postulated that oxidation of the sulfur to the sulfoxide **40** or sulfone **41** (Figure 5.2) may activate the targeted olefin toward reduction by reducing the

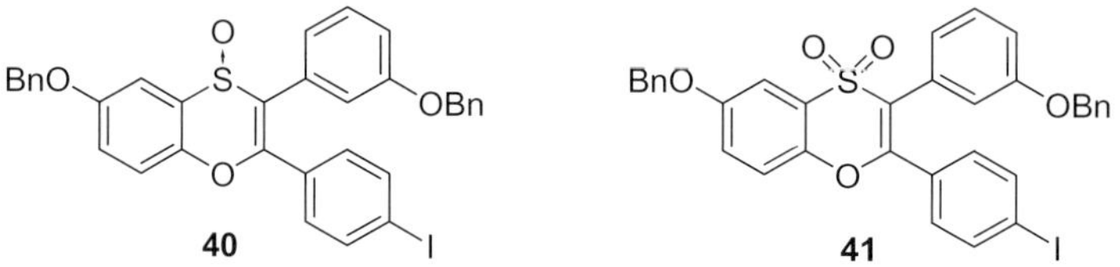

Figure 5.2 Potentially more reactive substrates for reduction.

electron density of the double bond and the chiral sulfoxide may offer a handle for chelation-controlled stereoselective reduction.

Prior literature indicated that olefins substituted with chiral sulfoxides could indeed be reduced by hydride or hydrogen with modest stereoselectivity, as summarized in Scheme 5.10. Ogura *et al.* reported that borane reduction of the unsaturated sulfoxide **42** gave product **43** in 87 : 13 diastereomer ratio and D_2O quench of the borane reduction mixture gave the product **43** deuterated at the α-position to the sulfoxide, consistent with the hydroboration mechanism [10a]. In another paper, Price *et al.* reported diastereoselective hydrogenation of *gem*-disubstituted olefin *rac*-**44** to **45** with excellent diastereoselectivity using a rhodium catalyst [10b].

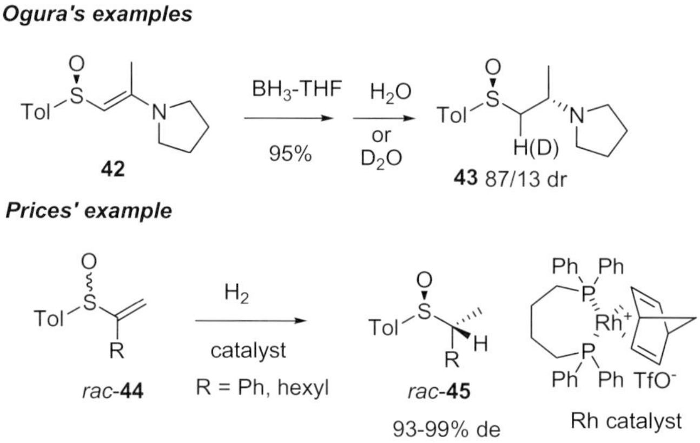

Scheme 5.10 Sulfoxide-directed olefin reduction.

To test the hypothesis, a model sulfoxide **46** was prepared from commercially available **47** via sodium periodate oxidation (Scheme 5.11). The goal was to find reaction conditions that would reduce the olefin double bond first, producing the saturated sulfoxide intermediate that could be further reduced from sulfoxide to sulfide **48**. The undesired reaction sequence would be reduction of sulfoxide occurring *before* that of the carbon–carbon double bond. Typical hydrogenation conditions with Pd and Pt catalysts did not reduce the olefin of **46** but only led to cleavage of the sulfur–oxygen bond of **46**, producing **47**. No reaction was observed using Prices' hydrogenation conditions (Scheme 5.10).

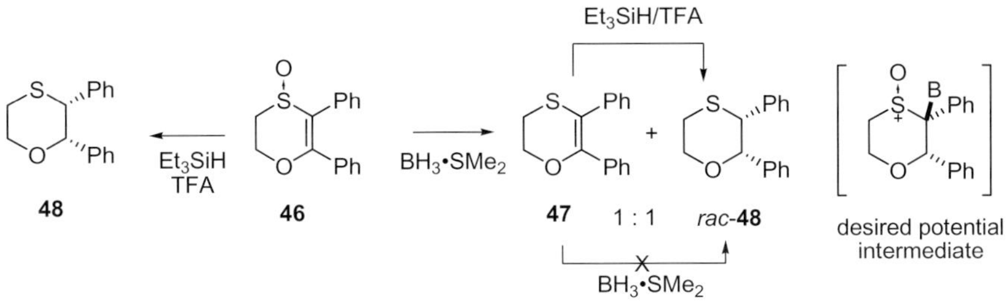

Scheme 5.11 Sulfoxide-directed olefin reduction – model compounds.

When **46** was treated with Et_3SiH in the presence of TFA, **48** was produced, meaning that both the olefin double bond and the sulfur–oxygen bond were reduced. In addition, **47** could also be reduced to **48** under the same conditions. Therefore, it was not clear from these results whether the reduction of the olefin in sulfoxide **46** was directed by the sulfoxide oxygen.

On the other hand, borane reduction of sulfoxide **46** gave a mixture of **47** and **48** in equal amounts and, more interestingly, no reaction of **47** occurred with borane under the same conditions. These results indicated that borane reduction of the olefin in **46** required the activation and potential direction by the sulfoxide oxygen.

With this encouraging result from the model system, a gram quantity of the racemic sulfoxide **40** was prepared by oxidation of benzoxathiin **16** with *m*CPBA and a small amount of chiral sulfoxide *(S)*-**40** with 94% ee was isolated by subsequent chiral HPLC separation (Scheme 5.12). When chiral sulfoxide *(S)*-**40** was treated with borane-dimethylsulfide, a clean reduction of the olefin and the sulfoxide was observed. More surprisingly, only the desired *cis*-diaryl dihydrobenzoxathiin **12** was observed in high yield and unchanged 94% ee. No *trans*-isomer or **16** was observed. With this proof of concept in hand, an efficient

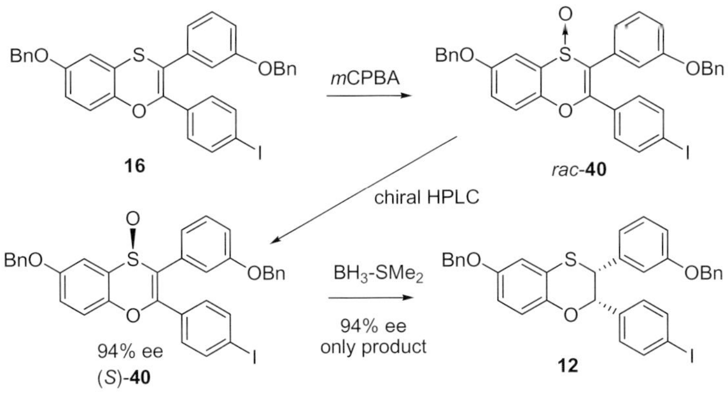

Scheme 5.12 Sulfoxide-directed olefin reduction – proof of concept.

preparation of the chiral sulfoxide *(S)*-**40** became the key reaction to set the *cis*-diaryl stereochemical centers. Mechanistic investigations of this novel reduction will be discussed in Section 5.2.

5.1.2.3.4 **Preparation of the Chiral Sulfoxide 40 and Its Reduction**

We initially screened a number of different oxidation methods and quickly came to the conclusion that the Kagan oxidation of sulfides to chiral sulfoxides was the method of choice due to its low cost and high efficiency [11, 12]. However, we also found that the Kagan oxidation needed to be optimized for this specific substrate. Typical Kagan oxidation is carried out with $Ti(Oi\text{-}Pr)_4$ and a dialkyl tartrate ligand. Our preliminary screening of ligands under the original Kagan conditions with 2 equiv of ligand, 1 equiv of water, 1 equiv of $Ti(Oi\text{-}Pr)_4$ in dichloromethane at 25 °C revealed that diisopropyl tartrate gave an ee of 38% but diethyl tartrate, BINOL, hydrobenzoin, TADDOL and *N,N*-dibenzyl tartramide all gave lower selectivities. Thus, the diisopropyl tartrate-catalyzed reaction was further optimized. Addition of Hünig's base to the reaction mixture was found to dramatically improve the ee [11d–f]. Using THF as the reaction solvent led to slightly better ee and also facilitated the isolation of product **40** when compared to ethyl acetate, chlorobenzene or dichloromethane. The reaction temperature had negligible effect on the ee within the tested range of 0–30 °C. The order of addition of the reagents was found to be critically important. The optimal sequence for adding the reagents was to mix diisopropyl tartrate, Hünig's base and water in THF followed by the addition of $Ti(Oi\text{-}Pr)_4$. This catalyst mixture was aged at ambient temperature overnight and the vinyl sulfide **16** was subsequently added to the catalyst mixture followed by addition of cumene hydrogenperoxide (CHP). The extended overnight age of the catalyst mixture was found to improve reproducibility for the selectivity of the oxidation. Without aging, the ee of **40** ranged from 80–90%. With aging the ee of **40** varied very little between 91 and 92% run to run. The catalyst loading was ultimately reduced to 15 mol% without compromising the ee or reproducibility. When the reaction was carried out under optimized conditions in THF, (*S*)-**40** precipitated out during the course of the reaction. At the end of the reaction, HPLC assay of the crude reaction slurry typically showed 95% assay yield of **40** with 92% ee. Crystallization significantly upgraded the purity of **40**. Upon filtration, crystalline **40** was typically isolated in 86% yield with 95% purity and 99% ee. This remarkably efficient direct isolation of **40** also avoided the tedious separation of TiO_2 normally associated with an aqueous work-up.

Reduction of sulfoxide (*S*)-**40** was carried out in toluene using 1.05 equiv of 1 M BH_3–THF at 10 °C, producing the *cis*-diaryl dihydrobenzoxathiin **12** with complete stereoselectivity. Upon work-up and crystallization from toluene and heptane, the desired product **12** was isolated in 88% yield and over 99% ee (Scheme 5.13). Borane dimethylsulfide led to a slower reaction. To the best of our knowledge, this synthetically useful reaction is unprecedented in the literature [13].[1] Investigation

1) The sulfoxide directed reduction of olefin is known, so is sulfoxide reduction by chloroborane, but not simultaneous reduction of both by borane.

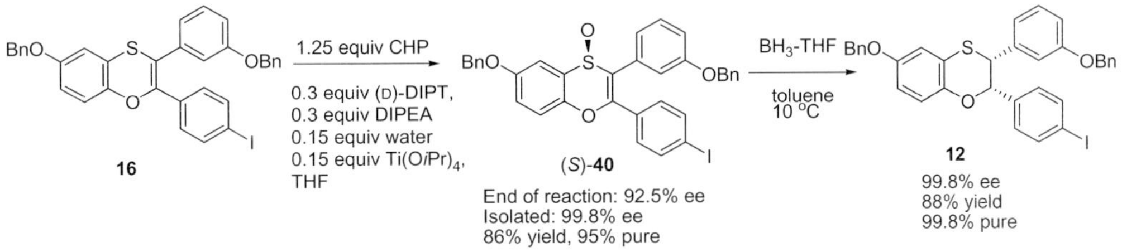

Scheme 5.13 Chiral sulfoxide preparation and olefin reduction with borane.

on the scope and mechanism of this novel reaction will be discussed in the chemistry development section later. Attempts to extend this novel reduction to other hydride reducing reagents have given poor results. Reduction of (*S*)-**40** with Et_3SiH in the presence of TFA also gave *cis*-diaryl-dihydrobenzoxathiin product **12**, but only in 20% ee favoring the desired enantiomer. DIBAL and $LiBHEt_3$ gave complex product mixtures.

5.1.2.4 **Installation of Pyrrolidinyl Ethanol**

For the installation of the pyrrolidinylethanol moiety **10** on the aryl group, we first tested Buchwald's Cu-catalyzed conditions with **10**, aryl iodide **12**, Cs_2CO_3, CuI and 1,10-phenanthroline at 110 °C in toluene to prepare the penultimate **49** [14a]. The reaction was very slow, giving only 5–10% conversion even after 2 days. The reaction was faster at higher temperatures but two impurities **50** and **51** were observed (Scheme 5.14). To find the optimal conditions, xylene and diglyme were tested as solvents, lithium, potassium and cesium carbonates were screened as bases and 2,2′-bipyridyl, TMEDA and 1-(2-dimethylaminoethyl)-4-methylpiperazine were examined as ligands. The optimized protocol was identified as 10 mol% of

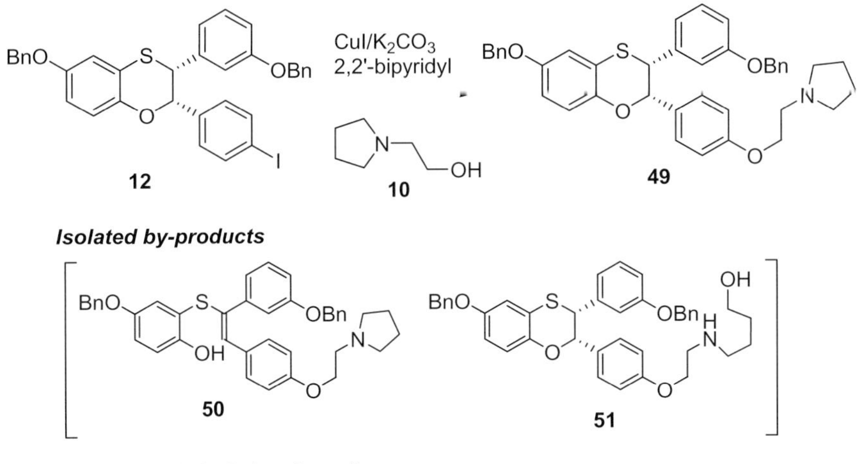

Scheme 5.14 Pyrrolidinylethanol installation.

CuI, 12 mol% of 2,2′-dipyridyl, in 10 vol of xylene:diglyme (9 : 1) at 140 °C with azeotropic removal of the water as it was formed. The azeotropic removal of water helped alleviate the problem of solids coating the reaction vessel walls, which led to stalling of the reaction. The reaction was complete in *less than 10 h*, typically with 96% assay yield and 92% isolated yield for **49** after aqueous work-up and subsequent crystallization [14b–d]. It was noteworthy that this catalytic system composed of the copper(I) salt with bipyridyl ligand was recently reported to be applicable to a wide range of Ullmann-type ether formations [14d].

5.1.2.5 **Final Deprotection and Isolation of Compound 1**

To remove the benzyl protecting groups on the two phenolic oxygens in **49**, we first tested the standard hydrogenolysis conditions. However, the hydrogenolysis of **49** was sluggish with a variety of heterogeneous Pd catalysts, requiring very high Pd loading. Catalyst poisoning from thiol side products was the probable cause of this sluggishness. Incomplete hydrogenolysis also produced two mono-benzyl intermediates which were very difficult to remove from **1** and hydrogenolysis under more forcing conditions resulted in lower yields, presumably due to cleavage of the dihydrobenzoxathiin ring. To find alternate debenzylation conditions, TMSI, $AlCl_3$ and various boron halides were examined. The TMSI reaction was found to be the most effective and provided **1** in the highest yield. However the benzyl iodide, by-product in this reaction, reacted with the desired product **1** to form an *N*-benzylated impurity and three other aryl ring-benzylated impurities (1 to 1.5% each, Scheme 5.15) which were especially difficult to remove from **1**. The *N*-benzylation side reaction became more extensive during work-up with aqueous Na_2CO_3, which necessitated the neutralization of the benzyl iodide before the work-up. After examining various benzyl iodide scavengers such as imidazole, *N*-methyl imidazole, DABCO, KSCN and thiourea, it was found that presence of both thiourea and *N*-methyl imidazole together during the TMSI debenzylation in acetonitrile suppressed the formation of these benzylation side products to below

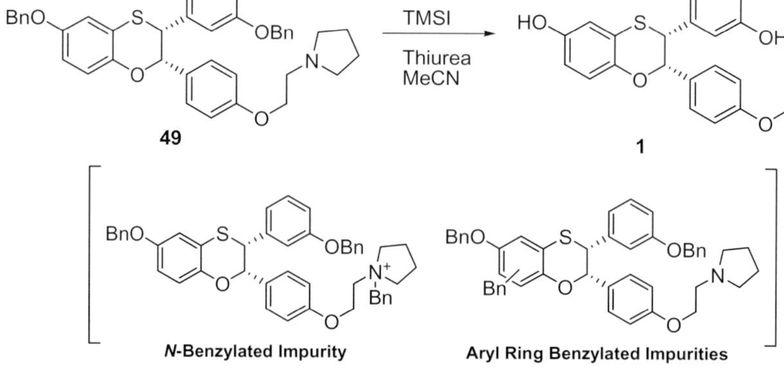

Scheme 5.15 Debenzylation and isolation of compound **1**.

0.4% each. After work-up, the final compound **1** was isolated as the HCl salt, first from ethanol as an ethanol solvate. Recrystallization from acetonitrile then gave pharmaceutically acceptable purity. The final product **1** was obtained in 81% yield and 99.4% purity.

5.1.2.6 Overall Synthesis Summary

In conclusion, we have developed an efficient asymmetric synthesis of the selective estrogen receptor modulator **1**. This process was scaled up in the pilot plant for the early steps in the synthetic route and in kilo laboratory scale for the later steps and finally 4 kg of the final product was made. The key process development and project support milestones are summarized below:

1) The project was supported with 4 kg SERM candidate **1** via the new synthesis.
2) A novel, unprecedented sulfoxide-directed borane reduction of the α,β-unsaturated sulfoxide to the saturated sulfide was discovered.
3) An efficient chiral sulfoxide preparation of **40** was developed which was the key to set the two chiral centers.
4) A new Ullmann ether protocol to install the pyrrolidinylethanol **10** was developed, obviating the need for the Mitsunobu reaction.
5) All chromatographic purifications were eliminated.
6) The new eight-step sequence gave **1** in an overall yield of 37% with high purity.

5.2 Chemistry Development

5.2.1 Mechanism of the Sulfoxide-Directed Olefin Reduction

The stereospecific sulfoxide-directed reduction of the neighboring olefin was unprecedented in the literature. The absolute configuration of the sulfoxide (*S*)-**40** and the reduced product **12** were unambiguously determined with single crystal X-ray crystallography. The oxygen on the sulfoxide and the two hydrogens were on the same face of the ring, indicating the directing effect of the sulfoxide [5]. This reaction was the key step in the efficient synthesis of **1**. We were interested in investigating the scope and understanding the mechanism of this novel reaction.

Initially, two plausible mechanisms were considered, as depicted in Scheme 5.16. The first was a hydroboration route (*a*), where the B–H bond was added across the olefin from the same face of S–O and upon aqueous work-up, the resulting C–B bond was replaced with a C–H bond. The *cis* B–H addition to the olefin led to the *cis*-stereochemistry of the two adjacent aryl substituents. The reduction of the sulfoxide oxygen occurs in the next step. The alternative mechanism was the borane reduction route (*b*), which was similar to 1,4-addition of hydride,

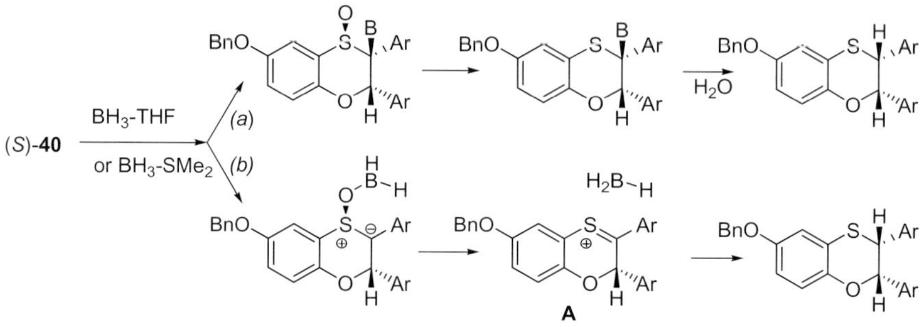

Scheme 5.16 Two plausible mechanistic pathways for the borane reduction.

leading to a sulfonium intermediate **A** after cleavage of the S–O bond. The intermediate **A** was then further reduced. The *cis*-stereochemistry of the product was made possible by the steric bias created by the first chiral center. To uncover the details of the reduction, a series of experiments was carried out to establish various characteristics of the reaction [5]. To establish the source of the two hydrogen atoms in the product, two deuterium labeling reactions were run, as summarized in Scheme 5.17. When (*S*)-**40** was reduced with BD_3–THF and the reaction was quenched with acetic acid, both of the newly formed C–H positions were found to be deuterated, so **12-*d*₂** was obtained. But when the BH_3–THF reduction was quenched with CD_3CO_2D, no deuterium was incorporated in the product **12**. These experiments established that both of the hydrogens were introduced from the borane source. This result was in contrast to the results reported by Ogura *et al.* (Scheme 5.10) [10a] where one of the hydrogens was found to come from the borane and the other from the quenching solvent. In that study, the orientation of the borane reduction was reported to be similar to hydroboration.

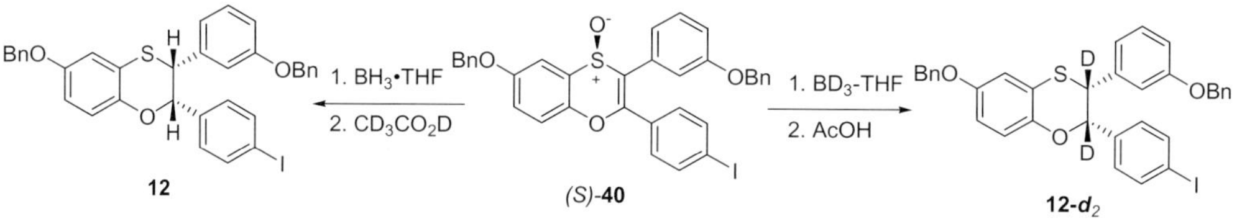

Scheme 5.17 Deuterium labeling studies.

To gain understanding of the interdependence between the olefin reduction and the sulfoxide reduction, the saturated sulfoxide **52** was prepared and treated with BH_3–THF. No reaction was observed under the similar conditions (Scheme 5.18). The unactivated vinyl sulfide **16** was also not reactive toward BH_3–THF. These results indicated that sulfoxide and olefin were reduced simultaneously, not independently. Again this phenomenon was unexpected and pointed to the unique nature of this reaction.

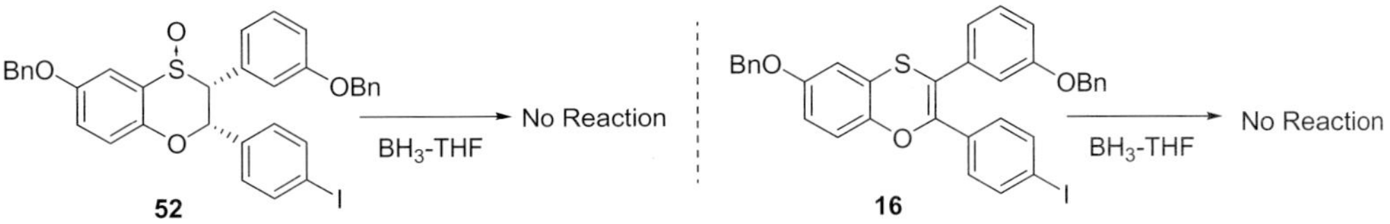

Scheme 5.18 Mechanistic investigation of the borane reduction – A.

To understand the interdependence of the creation of the two chiral centers relative to each other and to the sulfoxide, monosubstituted vinyl sulfoxides (*S*)-**53** and (*S*)-**54** were prepared and reduced with BH_3–THF under the same conditions (Scheme 5.19). Both the 2- and 3-phenyl substituted substrates gave the chiral products **54** and **55** with complete stereo specificities dictated by the configuration of the starting sulfoxides. These results again were unexpected and indicated that both hydrogens were delivered solely directed by the chiral sulfoxide. This was not consistent with the mechanism in which the chirality of the initially formed chiral center at the 3-postion dictates the chirality of the subsequently formed chiral center at the 2-position.

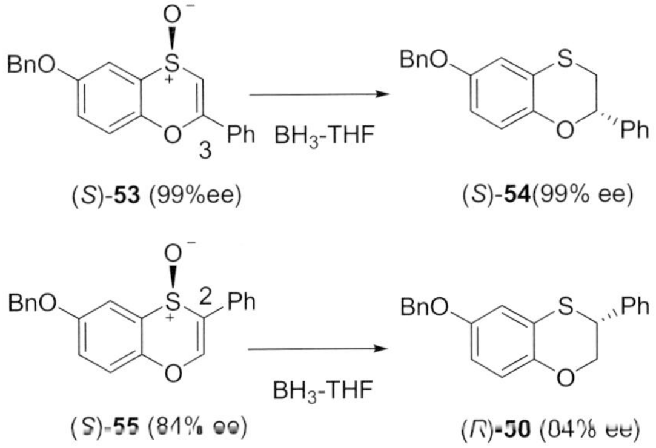

Scheme 5.19 Mechanistic investigation of the borane reduction – B.

The kinetics of the reaction was measured by NMR studies and the reaction was found to be first order relative to borane–dimethyl sulfide and to the substrate (Figure 5.3). This result was consistent with a bimolecular reaction in the rate-limiting step.

The deuterium kinetic isotope effect between BH_3–THF and BD_3–THF was obtained by measuring the reaction rate constants of the two reactions with the unsaturated sulfoxide (*S*)-**40** independently via React-IR. The $k(BH_3)/k(BD_3)$ is 1.4, consistent with hydrogen transfer not being the rate-limiting step [15, 16].

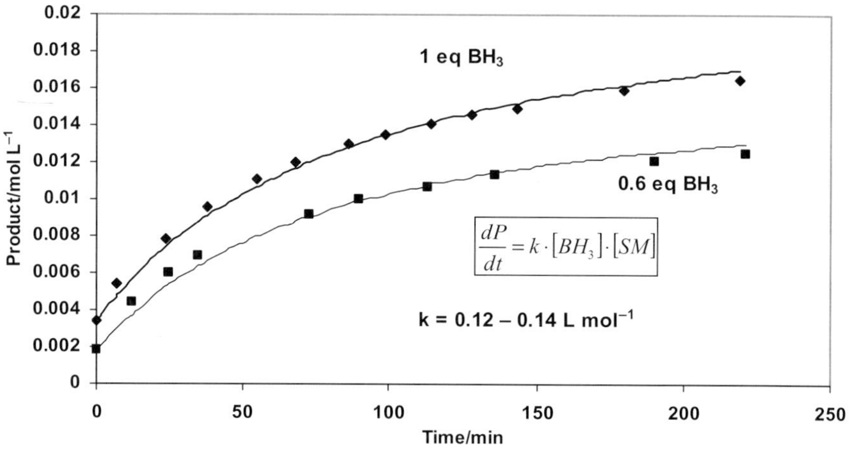

Figure 5.3 Kinetic data of the borane reduction.

Based on all the evidence collected, the mechanism of the sulfoxide-directed borane reduction was proposed as in Scheme 5.20. The rate-limiting step is proposed to be the transfer of BH_3 from either THF or SMe_2 to the sulfoxide in an SN_2-like mechanism on the boron center. The resulting S–O–BH_3 complex **B** then undergoes a hydrogen transfer from the borane to the β-carbon (C3) followed by another hydrogen transfer to the α-carbon (C2) with concurrent S–O bond cleavage. Both hydrogen transfers are directed by the sulfoxide to occur on the same face of the ring as the sulfoxide oxygen. The sequence of the two hydrogen transfers proposed was supported by the fact that the hydrogen content in **12-*d*$_2$** (Scheme 5.17) is higher on the C3 carbon than on the C2 carbon, as indicated by proton NMR, due to the contamination of BD_2H in the BD_3 used, presumably due to an H/D isotope effect.

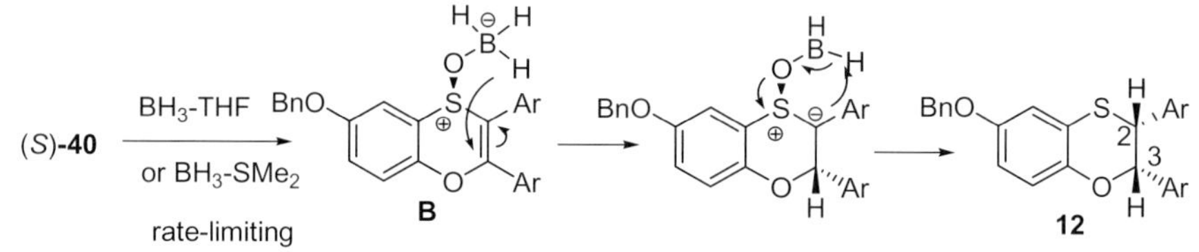

Scheme 5.20 Proposed mechanism of the borane reduction.

5.2.2
Application of the Sulfoxide-Directed Borane Reduction to Other Similar Compounds

The chiral 2,3-dihydro-1,4-benzoxathiins are useful compounds but there were no effective asymmetric methods to access these compounds in the literature [15, 17,

19]. The sequence of asymmetric oxidation of vinyl sulfide to chiral sulfoxide followed by stereospecific borane reduction should be applicable to analogous compounds. Thus, a series of 3-substituted 1,4-benzoxathiins were prepared with the same method as used in the synthesis of SERM **1**. As shown in Scheme 5.21, the oxidation of the sulfoxide was found to give very low ee with the diisopropyl tartrate ligand. After some screening, BINOL was found to give the best ee for this class of substrates [19]. Individual substrates gave variable ee depending on the steric and electronic nature of the substituents. The borane reduction still reliably gave the same ee product as the sulfoxides [11h, 12, 15f,g].

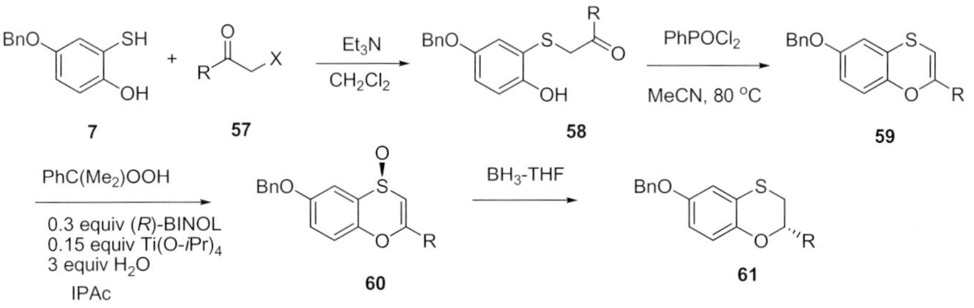

Entry	Product 60 and 61	R	Sulfoxide ee (%)	Dihydrobenzoxathiin ee (%)	Yield (%) (59 → 61)
1	a	4-$CH_3C_6H_5$	59	60	64
2	b	4-$CH_3OC_6H_5$	48	48	70
3	c	4-BrC_6H_5	26	26	68
4	d	CH_3	52	51	63
5	e	*t*-Bu	96	97	86

Scheme 5.21 Synthesis of analogous 1,4-benzoxathiins.

We next applied this reduction methodology to the preparation of **67**, a compound reported to be 500 times sweeter than sucrose [18d]. The synthesis starting from isovanillin is shown in Scheme 5.22. The phenol in intermediate **62** was protected as the electron-withdrawing *p*-chlorobenzoate (*p*CB) to facilitate the dehydrative cyclization to prepare benzoxathiin **64**. After mono-bromination of **64**, S_N2 displacement with mercaptophenol **7** gave the hydroxyketone **63**. The dehydrative cyclization cleanly furnished the desired **64** in 82% isolated yield. The compound **64** was oxidized using the typical procedure, but gave the sulfoxide in only 40% ee. It was found that the oxidation of free phenol **65** gave the corresponding sulfoxide in improved 78% ee. This difference in enantioselectivity was attributed to the different electronic effects of the *p*-chlorobenzoate **64** and the free phenol **65**. Borane reduction of the crude product furnished the desired dihydrobenzoxathiin **66** in 78% ee and 60% yield. Debenzylation via hydrogenolysis using

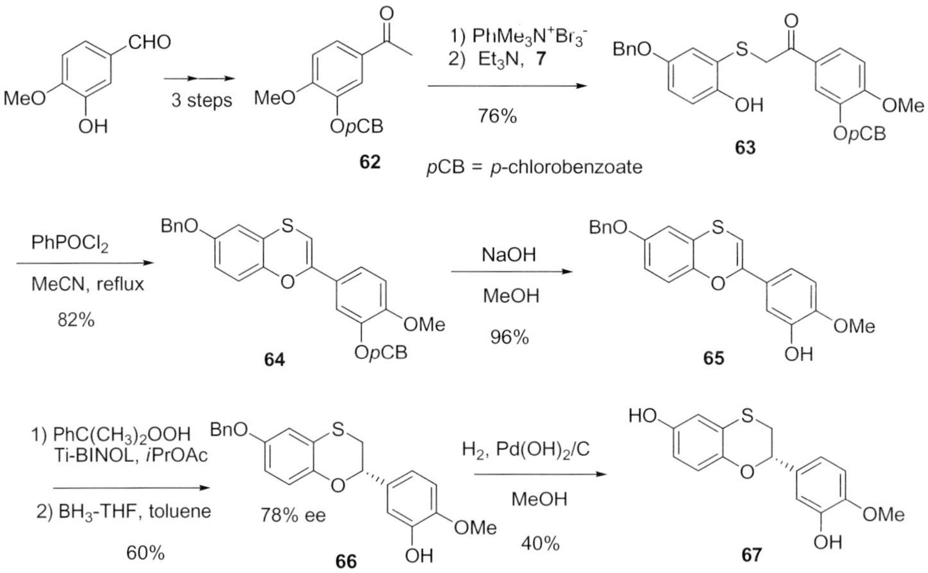

Scheme 5.22 Synthesis of the sweetener compound **67**.

Pearlman's catalyst provided the target compound **67** in 40% yield and 78% ee, completing the first asymmetric synthesis of this compound.

5.3 Conclusion

We have developed the efficient synthesis of the SERM drug candidate **1** and successfully demonstrated the process on a multiple kilogram scale to support the drug development program. A novel sulfoxide-directed borane reduction of vinyl sulfoxides was discovered. The mechanistic details of this novel reaction were explored and a plausible mechanism proposed. The sequence of asymmetric oxidation of vinyl sulfoxides followed by stereospecific borane reduction to make chiral dihydro-1,4-benzoxathiins was applied to the asymmetric synthesis of a number of other dihydro-1,4-benzoxathiins including the sweetening agent **67**.

Acknowledgments

The author would like to thank the SERM team members in Merck Process Research, who were the coauthors for the original publications reviewed in this chapter, for their original contributions to the work.

References

1 Recent Reviews of SERMs: (a) Peng, J., Sengupta, S., and Jordan, V.C. (2009) *Anti-Cancer Agents Med. Chem*, **9**, 481; (b) Oseni, T., Patel, R., Pyle, J., and Jordan, V.C. (2008) *Planta Med.*, **74**, 1656; (c) Jordan, V.C., and O'Malley, B.W. (2007) *J. Clin. Oncol.*, **25**, 5815; (d) Bai, C., Flores, O., and Schmidt, A. (2007) *Expert Opin. Drug Discov.*, **2**, 725.

2 Recent reports of SERM research (a) Kumar, S., Deshpande, S., Chandra, V., Kitchlu, S., Dwivedi, A., Nayak, V.L., Konwar, R., Prabhakar, Y.S., and Sahu, D.P. (2009) *Bioorg. Med. Chem.*, **17**, 6832; (b) Scanlan, T.S., and Iijima, T. (2009) U.S. Pat. US 2009124698 A1 20090514; (c) Barrett, I., Meegan, M.J., Hughes, R.B., Carr, M., Knox, A.J.S., Artemenko, N., Golfis, G., Zisterer, D.M., and Lloyd, D.G. (2008) *Bioorg. Med. Chem.*, **16**, 9554.

3 Tamoxifen new synthesis: Shiina, I., Sano, Y., Nakata, K., Suzuki, M., Yokoyama, T., Sasaki, A., Orikasa, T., Miyamoto, T., Ikekita, M., Nagahara, Y., and Hasome, Y. (2007) *Bioorg. Med. Chem*, **15**, 7599.

4 (a) DiNinno, F.P., Wu, J.Y., Kim, S., and Chen, H.Y. (2002) US Pat. 2002165226; (b) DiNinno, F.P., Chen, H.Y., Kim, S., and Wu, J.Y. (2002) PCT Int. Appl. WO 2002032377; (c) DiNinno, F.P., Kim, S., and Wu, J.Y. (2002) PCT Int. Appl. WO 2002032373; (d) Chen, H.Y., Kim, S., Wu, J.Y., Birzin, E.T., Chan, W., Yang, Y.T., Dahllund, J., DiNinno, F., Rohrer, S.P., Schaeffer, J.M., and Hammond, M.L. (2004) *Bioorg. Med. Chem. Lett.*, **14**, 2551; (e) Blizzard, T.A., Gude, C., Morgan, J.D., Chan, W., Birzin, E.T., Mojena, M., Tudela, C., Chen, F., Knecht, K., Su, Q., Kraker, B., Mosley, R.T., Holmes, M.A., Rohrer, S.P., and Hammond, M.L. (2007) *Bioorg. Med. Chem. Lett.*, **17**, 6295; (f). Blizzard, T.A., Gude, C., Chan, W., Birzin, E.T., Mojena, M., Tudela, C., Chen, F., Knecht, K., Su, Q., Kraker, B., Holmes, M.A., Rohrer, S.P., and Hammond, M.L. (2007) *Bioorg. Med. Chem. Lett.*, **17**, 2944.

5 Part of the process development and mechanism investigation has been reported. Song, Z.J., King, A.O., Waters, M.S., Lang, F., Zewge, D., Bio, M., Leazer, J.L., Javadi, G., Kassim, A., Tschaen, D.M., Reamer, R.A., Rosner, T., Chilenski, J.R., Mathre, D.J., Volante, R.P., and Tillyer, R. (2004) *Proc. Nat. Acad. Sci.*, **101**, 5776.

6 (a) Kim, S., Wu, J.Y., Chen, H.Y., and DiNinno, F. (2003) *Org. Lett.*, **5**, 685 and references cited therein; (b) Nair, V., Mathew, B., Menon, R.S., Mathew, S., Vairamani, M., and Prabhakar, S. (2002) *Tetrahedron*, **58**, 3235; (c) Nair, V., and Mathew, B. (2002) *Heterocycles*, **56**, 471; (d) Capozzi, G., Falciani, C., Menichetti, S., and Nativi, C. (1997) *J. Org. Chem.*, **62**, 2611; (e) Nair, V., and Mathew, B. (2000) *Tetrahedron Lett.*, **41**, 6919.

7 Davies, I.W., Marcoux, J., Corley, E.G., Journet, M., Cai, D.-W., Palucki, M., Wu, J., Larsen, R.D., Rossen, K., Pye, P.J., DiMichele, L., Dormer, P., and Reider, P.J. (2000) *J. Org. Chem.*, **65**, 8415.

8 Dormer, P.G., Kassim, A.M., Leazer, J.L., Lang, F., Xu, F., Savary, K.A., Corley, E.G., DiMichele, L., DaSilva, J.O., King, A.O., Tschaen, D.M., and Larsen, R.D. (2004) *Tetrahedron Lett.*, **45**, 5429.

9 (a) Lightfoot, A., Schnider, P., and Pfaltz, A. (1998) *Angew. Chem. Int. Ed. Engl.*, **37**, 2897; (b) Blackmond, D.G., Lightfood, A., Pfaltz, A., Rosner, T., Schnider, P., and Zimmermann, N. (2000) *Chirality*, **12**, 442; (c) Burk, M.J., Gross, M.F., and Martinez, J.P. (1995) *J. Am. Chem. Soc.*, **117**, 9375; (d) Pye, P.J., Rossen, K., Reamer, R.A., Tsou, N.N., Volante, R.P., and Reider, P.J. (1997) *J. Am. Chem. Soc.*, **119**, 6207; (e) Tang, W., Wu, S., and Zhang, X. (2003) *J. Am. Chem. Soc.*, **125**, 9570; (f) Schrems, M.G., Wang, A., and Pfaltz, A. (2008) *Chimia*, **62**, 506.

10 (a) Ogura, K., Tomori, H., and Fujita, M. (1991) *Chem. Lett.*, 1047; (b) Ando, D., Bevan, C., Brown, J.M., and Price, D.W. (1992) *J. Chem. Soc. Chem. Comm.*, 592.

11 Kagan oxidations: (a) Kagan, H.B. (2000) Asymmetric oxidation of sulfides, in *Catalytic Asymmetric Synthesis*, 2nd edn (ed. I. Ojima), John Wiley & Sons, Inc., New York, pp. 327; (b) Pitchen, P.,

Dunach, E., Deshmukh, M.N., and Kagan, H.B. (1984) *J. Am. Chem. Soc.*, **106**, 8188; (c) Brunel, J.-M., Diter, P., Duetsch, M., and Kagan, H.B. (1995) *J. Org. Chem.*, **60**, 8086; (d) Larsson, M.E., Stenhede, U.J., Sorensen, H., Unge, S.P.K.V., and Cotton, H.K. (1999) US Pat. 5948789; (e) Hogan, P.J., Hopes, P.A., Moss, W.O., Robinson, G.E., and Patel, I. (2002) *Org. Process Res. Dev.*, **6**, 225; (f) Cotton, H., Thomas Elebring, T., Larsson, M., Li, L., Sörensen, H., and von Unge, S. (2000) *Tetrahedron Asym.*, **11**, 3819; (g) Potvin, P.G., and Fieldhouse, B.J. (1999) *Tetrahedron Asym.*, **9**, 1661; (h) Kagan, H.B., and Luukas, T.O. (2004) Catalytic Sulfide Oxidations, in *Transition Metals for Organic Synthesis*, vol. **2**, 2nd edn (Ed. Beller, M., Bolm, C.), Wiley-VCH Verlag GmbH, Weinheim, Germany, pp. 479.

12 Other asymmetric sulfide oxidation methods: (a) Bolm, C. and Bienewald, F. (1995) *Angew. Chem. Int. Ed.*, **34**, 2640; (b) Davis, F.A., Reddy, R.T., Han, W., and Carroll, P.J. (1992) *J. Am. Chem. Soc.*, **114**, 1428; (c) Palucki, M., Hanson, P., and Jacobsen, E.N. (1992) *Tetrahedron Lett.*, **33**, 7111; (d) Rossi, C., Fauve, A., Madeslaire, M., Roche, D., Davis, F.A., and Reddy, R.T. (1992) *Tetrahedron Asym.*, **3**, 629; (e) Matsugi, M., Hashimoto, K., Inai, M., Fukuda, N., Furuta, T., Minamikawa, J., and Otsuka, S. (1995) *Tetrahedron Asym.*, **6**, 2991; For more recent reviews: (f) Legros, J., Dehli, J.R., and Bolm, C. (2005) *Adv. Syn. Catal.*, **347** (1), 19; (g) Bryliakov, K.P., and Talsi, E.P. (2008) *Curr. Org. Chem.*, **12**, 386.

13 (a) Brown, H.C., and Ravindran, N. (1973) *Synthesis*, 42; (b) Brown, H.C., and Murray, K.J. (1986) *Tetrahedron*, **42**, 5497.

14 Ullmann-type ether formations (a) Wolter, M., Nordmann, G., Job, G.E., and Buchwald, S.L. (2002) *Org. Lett.*, **4**, 973; more recent reports: (b)Shafir, A., Lichtor, P.A., and Buchwald, S.L. (2007) *J. Am. Chem. Soc.*, **129**, 3490; (c) Zhang, H., Ma, D., and Cao, W. (2007) *Synlett*, 243; (d) broad scope of the copper(I) bipyridyl complex in Ullmann-type ether formation was reported more recently: Niu, J., Zhou, H., Li, Z., Xu, J., and Hu, S. (2008) *J. Org. Chem.*, **73**, 7814.

15 (a) Waters, M.S., Onofiok, E., Tellers, D.M., Chilenski, J.R., and Song, Z.J. (2006) *Synthesis*, **20**, 3389; (b) Song, Z.J., and Waters, M.S. (2005) U.S. Pat. Appl. Publ. US 2005148781 A1 20050707.

16 Isotope effects in borane reductions (a) Hawthorn, M.F., and Lewis, E.S. (1958) *J. Am. Chem. Soc.*, **80**, 4296; (b) Lewis, E.S., and Grinstein, R.H. (1962) *J. Am. Chem. Soc.*, **84**, 1158; (c) White, S.S., Jr., and Kelly, H.C. (1970) *J. Am. Chem. Soc.*, **92**, 4203; (d) Corey, E.J., Link, J.O., and Bakshi, R.K. (1992) *Tetrahedron Lett.*, **33**, 7107; (e) Linney, L.P., Self, C.R., and Williams, I.H. (1994) *J. Chem. Soc., Chem. Commun.*, 1651.

17 (a) Sasaki, T., Takahashi, T., Nagase, T., Mizutani, T., Ito, S., Mitobe, Y., Miyamoto, Y., Kanesaka, M., Yoshimoto, R., Tanaka, T., Takenaga, N., Tokita, S., and Sato, N. (2009) *Bioorg. Med. Chem. Lett*, **19**, 4232; (b) Diaz-Gavilan, M., Conejo-Garcia, A., Cruz-Lopez, O., Nunez, M.C., Choquesillo-Lazarte, D., Gonzalez-Perez, J.M., Rodriguez-Serrano, F., Marchal, J.A., Aranega, A., Gallo, M.A., Espinosa, A., and Campos, J.M. (2008) *ChemMedChem*, **3**, 127.

18 (a) Tegeler, J.J., Ong, H.H., and Profitt, J.A. (1983) *J. Heterocyclic Chem.*, **20**, 867; (b) Capozzi, G., Lo Nostro, P., Menichetti, S., Nativi, C., and Sarri, P. (2001) *Chem. Commun.*, 551; (c) Melchiorre, C., Brasili, L., Giardina, D., Pigini, M., and Strappaghetti, G. (1984) *J. Med. Chem.*, **27**, 1535; (d) Arnoldi, A., Bassoli, A., Merlini, L., and Ragg, E. (1993) *J. Chem. Soc., Perkin Trans 1*, 1359.

19 Komatsu, N., Hashizume, M., Sugita, T., and Uemura, S. (1993) *J. Org. Chem.*, **58**, 4529.

6
HIV Integrase Inhibitor: Raltegravir

Guy R. Humphrey and Yong-Li Zhong

Human immunodeficiency virus type 1 (HIV-1) is the etiological agent of the acquired immunodeficiency syndrome (AIDS). It is currently estimated that 33 million people are living with HIV/AIDS worldwide, with infection and death rates of about 4 million and 3 million per year, respectively [1, 2]. Three enzymes have been identified and targeted to arrest the HIV life cycle: reverse transcriptase (RT), protease (PR) and integrase (IN). Most oral drug treatments of HIV infection inhibit the first two of these enzymes. Until recently multidrug cocktails consisting of a protease inhibitor or a non-nucleoside reverse transcriptase inhibitor (NNRTI) in combination with two nucleoside reverse transcriptase inhibitors were the standard HIV therapy (HAART). Significant limitations on the effectiveness of these HAART combinations, due to drug-related toxicities and the emergence of resistant viruses, is a major issue [3]. Drugs with novel modes of action, such as viral entry inhibitors and IN inhibitors, have been sought in order to offer antiretroviral treatment to experienced HIV patients who harbor drug-resistant virus or suffer toxicities with HAART. IN inhibitors have been of particular interest to HIV/AIDS researchers because, unlike RT and PR, IN has no human homolog, and thus inhibitors of IN might be better tolerated at high doses [4].

Raltegravir (**1**; Figure 6.1), discovered and developed in the Merck research laboratories, was approved under the trademark ISENTRESS® in October 2007 as the first commercially available antiretroviral agent to target IN [5]. Although Raltegravir was originally indicated for combination therapy with other antiretroviral agents in treatment-experienced adults, the FDA recently approved an expanded indication to include treatment-naive adult patients [6]. In the first part of this

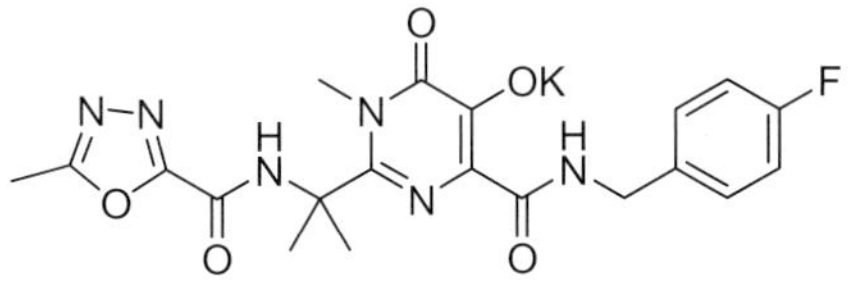

1 potassium salt

Figure 6.1 Structure of Raltegravir (**1**).

The Art of Process Chemistry. Edited by Nobuyoshi Yasuda

ISBN: 978-3-527-32470-5

chapter the original chemical synthesis of Raltegravir will be described along with details of the development of a new, highly efficient manufacturing synthesis capable of consistently supplying multi-ton quantities of high purity bulk drug. A key driver throughout the development work was the need for the most cost effective process to enable worldwide patient access. In the second part, further development of the key thermal rearrangement to pyrimidinones is discussed.

6.1 Project Development

6.1.1 Medicinal Chemistry Route

The initial retrosynthetic analysis of **1** resulted in the cleavage of the two amide bonds and a C–N bond leading to the four components; oxadiazole carbonyl chloride **2**, methyl iodide, 4-fluorobenzylamine (4-FBA) and the densely functionalized hydroxypyrimidinone **3** (Scheme 6.1). These synthetic disconnections were reasonable and should be applicable for long term route development.

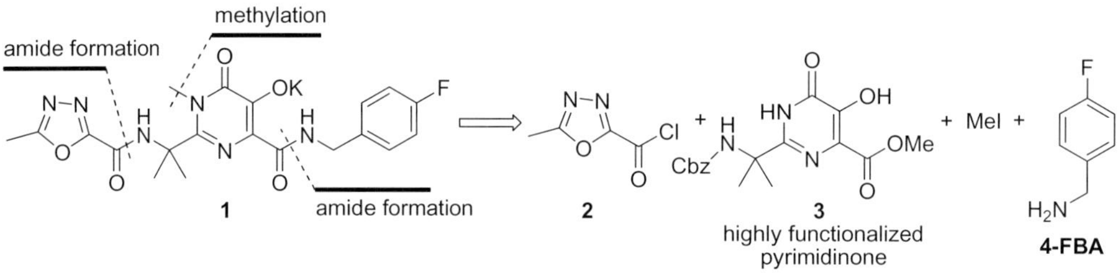

Scheme 6.1 Retrosynthetic analysis of **1**.

The medicinal route for the preparation of **1** started from a Strecker reaction of commercially available and inexpensive acetone cyanohydrin **4** with ammonia (Scheme 6.2). Cbz-protection of the aminonitrile **5**, under Schotten–Baumann conditions, afforded intermediate **6**. Hydroxylamine addition to the nitrile group provided the amidoxime **7**. Two-component coupling reaction between **7** and dimethyl acetylenedicarboxylate (DMAD) gave a *Z*/*E* mixture of adduct **8**, which was thermally rearranged to the key intermediate hydroxypyrimidinone **3**. Selective hydroxy group protection of **3** using benzoic anhydride, followed by chromatographic purification afforded benzoate **9**. *N*-Methylation of the resulting protected pyrimidinone **9** using Me_2SO_4 and LiH in 1,4-dioxane, followed by another chromatographic purification, to remove the 6-*O*-methylated by-product, gave intermediate **10**. The Cbz-protecting group was cleaved by Pd-mediated hydrogenation

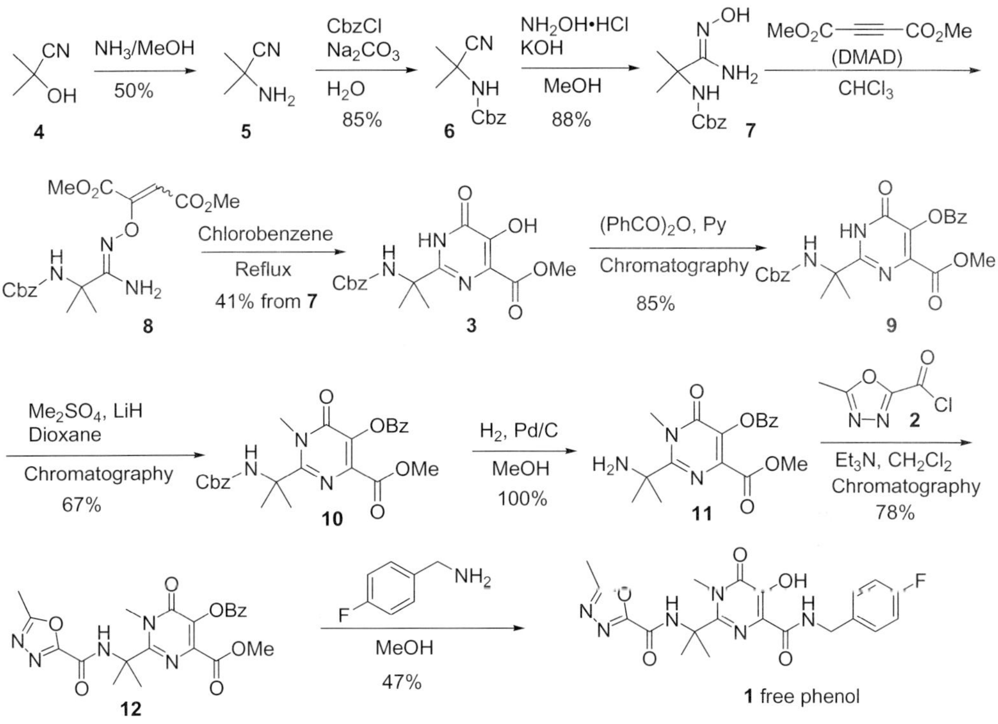

Scheme 6.2 Synthetic route of medicinal chemistry to free phenol **1**.

to obtain free amine **11**. Coupling of the resulting amine **11** with oxadiazole carbonyl chloride **2**, followed by chromatographic purification, afforded penultimate **12**. Amidation of the methyl ester **12** with 4-FBA completed the synthesis of **1**. The synthesis was carried out in ten linear steps in approximately 3% overall yield [5].

After careful scrutiny of the Medicinal Chemistry route, we evaluated the advantages and shortcomings for the synthesis of **1**:

6.1.1.1 Advantages of the Medicinal Chemistry Route

1) All the starting materials were inexpensive.
2) The highly functionalized hydroxypyrimidinone **3**, the key intermediate for the synthesis of **1**, was rapidly assembled by a two-component coupling reaction, followed by a thermal rearrangement.

6.1.1.2 Problems with the Medicinal Chemistry Route

1) Several low yielding steps, especially in the Strecker reaction, the thermal rearrangement, unselective *N*-methylation and the final amidation.

2) Multiple use of halogenated solvents such as chloroform and dichloromethane, and highly toxic and expensive 1,4-dioxane, which was not desirable for large scale due to environmental concerns.
3) Multiple chromatographic purifications.

6.1.2
Process Development

6.1.2.1 First Generation Manufacturing Process for the Synthesis of 1

Although all the above issues for the long term and manufacturing preparation of our target **1** needed to be addressed, we recognized that the *key* challenges were the efficient construction of the hydroxypyrimidinone **3** and the optimization of the selective *N*-methylation. [7] Initial process development for **1** is discussed as outlined below, focusing on these two major issues.

1) Route selection for the construction of hydroxypyrimidinone **3**.
   - Condensation of dihydroxyfumarate derivatives with amidines.
   - Michael addition of amidoxime **7** to DMAD, followed by thermal rearrangement.
2) Optimization of the selected route for the synthesis of **3**.
3) Optimization of selective *N*-methylation of intermediate **9**.
4) Amidation with 4-FBA prior to installing oxadiazole carboxamide.
5) Preparation of 5-methyl-1,3,4-oxadiazole-2-carbonyl chloride **2**.
6) Installation of oxadiazole carboxamide as the final step.

6.1.2.1.1 Route Selection for Synthesis of 3

Literature reports on synthetic methods for the construction of the pyrimidinone core were very limited. Most of the synthetic strategies toward the densely functionalized core fell into two methodologies, which start from the same amidoxime **13** (Scheme 6.3). Route **A** is a three-step sequence that involves hydrogenation of **13** to prepare amidine **14**. Claisen condensation of commercially available α-benzyloxy acetate and methyl *tert*-butyl oxalate provides the dihydroxyfumarate

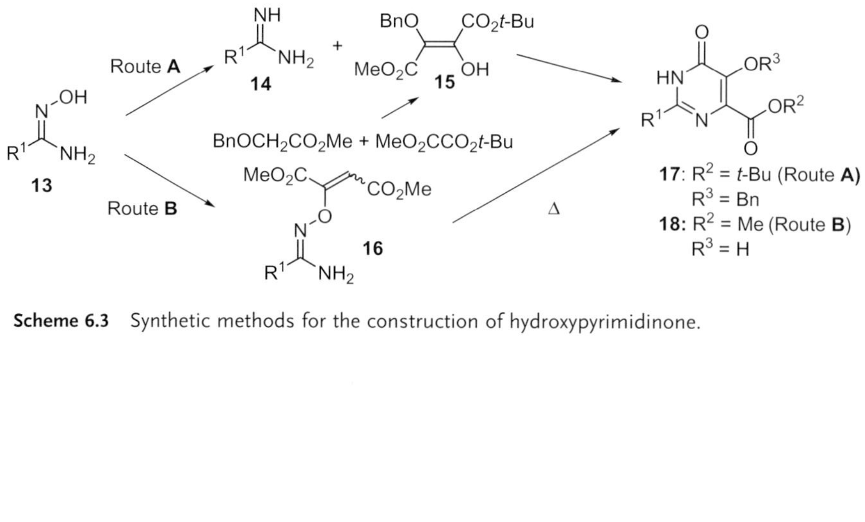

Scheme 6.3 Synthetic methods for the construction of hydroxypyrimidinone.

derivative **15**. Finally, condensation between **14** and **15** furnishes the benzyl-protected pyrimidinone **17** [8]. On the other hand, route **B** is a two-step approach starting with a Michael reaction between **13** and DMAD to afford the intermediate **16**, followed by a thermal rearrangement to hydroxypyrimidinone **18** [9].

Route A: Condensation of dihydroxyfumarate derivatives with amidines Recently, Dreher *et al.* [8d] developed and optimized the condensation reaction of amidines **14** with **15**, which led to the benzyl pyrimidinones **17** in up to 96% yield. This was an attractive possibility for an efficient construction of the hydroxypyrimidinone core. Due to the incompatibility of the *N*-Cbz group for the hydrogenation of **13** to **14**, an alternative protecting group was sought for our route development. Towards this end, Boc- or Nosyl-protection of aminonitrile **5**, followed by hydroxyamine addition to the nitrile group afforded amidoximes **19** and **20** in 90% yield, respectively (Scheme 6.4). Hydrogenation of **19** and **20** gave the corresponding amidines **21** and **22** in 85% yield. Optimized condensation of amidines **21** or **22** with fumarate derivative **15** gave the desired functionalized pyrimidinones **23** and **24** in only 50% and 55% yield, respectively. While the condensation approach may be of use for the preparation of a wide variety of pyrimidinones, the moderate yield, increased number of steps, stability concerns with **15** and the overall poor atom-economy, led us to abandon this route for our long term synthesis development.

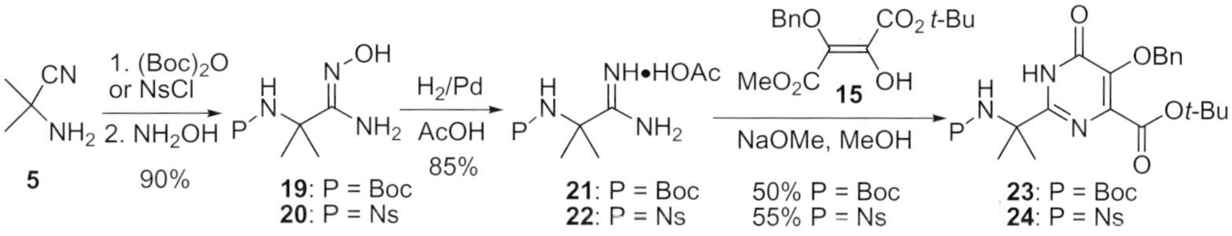

Scheme 6.4 Condensation of dihydroxyfumarate derivatives with amidines to pyrimidinone core.

Route B: Michael addition of amidoxime 7 to DMAD, followed by thermal rearrangement Michael addition of amidoxime **7** to DMAD, followed by thermal rearrangement to construct the hydroxypyrimidinone core, was the route originally used by the medicinal chemists. Although the initial yield for the thermal rearrangement step was only 41%, the high atom-economy, simplicity, low cost and brevity of the chemistry were very attractive for the long term route development. We decided to investigate this route further with a view to gaining a full understanding of the chemistry and optimizing with respect to yield and operability at scale.

6.1.2.1.2 Optimization of the Selected Route for the Synthesis of 3

Solvent effects for the thermal rearrangement Solvent effects on the thermal rearrangement were evaluated using a microwave technique [9e].

Table 6.1 Solvent effects for the microwave-assisted rearrangement.

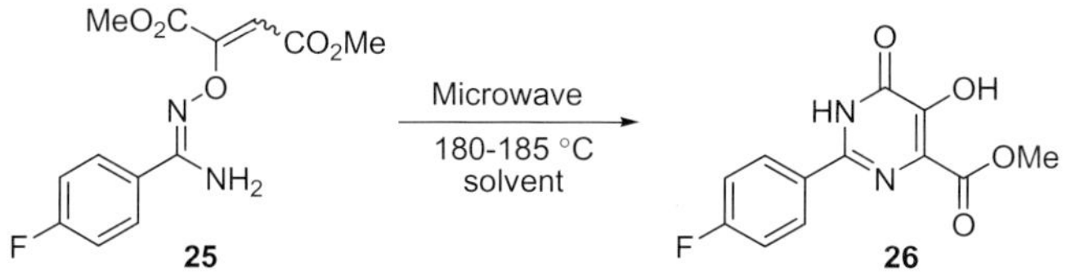

Entry	Solvent	Assay yield (%)
1	neat	52
2	1,2-dichlorobenzene	62
3	DME	54
4	1,2-DCE	50
5	1,4-dioxane	66
6	DMF	38
7	IPA	35
8	Toluene	47
9	MeCN	48
10	*o*-Xylene (or xylenes)	68

Due to some stability concerns with the *N*-Cbz group of **8** at high temperatures, compound **25** was used as a model substrate for the reaction. Substrate **25** was irradiated for 2 min (internal temperature reached 185 °C) in a variety of solvents and all thermal reactions reached >95% conversion (Table 6.1). Both aprotic polar solvents (entries 6 and 9) and protic polar solvent (entry 7) gave poor assay yields of product **26**. With nonpolar solvents (entry 10) such as *o*-xylene and xylenes, the rearrangement reaction provided the highest assay yield and proved to be the best solvent choice [9e].

E*- and *Z*-Amidoxime DMAD adduct issues for the thermal rearrangement** The thermal rearrangement of **8** to hydroxypyrimidinone **3** was investigated in detail. Addition of amidoxime **7** to DMAD typically affords a mixture of *E*/*Z* adducts **8**. Initial experiments using a mixture of *E*- and *Z*-adducts **8** in *o*-xylene at 125 °C, indicated that one of the isomers was consumed faster than the other. At this point, the two isomers were separated using chromatography and the structure of the *Z*-adduct **8*Z was unambiguously determined by X-ray crystallography (Figure 6.2).

The *E*- and *Z*-adducts were separately subjected to the thermal rearrangement conditions, as shown in Scheme 6.5. The thermal rearrangement of the *Z*-adduct **8*Z*** occurred at a lower temperature (125 °C) and gave a 72% assay yield of product **3**. On the other hand, thermal rearrangement of the *E*-adduct **8*E*** required higher temperature (135 °C) and gave a lower assay yield. Both of the reactions generated about 5% of imidazole **27** as a by-product. In order to improve the assay yield and

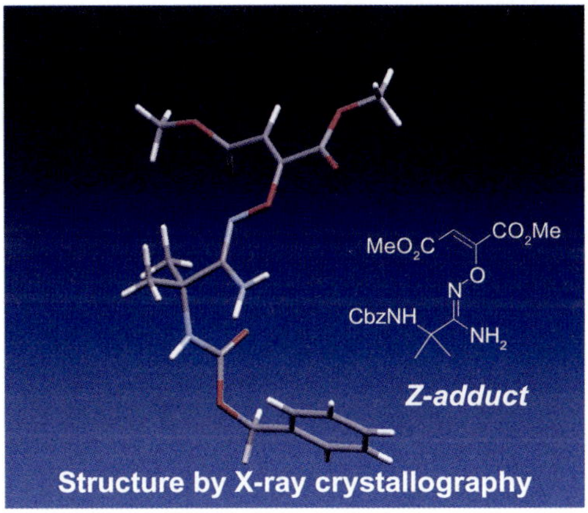

Figure 6.2 X-ray crystallography of *Z*-adduct **8*Z***.

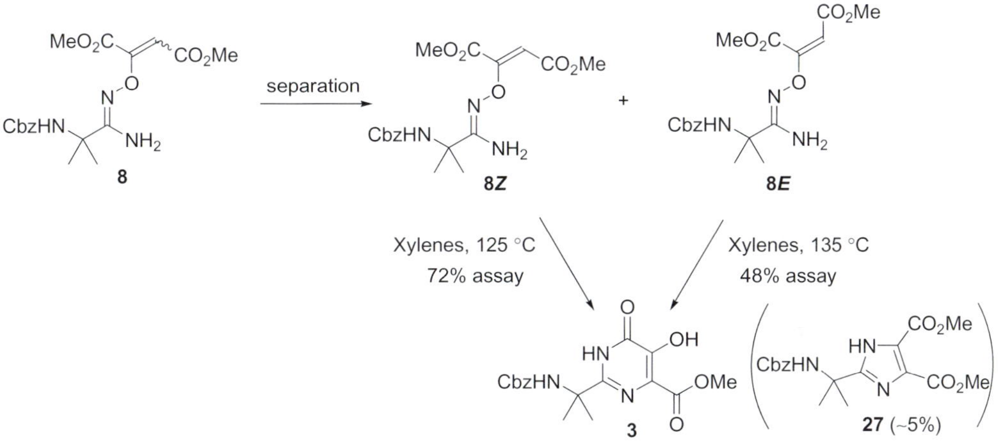

Scheme 6.5 Thermal rearrangement of *E*- and *Z*-amidoxime–DMAD adducts.

minimize the reaction time and temperature, the *Z*-selective DMAD adduct formation was investigated.

The solvent and temperature effects for the Michael addition of amidoxime **7** to DMAD were probed because the reaction itself occurs without any other catalysts. As shown in Table 6.2, the reaction gave a high ratio of **8*E*** in strongly aprotic polar solvents such as DMF and DMSO (entry 1 and 2). **8*E*** was also found as the major product in MeCN (entry 3), dichloromethane (entry 4), and xylenes (entry 5). To our delight, the desired **8*Z*** was obtained as the major component in methanol (entry 6). The stereoselectivity of **8*Z*** versus **8*E*** was better at low temperature (entry 7). A similar result was observed when the reaction was run in THF or dichloroethane in the presence of a catalytic amount of DABCO (entries 9 and 10).

Table 6.2 Optimization of stereoselective amidoxime addition to DMAD.

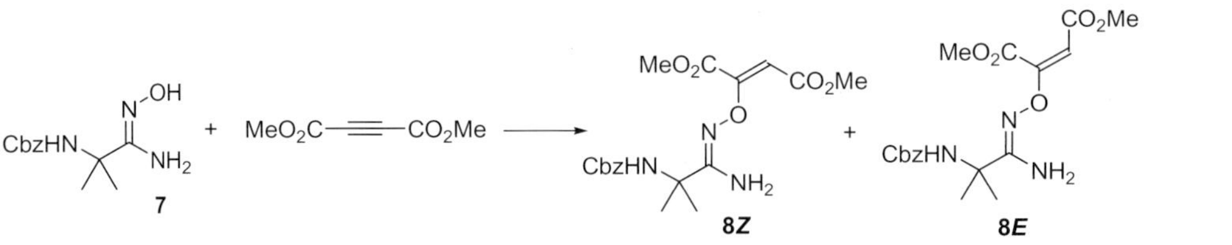

Entry	Solvent	Additive (mol%)	*T* (°C)	Conversion (%)	*Z*/*E* ratio
1	DMF	N/A	rt	100	2:98
2	DMSO	N/A	rt	100	1:99
3	MeCN	N/A	rt	100	20:80
4	DCM	N/A	rt	70	31:69
5	Xylenes	N/A	80	100	21:79
6	MeOH	N/A	rt	100	65:35
7	MeOH	N/A	−10	100	75:25
8	AcOH	N/A	rt	30	50:50
9	THF	DABCO (15)	−70	100	79:21
10	1,2-DCE	DABCO (5)	rt	100	74:26

Isomerization of **8*E*** to **8*Z*** was investigated under typical isomerization conditions of double bonds. Unfortunately, all attempts were unsuccessful.

The ratio of **8*Z*** to **8*E*** is slightly better in the presence of DABCO in THF (Table 6.2; entry 9) than in pure methanol (Table 6.2; entry 7). Since the DABCO must be removed prior to the thermal rearrangement and the minimal impact on overall yield, we decided to run the Michael addition in methanol to afford a mixture of *Z*- and *E*-adducts **8** in quantitative yield. The resulting solution of adducts **8** was solvent-switched to xylenes and heated at 125 °C for 2 h, and at 135 °C for 4 h to give a 62% assay yield of desired product **3**. The reaction mixture was concentrated and hydroxypyrimidinone **3** was directly crystallized in 54% isolated yield as a white crystalline solid.

Since the key thermal rearrangement was optimized, we turned our attention to earlier steps. The synthesis of amidoxime **7** was optimized from acetone cyanohydrin **4** (Scheme 6.6). The original Strecker reaction was carried out with ammonia

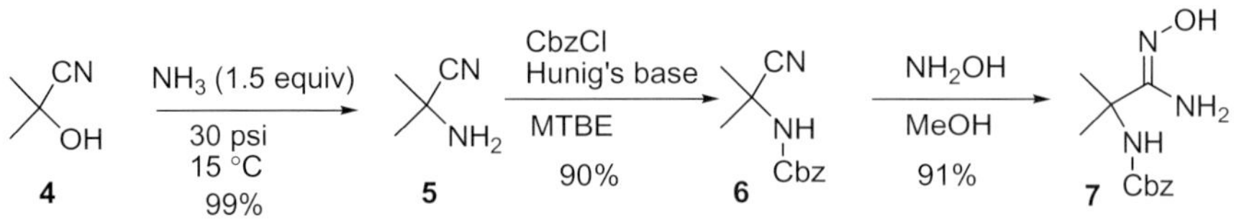

Scheme 6.6 Optimized synthesis of amidoxime **7**.

in methanol solution. However, under these conditions, the reaction did not reach completion and a large amount of product was lost in the work-up. Under our optimized conditions, the Strecker reaction was accomplished in 1.5 equiv of liquid ammonia at 30 psi, 15 °C and gave aminonitrile **5** in 99% yield. *N*-Cbz-protection of the aminonitrile **5** with benzyl chloroformate (Cbz-Cl) in the presence of Hünig's base afforded intermediate **6** in 90% assay yield. Intermediate **6** was treated with hydroxyamine in methanol to provide amidoxime **7**, as a crystalline solid, in 91% isolated yield. The overall isolated yield of the three-step sequence to amidoxime **7** from **4** was increased from 37% to 81%.

6.1.2.1.3 Optimization of the *N*-Methylation Selectivity

In the original Medicinal Chemistry route, protection of the phenolic OH with a benzoate was carried out prior to *N*-methylation. In order to simplify the process, the direct *N*-methylation of hydroxypyrimidinone **3** was investigated. To our delight, methylation of **3** gave a mixture of the desired *N*-methyl product **31** and the undesired *O*-methyl by-product **32** as a 70 : 30 mixture (Scheme 6.7; path b). Surprisingly, methyl ethers **28–30** were not observed at all (Scheme 6.7; path a).

With proof-of-concept on the direct methylation in hand, the optimization of the methylation with respect to regioselectivity, yield and isolation conditions was initiated to eliminate the use of 1,4-dioxane and LiH (Table 6.3). Under similar reagent/base conditions (Me_2SO_4, Li*Ot*-Bu), better selectivity was observed in non-polar solvents (entries 2 and 3) than in polar (entry 1) and aprotic solvents (entries 4 and 5). MeI was a better methylation reagent than dimethyl sulfate under otherwise the same reaction conditions (entries 5 and 7).

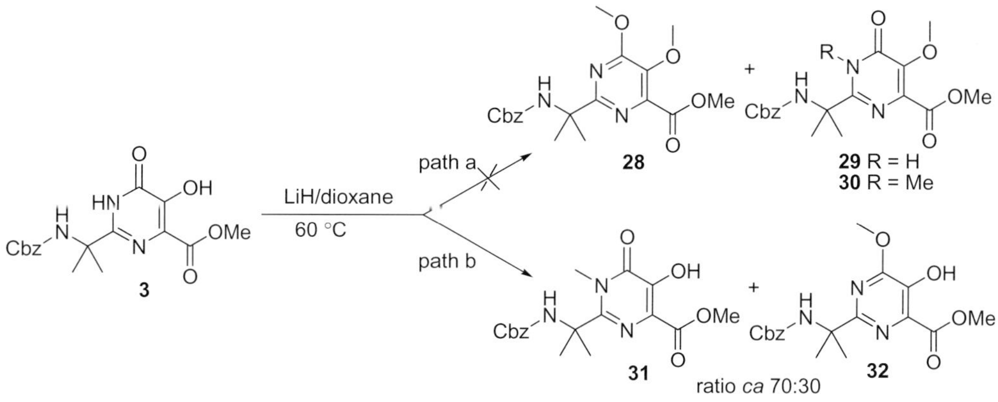

Scheme 6.7 Direct *N*-methylation of hydroxypyrimidinone **3**.

Interestingly, the selectivity was significantly improved by addition of 1 equiv of $MgBr_2{\cdot}OEt_2$ to the reaction (entry 6). A possible mechanism involving chelation between the hydroxypyrimidinone **3** and $Mg(OMe)_2$ was proposed (**33** and **34** in Scheme 6.8). Encouraged by the above results, we found that the best performance

Table 6.3 Direct methylation of hydroxypyrimidinone 3.

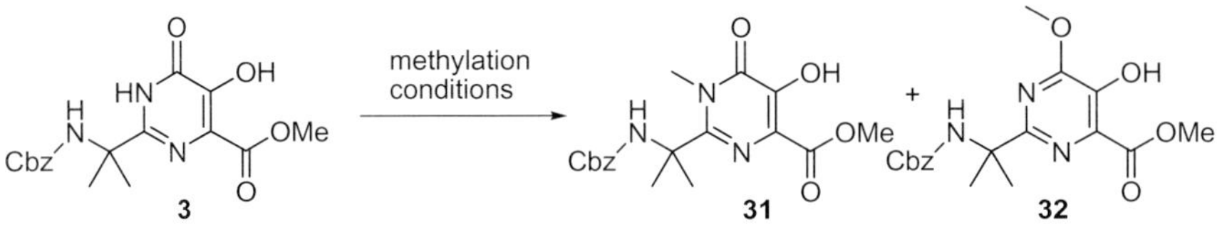

Entry	Reagent	Base	Solvent	Additive	*T* (°C)	Conversion (%)	31:32 ratio
1	Me_2SO_4	LiO*t*-Bu	MeOH	N/A	50	25	16:84
2	Me_2SO_4	LiO*t*-Bu	Toluene	N/A	50	40	48:52
3	Me_2SO_4	LiO*t*-Bu	THF	N/A	50	74	47:53
4	Me_2SO_4	LiO*t*-Bu	DMSO	N/A	50	100	12:88
5	Me_2SO_4	LiO*t*-Bu	DMF	N/A	80	100	9:91
6	Me_2SO_4	LiO*t*-Bu	DMF	$MgBr_2 \cdot OEt_2$	80	86	48:52
7	MeI	LiO*t*-Bu	DMF	N/A	80	65	36:64
8	MeI	$Mg(OMe)_2$	TFH	N/A	20	16	52:48
9	MeI	$Mg(OMe)_2$	DMF	N/A	35	58	64:36
10	MeI	$Mg(OMe)_2$	NMP	N/A	20	70	63:37
11	MeI	$Mg(OMe)_2$	DMAc	N/A	20	68	61:39
12	MeI	$Mg(OMe)_2$	DMSO	N/A	35	56	78:22
13	MeI	$Mg(OMe)_2$	DMSO	N/A	60	100	78:22

of the methylation was using MeI as methylation reagent, $Mg(OMe)_2$ as base and DMSO at 60 °C (entry 13). Under our optimized conditions, the methylation afforded the desired product **31** in 78% assay yield and 70% isolated yield after crystallization.

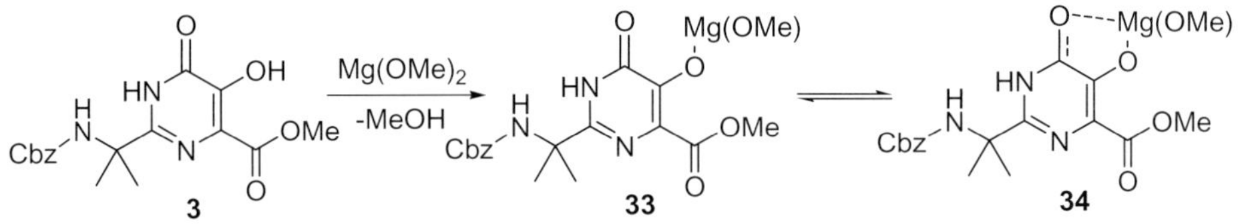

Scheme 6.8 Proposed chelation of hydroxypyrimidinone **3** with Mg^{2+}.

6.1.2.1.4 **Amidation with 4-FBA Prior to Installing the Oxadiazole Carboxamide**

The Medicinal Chemistry route introduced the oxadiazole fragment prior to installation of the 4-FBA (Scheme 6.1). The overall yield for these two steps was only 37%. The oxadiazole required a two-step synthesis and was a much more expensive reagent than 4-FBA. In order to improve the chemical yield, reduce cost and improve the overall process robustness, we investigated the amidation with 4-FBA prior to installing the oxadiazole moiety.

Treatment of intermediate **31** with 2.2 equiv of 4-FBA in EtOH at 72 °C afforded **35** as a white crystalline solid in 90% isolated yield (Scheme 6.9). Hydrogenation in the presence of 5% of Pd/C and 1 equiv of MsOH, efficiently removed the Cbz-protected group. MsOH was used to prevent fluoride reduction resulting in low levels of the des-fluoro by-product. Catalyst filtration, followed by neutralization of the crude reaction mixture with NaOH, afforded free amine **36** as a white crystalline product in 99% isolated yield. Free-amine **36** was isolated as a dihydrate which necessitated drying prior to coupling with oxadiazole chloride **2**.

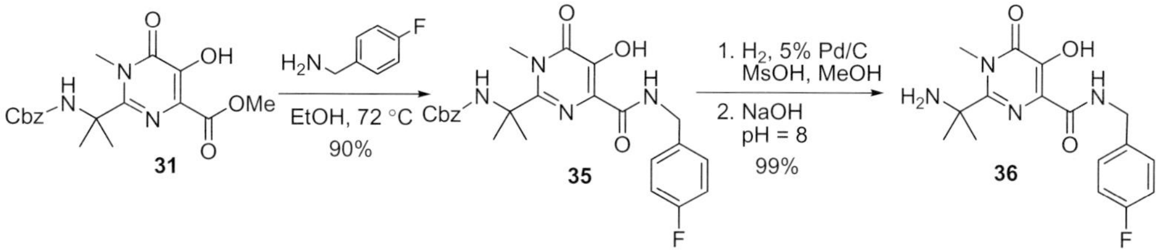

Scheme 6.9 Amidation with 4-FBA prior to install the oxadiazole moiety.

6.1.2.1.5 **Preparation of 5-Methyl-1,3,4-Oxadiazole-2-Carbonyl Chloride (2)**

The original route for the synthesis of potassium 5-methyl-1,3,4-oxadiazole-2-carboxylate (**39**) via an intramolecular condensation of ethyl (2-acetylhydrazinyl)-(oxo)acetate (**37**) suffered from moderate yields (Scheme 6.10).

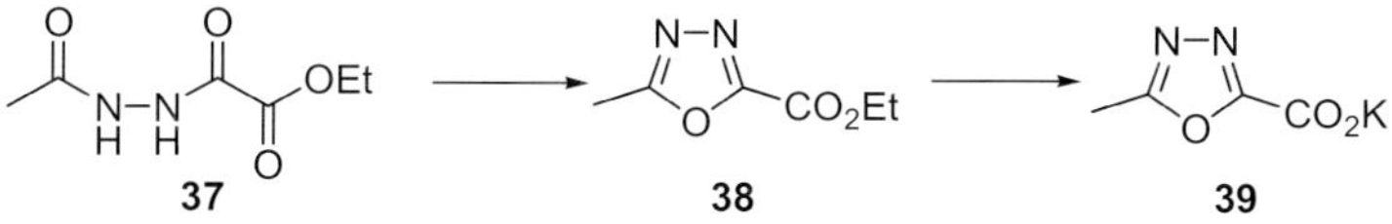

Scheme 6.10 The original preparation of potassium salt of oxadiazolyl acid.

A modified literature procedure [10] provided a long-term high yield manufacturing process to the oxadiazole fragment, as depicted in Scheme 6.11. Amidation of tetrazole **40** with ethyl oxalyl chloride (**41**) afforded intermediate **42** which was

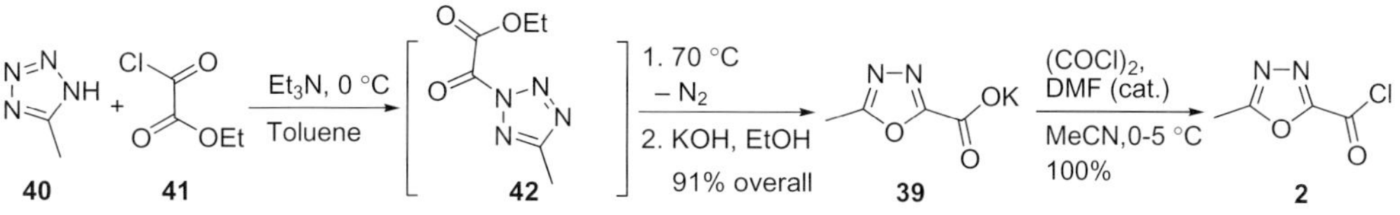

Scheme 6.11 Preparation of oxadiazole carbonyl chloride fragment **2**.

heated at 70 °C to afford the oxadiazole ester **38** in high yield. Hydrolysis with KOH gave potassium salt **39** as a crystalline solid, in 91% overall yield. Treatment of **39** with oxalyl chloride in the presence of a catalytic amount of DMF provided 5-methyl-1,3,4-oxadiazole-2-carbonyl chloride (**2**) in quantitative yield.

6.1.2.1.6 Installation of Oxadiazole Carboxamide as the Final Step (End Game)

With the free amine **36** and carbonyl chloride **2** in hand, the coupling of these two fragments was achieved in the presence of 4-methylmorpholine (NMM) to afford the desired free phenol **1** and the ester **43** in about 10 : 1 ratio. The reaction mixture was directly treated with aqueous $MeNH_2$ to cleave the ester of **43**. Acidification with HCl provided free phenol **1** in 88% isolated yield after crystallization and filtration. Raltegravir (**1**) was isolated as a crystalline potassium salt (1 : 1 molar ratio) by addition of KOEt to a solution of the free phenol **1** in a 93% yield, and 99.5% purity (Scheme 6.12).

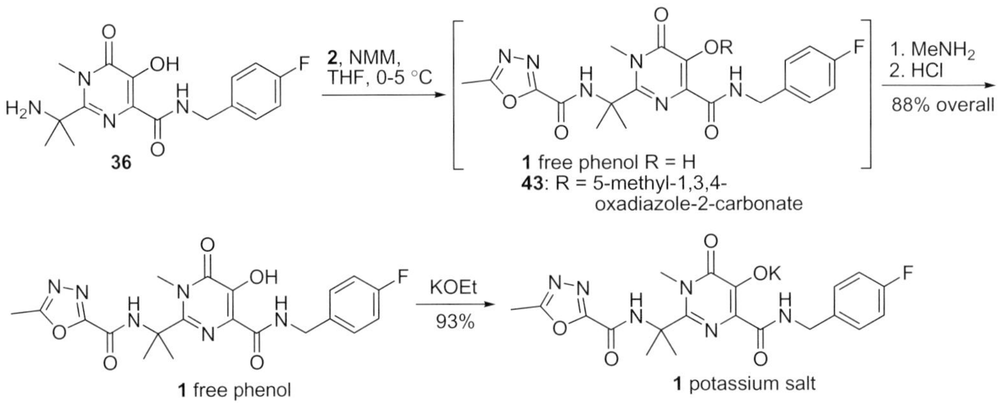

Scheme 6.12 Preparation of Raltegravir (**1**).

6.1.2.1.7 Summary of the First Generation of Manufacturing Process Development

Key accomplishments for the first generation manufacturing synthesis of Raltegravir **1** are summarized below:

1) An efficient and practical synthesis of Raltegravir (**1**) was developed via a key thermal rearrangement of amidoxime DMAD adducts to construct the key highly functionalized hydroxpyrimidione **3**.
2) The thermal rearrangement of the amidoxime DMAD *Z*-adduct **8*Z*** was more robust to be converted to the desired product **3** than the corresponding *E*-adduct **8*E***. This discovery opened a new door for us to revisit this old chemistry.

3) Raltegravir (**1**) was prepared in nine linear chemical steps with 22% overall yield.
4) Key problems of the original route were effectively addressed.
   - The low yielding steps were significantly improved.
   - All halogenated, highly toxic and expensive solvents were eliminated from the process.
   - Undesired base LiH was substituted by $Mg(OMe)_2$.
5) No chromatographic operations throughout the process.

The first generation process was rapidly and successfully scaled up to provide multi-ton quantities of bulk drug to adequately supply needs through phase III, drug filing and launch.

Since Raltegravir was a critical new tool to address the worldwide HIV/AIDS epidemic, in order to further increase global patient access, an even more efficient, economical and environmentally friendly synthesis was desired.

6.1.2.2 Second Generation Manufacturing Process for the Synthesis of 1

Work towards a second generation manufacturing route began with a close examination of previously investigated synthetic routes, brainstorming new ideas and a full evaluation of the current route and potential areas for improvements. The construction of the pyrimidinone **3** via the cyclo-isomerization chemistry used cheap and readily available raw materials and provided the complex core in a single step with high efficiency. The chemistry was robust and simple to perform with a direct crystallization to afford the product **3** with good purity, as previously described. Thus the majority of the synthesis redesign was targeted at the chemical steps from pyrimidinone **3** to the final product **1**. [7]

Evaluation of each step of the existing route with respect to yield, productivity, efficiency, waste generation (PMI: process (product) mass intensity) and solvent use was carried out and a number of potential areas for improvement were identified as summarized below:

1) Excess (2.2 equiv) of expensive oxadiazole **2** used in final coupling.
2) Free amine **36** was isolated as a hygroscopic hydrate which was very difficult to dry in the solid state and required azeotropic drying using THF (expensive and time consuming).
3) Excess 4-FBA (2.3 equiv) was used in the amidation.
4) Methylation selectivity was only 78 : 22, resulting in a moderate isolated yield (70%).
5) Volume productivity on many steps was low.
6) Waste production (PMI) was high on many steps especially the methylation and final coupling.

6.1.2.2.1 Use of a Protecting Group for the Final Coupling

During the final coupling reaction between free amine **36** and oxadiazole **2** unselective acylation of the phenolic OH occurs to give the bis-acylated compound **43**.

Hydrolysis of **43** using $MeNH_2$ or KOH furnishes the coupled phenol **1**. Attempts to overcome the unselective acylation issue using alternative coupling reagents were unsuccessful. Thus, in order to reduce the need for 2.2 equiv of expensive oxadiazole **2**, to obtain high conversion/yield, protection of the OH group in amine **36** was desired. Selection of the "right" protecting group was critical. The group needed to be easily and economically installed in essentially quantitative yield and to provide a stable, isolable intermediate that could be stored and used directly in the coupling reaction. Ideally, the intermediate would be crystalline and non-hygroscopic, thus obviating the tedious, expensive drying step of amine **36** in the existing process. Lastly, the protecting group would ideally be cleaved under the same conditions (or milder) as used in the current process. The oxadiazole side chain in **1** is unstable under basic conditions and care is needed to avoid decomposition. Consequently, a number of acyl protecting groups were evaluated including acetate, propionate, benzoate and pivalate. A combination of positive attributes led us to select the pivalate ester derivative **44** for further development.

6.1.2.2.2 **Preparation and Properties of Free-Amine Pivalate Ester (FAPE) 45**

Pivalate ester **44** was prepared in excellent yield (99%) and purity using a highly optimized through-process (Scheme 6.13). Cbz-amide **35** was acylated with pivaloyl chloride in the presence of triethylamine and a catalytic amount (0.01 mol%) of DMAP in ethyl acetate at room temperature. Without DMAP the reaction was sluggish and typically stalled at 90–95% conversion. Attempted hydrogenation of **44** using the previously employed conditions (Pd/C, MsOH, MeOH) was complicated due to the insolubility of the MsOH salt of the amine **45** in the reaction medium, making catalyst filtration impossible. After screening several different acids, glycolic acid was identified as a suitable counterion with excellent solubility properties.

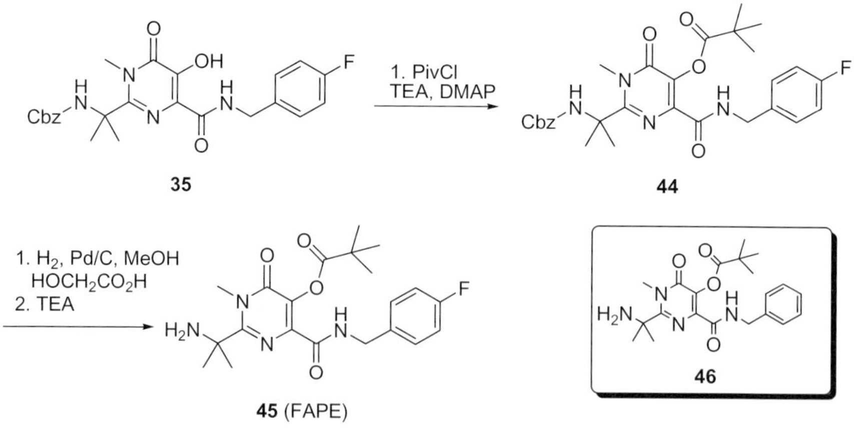

Scheme 6.13 Synthesis of amide **45**.

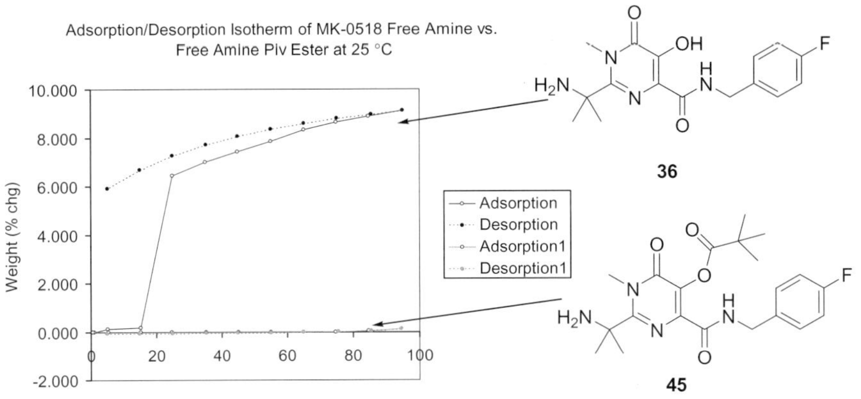

Figure 6.3 Hygroscopicity of amine **36** versus FAPE **45**.

Glycolic acid proved to be an essential additive allowing complete conversion to the amine without formation of the des-fluoro impurity **46** and maintaining solubility of the free amine in the reaction mixture.[1] In practice, after pivalate ester formation, the crude reaction mixture was washed with water to remove the $NEt_3 \bullet HCl$. Methanol, glycolic acid and Pd/C were added and the mixture hydrogenated at 5 psi of hydrogen for 2–3 h at 20–25 °C . On complete conversion, the catalyst was filtered and the pH of the solution was adjusted to pH 9 with triethylamine to crystallize free amine pivalate ester (FAPE) **45**, as the free base. The isolated solid was a high-melting, crystalline solid that proved stable to storage in the solid state and gratifyingly was non-hygroscopic between 0 and 95% relative humidity (Figure 6.3.) and could be dried easily using conventional filter drying or drying oven technology.

6.1.2.2.3 **Optimized Coupling of FAPE 45 with 2 to Provide 1**

With crystalline, dry, non-hygroscopic FAPE **45** in-hand, re-optimization of the final amide coupling was carried out (Scheme 6.14). In the original manufacturing route a mixture of solvents was required. THF was used to azeotropically dry the amine **36**, acetonitrile was used to prepare the acyl chloride **2** and IPA/water used to crystallize the free phenol **1**. The volume productivity was low while solvent waste generation was high and the multi-solvent waste mixture produced was difficult to recycle. In order to simplify the process and increase productivity, the coupling and isolation was optimized to a single organic solvent. Thus acyl chloride preparation was carried out in acetonitrile at 0–5 °C. A slurry of FAPE **45** and 4-NMM in acetonitrile was added to **2** at −10 °C. On complete reaction aqueous potassium hydroxide was added to convert **47** to free phenol **1**. Acetic acid was then added, followed by addition of water to crystallize the free phenol **1** in 99% overall isolated yield. When compared to the original process to couple **2** with **36** the new process is less complex from a process standpoint, uses approximately

1) Des-F **47** results in the formation of des-F **1**, an impurity not rejected by crystallization.

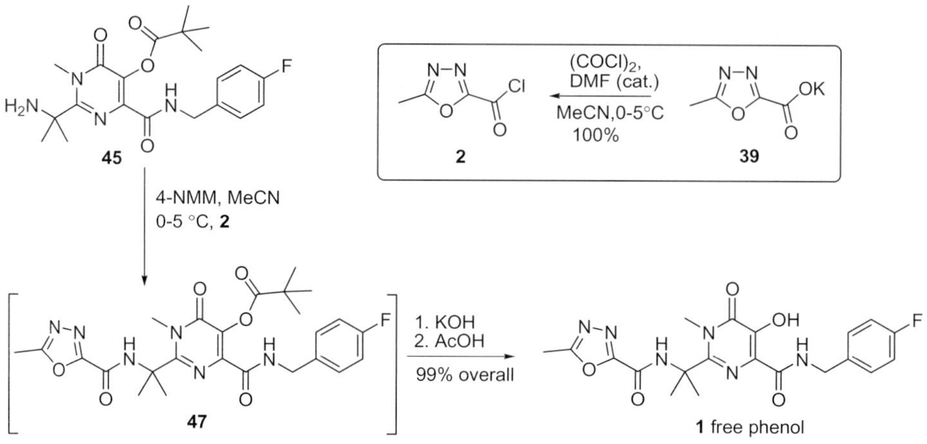

Scheme 6.14 Improved coupling of FAPE **45** with **2**.

one fifth the total solvent volume, and is about three times more productive. The new process also uses only 1.15 equiv of **2** (vs. 2.2 equiv), does not employ any solvent switches or time-consuming concentrations and provides free phenol **1** in almost quantitative yield.

6.1.2.2.4 **Improving the *N*-Methylation Selectivity**

With the identification of FAPE **45** as the key coupling intermediate and optimization of the final coupling complete we turned our attention to the remaining key chemistry challenge, namely, the selective *N*-methylation of pyrimidinone **3**. Considerable development work on the methylation reaction was carried out in the first generation route development. Under optimized conditions, use of $Mg(OMe)_2$ and MeI in DMSO at 60 °C provided a 78 : 22 selectivity for the *N*-Me versus *O*-Me products. Carefully controlled crystallization afforded pure *N*-Me compound **31** in 70% overall isolated yield. Apart from the moderate yield, the reaction suffered from lower than ideal productivity and very high solvent use, resulting in the production of a large amount of waste. In fact the PMI for this single step was over 100.

Since, based on the historical data, complete kinetic regioselectivity was unlikely to be achieved, our attention was focused on the possibility of a demethylation followed by an *N*-methylation recycle (Scheme 6.15). Unfortunately, the attempted

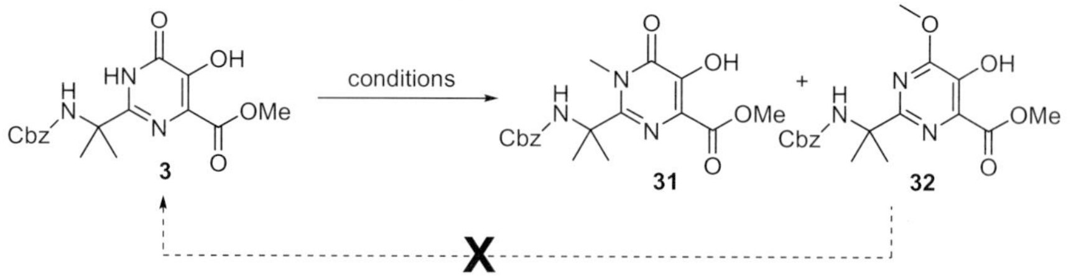

Scheme 6.15 Attempted demethylation of *O*-Me **32**.

demethylation of *O*-Me **32** under a variety of conditions resulted in extensive decomposition, due, in part, to the lability of the methyl ester functionality. In order to stabilize the intermediate to the relatively harsh demethylation conditions, we decided to look at reversing the step order and preparing amide **48** prior to methylation.

Amide **48** was prepared from hydroxypyrimidinone **3** using 1.2 equiv of 4-FBA in the presence of 1 equiv of triethylamine in methanol at reflux for 6 h. Addition of acetic acid and water to induce crystallization provided amide **48** in 99% isolated yield after filtration and drying (Scheme 6.16).

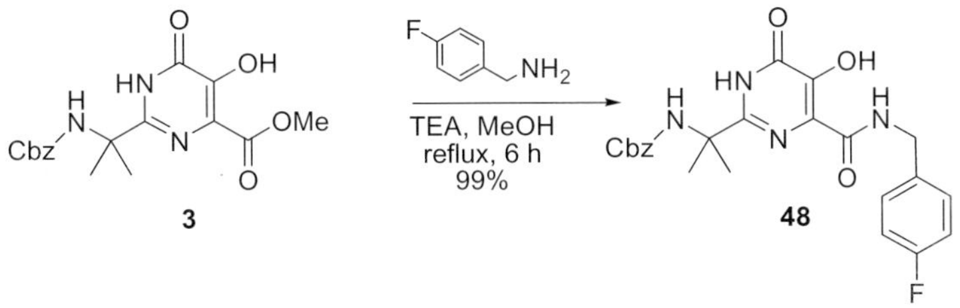

Scheme 6.16 Preparation of amide **48**.

Methylation of amide **48** under the previously optimized conditions [MeI, $Mg(OMe)_2$, DMSO] used for ester **3** provided similar *N*-Me to *O*-Me regioselectivity (78 : 22) as expected. With increased stability of amide **48** versus ester **3**, higher temperatures and longer reaction times were probed. Gratifyingly, a key observation was made from some initial probe experiments. Use of extended reaction time (20 h) at 65 °C gave a 99 : 1 selectivity for the *N*-methylated product **35**. In order to explore the *O*-Me to *N*-Me recycle strategy further, design of experiments (DOE) and extensive use of high throughput screening was utilized. DOE evaluation of a number of variables, including reaction concentration, temperature, time, and equivalents of reagents, showed that higher reaction temperature, higher concentration and higher equivalents of reagents all led to higher *N*-Me selectivity (Scheme 6.17). Under optimum conditions [3 equiv of $Mg(OMe)_2$, 5 equiv of MeI,

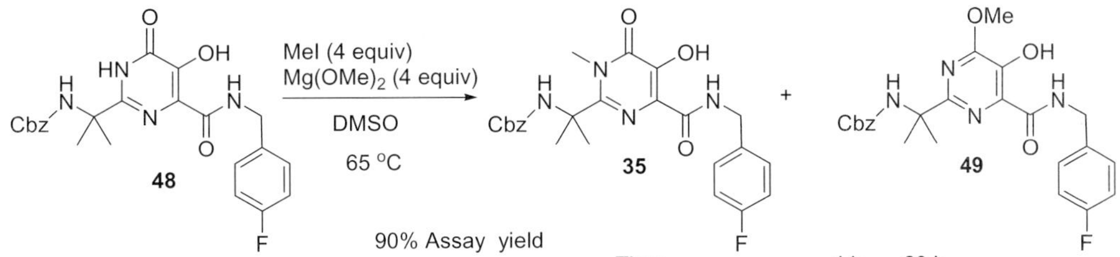

Time	4 h	20 h
Conversion	95%	99%
N-Me *vs* *O*-Me	80/20	99/1

Scheme 6.17 Methylation of **48** and conversion of *O*–Me to *N*–Me *in situ* using MgI_2.

20 h at 65 °C at a reaction concentration of 1.0 M] *N*-methyl amide **35** was obtained in 90% assay yield (99 : 1, *N*-Me vs. *O*-Me).

With successful proof of concept on the *in situ* conversion of *O*-Me **49** to *N*-Me **35**, we turned our attention to the use of high-throughput reaction screening to select the base, methylating reagent and solvent with respect to our long-term goals of productivity, economics, safety and environmental impact. Magnesium hydroxide [$Mg(OH)_2$] was selected as the base for optimization due to ready availability and very low cost. Trimethylsulfoxonium iodide [$Me_3S(O)I$, m.p. 208–212 °C] was selected as the methylation source due to relative cost and ease of handling when compared with methyl iodide. Both DMSO and DMF were suitable as reaction solvents, however NMP was chosen for intrinsic safety reasons.

The optimized methylation conditions (Scheme 6.18) provided >99% conversion and 92% isolated yield of **35** after *in situ* crystallization, filtration and drying. Addition of at least 1 equiv of water was essential for complete conversion of the *O*-Me to *N*-Me product. Under these reaction conditions MeI is released at the reaction temperature, resulting in an initial 4 : 1 mixture of **35** : **49**. *In situ*, iodide-promoted, demethylation of **49** followed by remethylation recycled the undesired *O*-methyl isomer **49** to **35** in a single-pot reaction. The reaction was generally complete in about 3–6 h at 100 °C.

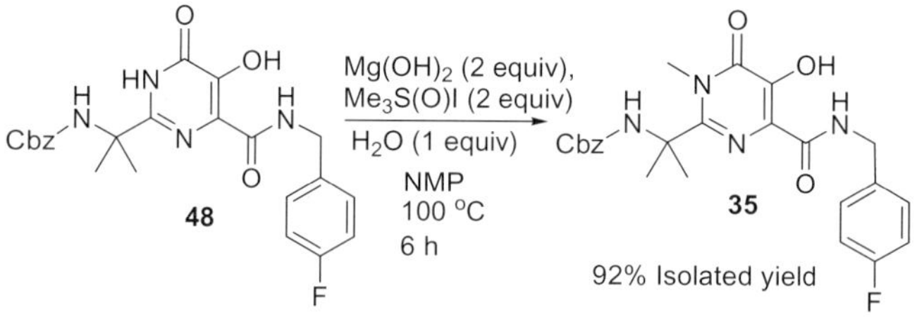

Scheme 6.18 Optimized methylation of **48** using $Mg(OH)_2$ and $Me_3S(O)I$.

6.1.2.2.5 Summary of Second Generation Manufacturing Route

Key developments for the second generation chemistry are summarized:

1) Identification of the pivalated amine intermediate (FAPE, **45**) as a non-hydroscopic coupling partner.
2) Improvement in productivity and yield for the final coupling reaction.
3) Discovery of *in situ* demethylation–remethylation conditions to isomerize the *O*-methyl to the desired *N*-methylpyrimidinone.
4) New high yield amidation with 4-FBA.

Process development in the conversion of hydroxypyrimidinone **3** to Raltegravir (**1**) resulted in an increase in overall yield from a poor 20% in the original medicinal chemistry synthesis to 51% in the first generation manufacturing route and finally to 84% for the second generation manufacturing route (Scheme 6.19). As well as reducing the overall cost to produce Raltegravir, the yield improvement, coupled with a three- to five-fold increase in productivity for each step, resulted in a reduction of organic and aqueous waste generation by 65%. The second

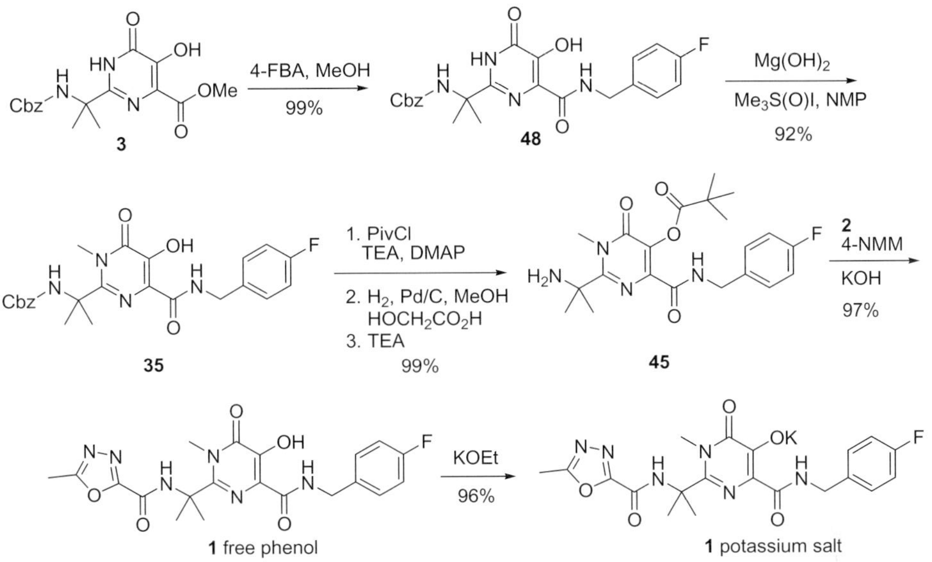

Scheme 6.19 Second generation manufacturing route.

generation manufacturing route was successfully demonstrated at metric ton scale.

6.2 Further Chemistry Development

6.2.1 Development of Microwave-Accelerated Thermal Rearrangement

During the course of our studies on the solvent effect on the high-temperature hydroxypyrimidinone formation we found that the thermal rearrangement, promoted by microwave irradiation, proceeded rapidly (<5 min) to afford moderate to good isolated yields of the desired product. Extension of this microwave protocol to explore the substrate scope was carried out. To a solution of amidoxime in methanol (10 vol) 1.05 equiv of DMAD was added dropwise at −10 °C and the mixture slowly warmed to ambient temperature (>98% conversion) over 6 h. The amidoxime–DMAD adducts were formed in moderate to good selectivity for the *Z*-adduct. The reaction mixture was solvent-switched to *o*-xylene (5 vol) at 25–40 °C. The xylene solution of amidoxime adducts was irradiated with microwaves.[2] The resulting slurry was stirred at room temperature for 1 h and the crystalline solid product filtered, washed and dried under vacuum to afford the corresponding hydroxypyrimidinone (Table 6.4). A variety of substrates, including

2) ETHOS D, Millestone, at 80% of a total output of 1000 W with temperature control set to 185 °C for 1–2 min.

Table 6.4 Scope of selectivity of amidoxime addition to DMAD and hydroxypyrimidinone formation via microwave-accelerated thermal rearrangement.

Entry	Starting material	*Z*-adduct/*E*-adduct	*Z*-adduct/ *E*-adduct (ratio)	Product	Conditions	Isolated yield (%)
	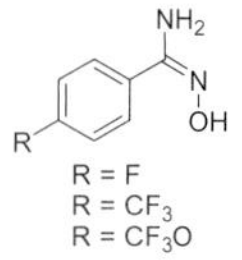	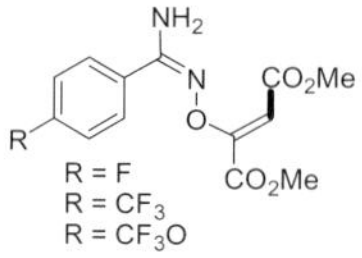		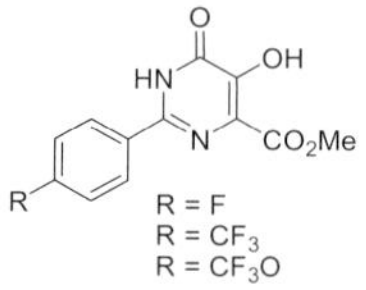		
1	R = F	R = F	90:10	R = F		60
2	R = CF_3	R = CF_3	87:13	R = CF_3	rt to 185 °C over 85 s	61
3	R = CF_3O	R = CF_3O	88:12	R = CF_3O		48
4	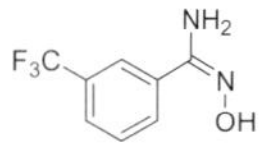	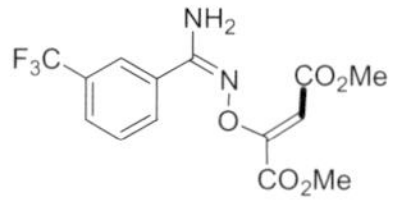	86:14	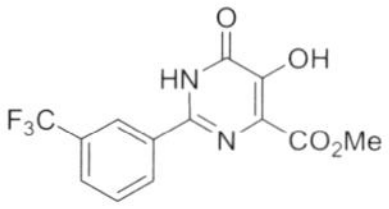	rt to 185 °C over 160 s	50
5	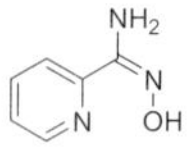	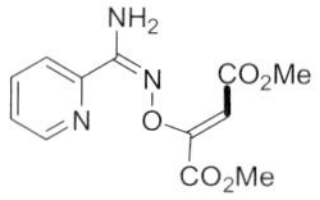	91:9	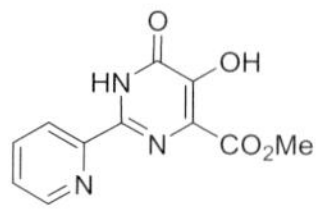	rt to 185 °C over 85 s	50
6	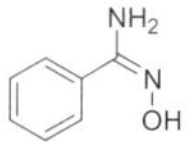	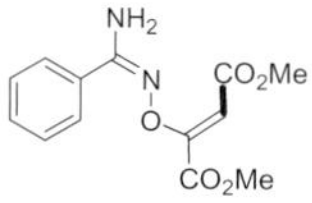	90:10	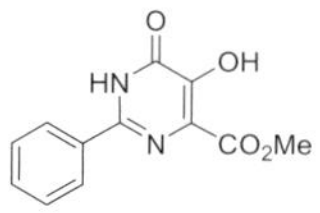	rt to 182 °C over 85 s	50
7	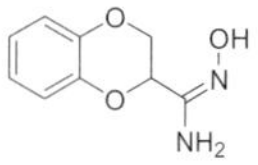	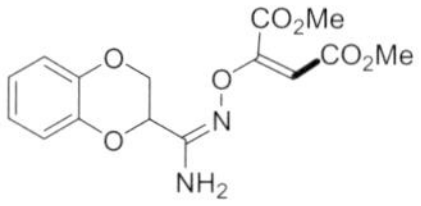	81:19	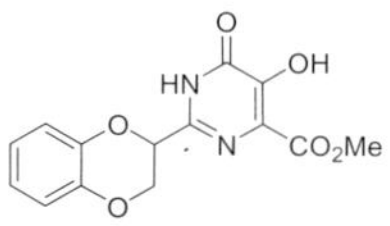	rt to 185 °C over 85 s	59
	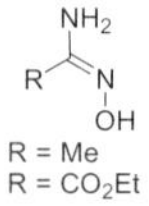	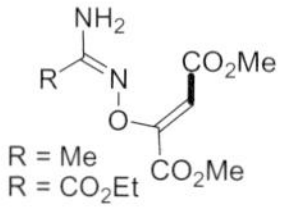		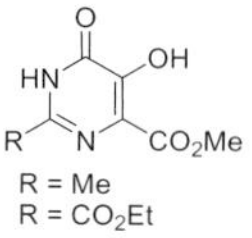		
8	R = Me	R = Me	89:11	R = Me	rt to 185 °C over 85 s	48
9	R = CO_2Et	R = CO_2Et	77:23	R = CO_2Et		50
10	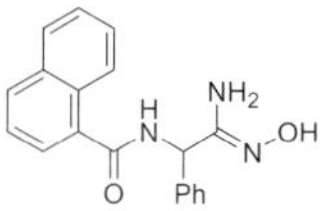	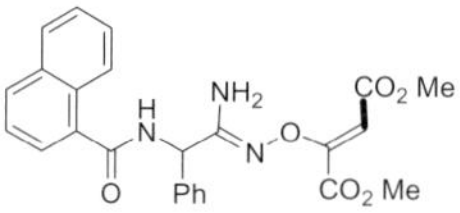	67:33	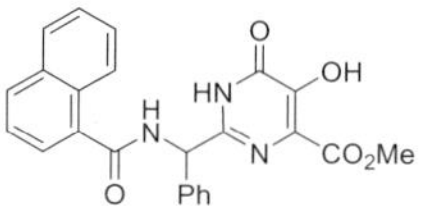	rt to 185 °C over 120 s	39
11	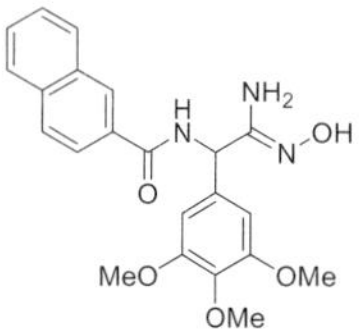	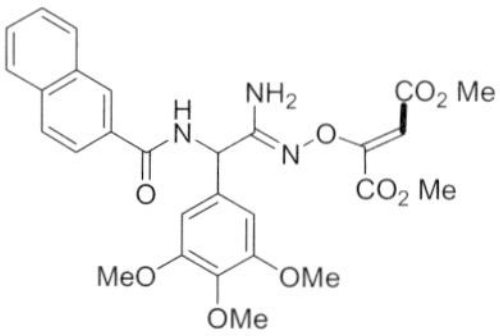	67:33	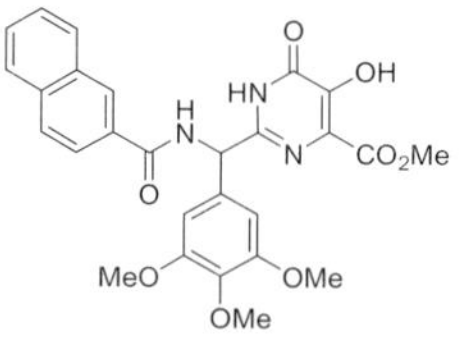	rt to 170 °C over 300 s	67

aromatic (entries 1–4, 6), pyrimidine ring (entry 5), and functionalized aliphatics (entries 7–9) were effectively cyclized to the hydroxypyrimidinones. *N*-Protected α-amino amidoxime–DMAD adducts (entries 10–11) were also converted to hydroxypyrimidinone in moderate yield.

6.2.2
Mechanistic Studies on the Thermal Rearrangement

As described previously, the two-component coupling reaction between amidoxime **50** and DMAD generated a mixture of *Z*- and *E*-adducts **51**, which was heated in xylenes to afford hydroxypyrimidinone **55** (Scheme 6.20). The previously proposed mechanism involved tautomerization of **51** to **52**, followed by a Claisen [3,3]-rearrangement to yield intermediate **53**. Subsequent tautomerization of the intermediate **53** to **54**, followed by cyclization would afford **55** [9a,f].

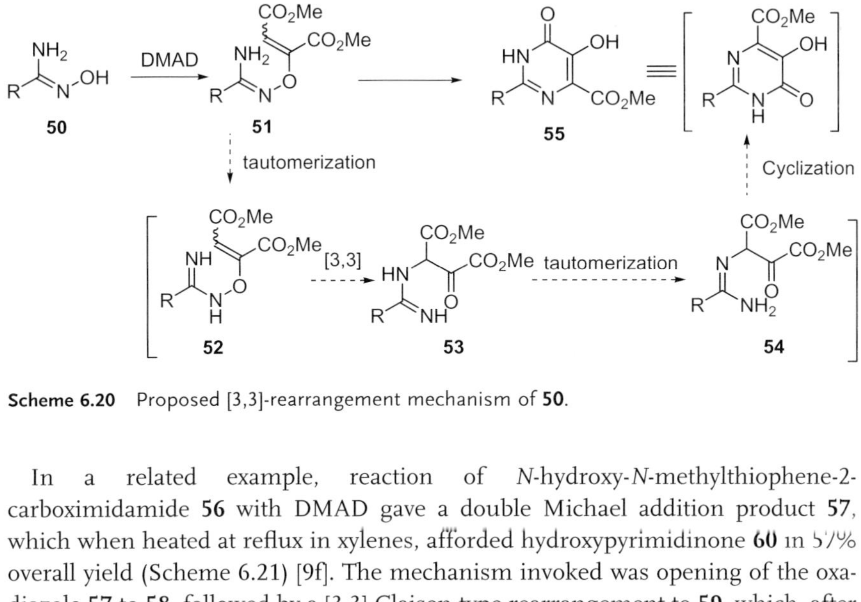

Scheme 6.20 Proposed [3,3]-rearrangement mechanism of **50**.

In a related example, reaction of *N*-hydroxy-*N*-methylthiophene-2-carboximidamide **56** with DMAD gave a double Michael addition product **57**, which when heated at reflux in xylenes, afforded hydroxypyrimidinone **60** in 57% overall yield (Scheme 6.21) [9f]. The mechanism invoked was opening of the oxadiazole **57** to **58**, followed by a [3,3]-Claisen-type rearrangement to **59**, which, after tautomerization and cyclization, afforded **60**.

In another example using the isomeric amidoxime substrate **61**, the formation of the expected [3,3]-rearrangement product **63** was not observed (Scheme 6.22). Instead the *Z*-adduct **62*Z*** cyclized to oxadiazoline **64**. Interestingly, the *E*-adduct **62*E*** rearranged to hydroxypyrimidinone **60** and imidazole **66** instead of **63**. The rearrangement of the substrate **62*E*** was proposed to occur via intermediate **65** via a [1,3]-sigmatropic rearrangement which, after cyclization, led to the observed products **60** and **66**.

Since the exact mechanism of the rearrangement of unsubstituted amidoxime DMAD adduct **51** was unclear, we decided to undertake our own studies. These

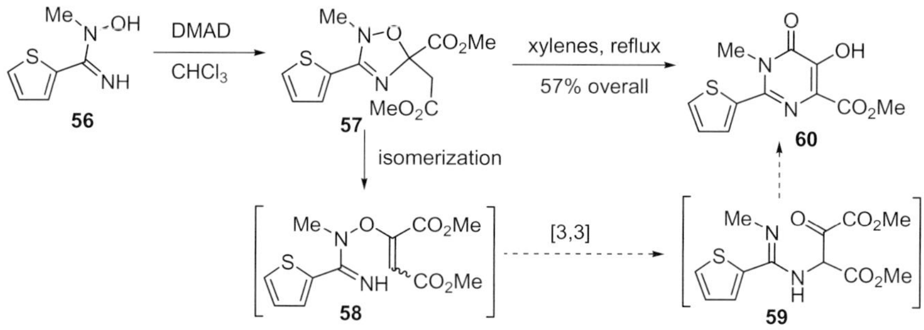

Scheme 6.21 Evidence for [3,3]-rearrangement mechanism of **56**.

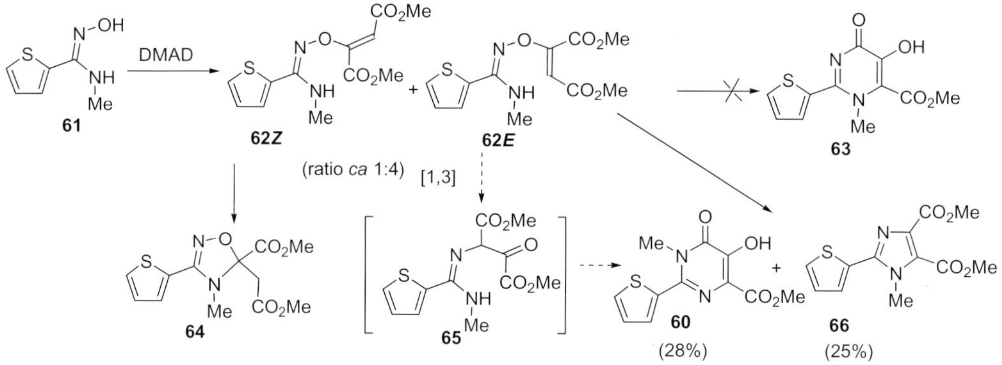

Scheme 6.22 Proposed [1,3]-rearrangement mechanism of **62**.

studies began with a ^{15}N-labeled experiment (Scheme 6.23) [9i]. Treatment of aminonitrile **6** with (^{15}N)-hydroxylamine afforded (^{15}N)-amidoxime **7***, which was converted to a mixture of Michael adducts **8*Z***/**8*E*** (65/35) at room temperature. Thermolysis of **8*Z***/**8*E*** provided the hydroxypyrimidinone **3***, in which the ^{15}N label was unexpectedly found to be exclusively at the position *ortho* to the ester substituent via **67***. Since this outcome is not consistent with a [3,3]-sigmatropic

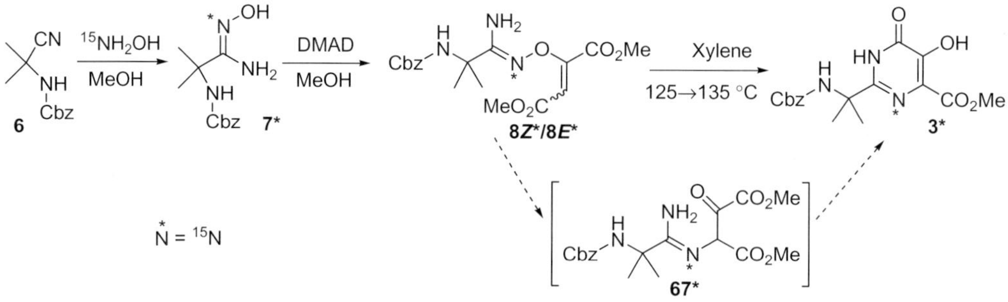

Scheme 6.23 ^{15}N-labeled vinyloxyamidine rearrangement. *All energies are referenced to **68*Z*/*E***. Pictures are for structures in the stepwise rearrangement of **68*E***. (N.D. = not determined.).

rearrangement mechanism, a [1,3]-rearrangement or a diradical mechanism must have occurred.

Computational mechanistic studies on this reaction were carried out in collaboration with Professor Kendall Houck. The results are summarized in Scheme 6.24. Three main reaction pathways were identified. Path **C** involves tautomerization of **68*Z*/*E*** to **69*Z*/*E***. Two transition states (**TS**) from **69*Z*/*E*** to product **72** via **71** and **70** were identified. The energies of **TS5** (36.5/37.8; *Z* /*E* respectively) via [3,3] sigmatoropic rearrangement are much lower energy than **TS4** via [1,3] sigmatropic rearrangement. **TS5** was identified as the lowest transition state in the gas phase, however, tautomerization of **68*Z*/*E*** to **69*Z*/*E*** was never observed in toluene even though the activation energy is only ~12 kcal mol^{-1}. Therefore, the reaction path **C** was eliminated. Path **A**, via a tightly hydrogen-bound polar radical pair (PRP) was identified as the second lowest transition state (**TS1**, 40.4/43.1). Significantly, **TS1-*Z*** is 3 kcal mol^{-1} lower in energy than **TS1-*E***, which is in excellent agreement with the experimentally favored reactivity of **8*Z*** in comparison with **8*E***. Finally,

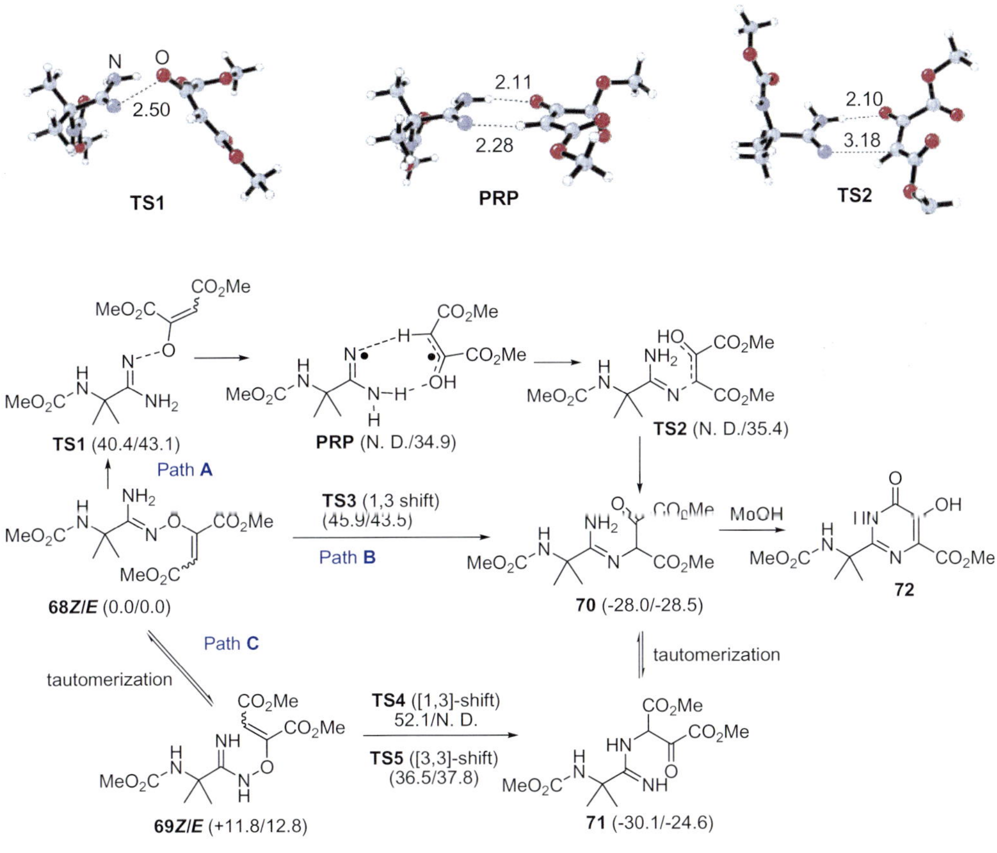

Scheme 6.24 Corrected B3LYP energies and relevant transition structures and intermediates in the possible rearrangement mechanisms of amidoximes–DMAD adducts **68*Z*/*E***.

path **B** goes through direct [1,3]-sigmatropic rearrangement. The activation energies of the transition state (**TS-3**) are 45.9 and 43.5 kcal mol^{-1} for the *Z* and *E* forms, respectively. The geometries of **TS1** and **TS3** are quite similar, being dissociative in nature, these differ most significantly in the orientation of the migrating amidino moiety in relation to the vinyloxy moiety. Given the significant error bars associated with computed barriers, the calculations do not definitely differentiate path **A** and path **B**. However, the relative reactivities computed for **68*Z*/*E***, which are similar to the experimental results, and the intramolecular trapping of a radical pair (see below) are consistent only with the radical pair mechanism.[3)]

An experimental probe for the presence of radical intermediates resulting from thermally induced homolytic cleavage of the *N–O* bond was derived by incorporating an alkene into a model substrate to act as a potential intramolecular radical trap (Scheme 6.25) [11]. In a control experimental, thermal reaction of **73** gave the desired product **74** in 66% isolated yield. On the other hand, thermal rearrangement of the unsaturated compound **75** under our typical conditions gave the desired hydroxypyrimidinone **76** in only 38% isolated yield. When the vinyl amidoxime mixture **75*Z*/*E*** was heated in *o*-xylene at 125 °C in the presence of a

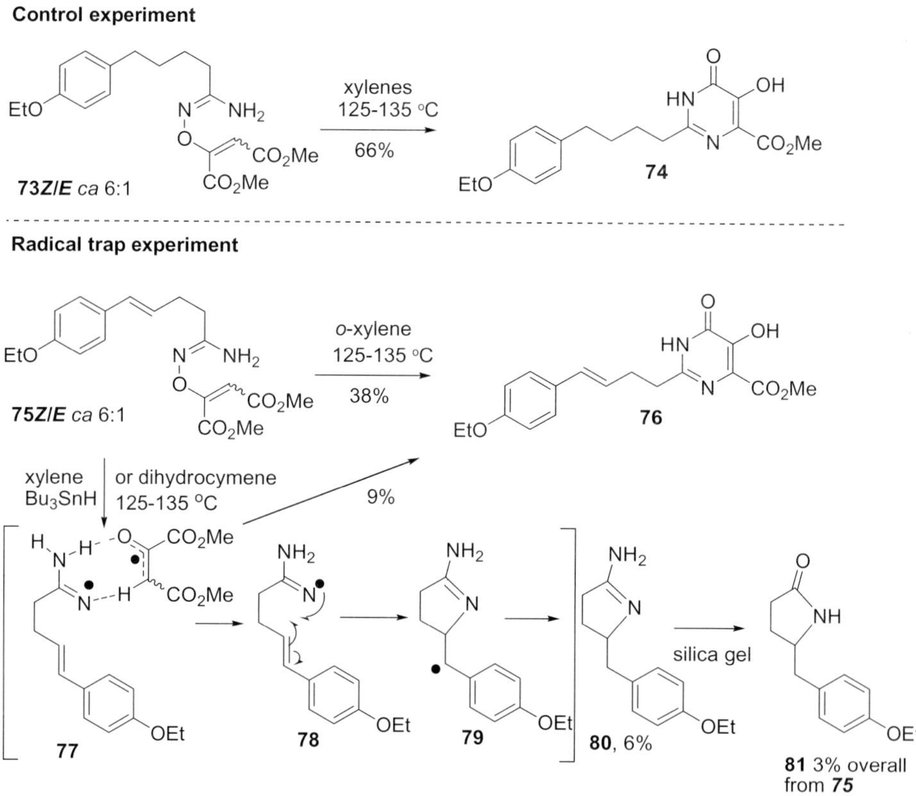

Scheme 6.25 Trapping of proposed radical intermediate.

3) See Ref. [9i] for a more detailed discussion of the computational results.

hydrogen atom source (Bu_3SnH or dihydrocymene), **76** was formed in only 9% isolated yield along with **81** (the product of hydrolysis of **80**) in 3% overall yield. This result strongly supported the PRP reaction pathway A.

Further evidence consistent with the polar radical pair mechanism was provided by a crossover experiment (Scheme 6.26). A 1:1 mixture of labeled **8*Z****/8*E****** and unlabeled **8*Z*/8*E*** was heated in xylene at 125 °C for 2 h and at 135 °C for 4 h to afford hydroxypyrimidinones **3***** and **3**. Analysis of the products by high resolution mass spectrometry showed no crossover between the labeled and unlabeled fragments. This result reinforces the computational results discussed previously wherein PRP recombines to give product within the solvent cage (Scheme 6.24).

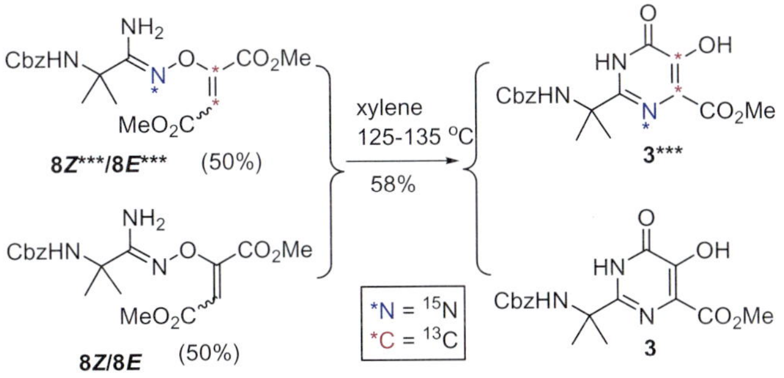

Scheme 6.26 ^{15}N- and ^{13}C-labeled vinyloxyamidine rearrangement.

Thus experimental and computational investigations have provided some evidence for the intermediacy of a polar radical-pair in the assembly of the pyrimidinone core of Raltegravir **1**.

6.3 Conclusion

A practical, highly efficient manufacturing route for the synthesis of HIV integrase inhibitor Raltegravir **1** was developed. A more than ten-fold increase in overall yield, from 3% in the original medicinal chemistry synthesis to 35% in the second generation manufacturing route, was realized through innovative chemistry development, reaction understanding and heavy optimization. The result of these developments was a highly efficient, productive, robust, economical and environmentally friendly synthesis capable of supplying the expected high-volume demand.

Acknowledgments

The authors wish to thank Remy Angelaud, David Askin, Kevin M. Belyk, Spencer Dreher, Tony Hudgens, Amar J. Mahajan, Peter E. Maligres, Danny Manchino, Ross A. Miller, Dermot O'Brien, Michael Palucki, Paul Phillips, Vanessa M. Pruzinsky, Philip J. Pye, Robert A. Reamer, Mary Stanik, Dietrich Steinhubel, Steve Weissman, and Timothy J. Wright for their contributions to the Raltegravir synthesis development effort.

References

1 UNAIDS/WHO (2007). AIDS epidemic update: December 2007. UNAIDS/07.27E. ISBN 92 9 173621 8.

2 Evering, T.E., and Markowitz, M. (2007) *Drugs Today*, **43**, 865.

3 Tozzi, V., Zaccarelli, M., Bonfigli, S., Lorenzini, P., Liuzzi, G., Trotta, M.P., Forbici, F., Gori, C., Bertoli, A., Bellagamba, R., Narciso, P., Perno, C.F., and Antinori, A. (2006) *Antivir. Ther.*, **11**, 553–560.

4 Havlir, D.V. (2008) *N. Engl. J. Med.*, **359**, 416–441.

5 Summa, V., Petrocchi, A., Bonelli, F., Crescenzi, B., Donghi, M., Ferrara, M., Fiore, F., Gardelli, C., Gonzalez Paz, O., Hazuda, D.J., Jones, P., Kinzel, O., Laufer, R., Monteagudo, E., Muraglia, E., Nizi, E., Orvieto, F., Pace, P., Pescatore, G., Scarpelli, R., Stillmock, K., Witmer, M.V., and Rowley, M.J. (2008) *J. Med. Chem.*, **51**, 5843–5855.

6 Department of Health and Human Services (2009) Panel on Antiretroviral Guidelines for Adults and Adolescents. Guidelines for the use of antiretroviral agents in HIV-1 infected adults and adolescents. December 1, 1–161.

7 Humphrey, G.R., Pye, P.J., Zhong, Y.-L., Angelaud, R., Askin, D., Belyk, K.M., Hudgens, T., Mahajan, A.J., Maligres, P.E., Manchino, D., Miller, R.A., O'Brien, D., Phillips, P., Pruzinsky, V.M., Reamer, R.A., Stanik, M., Weissman, S., and Wright, T.J. "Development of a 2nd Generation, Highly Efficient Manufacturing Route for the HIV Integrase Inhibitor Raltegravir Potassium", submitted.

8 (a) Johnson, T.B., and Caldwell, W.T. (1929) *J. Am. Chem. Soc.*, **51**, 873–880; (b) Budesinsky, I., Jelinek, V., and Prikryl, J. (1962) *J. Collect. Czech. Chem. Commun.*, **27**, 2550–2560; (c) Sunderland, C.J., Botta, M., Aime, S., and Raymond, K.N. (2001) *Inorg. Chem.*, **40**, 6746–6756; (d) Dreher, S.D., Ikemoto, N., Gresham, V., Liu, J., Dormer, P.G., Balsells, J., Mathre, D., Novak, T.J., and Armstrong, J.D., III (2004) *Tetrahedron Lett.*, **45**, 6023–6025.

9 (a) Culbertson, T.P. (1979) *J. Heterocycl. Chem.*, **16**, 1423; (b) Summa, V., Petrocchi, A., Matassa, V.G., Taliani, M., Laufer, R., Franasco, R.D., Altamura, S., and Pace, P. (2004) *J. Med. Chem.*, **47**, 5336–5339; (c) Stansfield, I., Avolio, S., Colarusso, S., Gennari, N., Narjes, F., Pacini, B., Ponzi, S., and Harper, S. (2004) *Bioorg. Med. Chem. Lett.*, **14**, 5085–5088; (d) Wagner, E., Becan, L., and Nowakowska, E. (2004) *Bioorg. Med. Chem. Lett*, **12**, 265–272; (e) Zhong, Y.-L., Zhou, H., Gauthier, D.R., Jr., and Askin, D. (2006) *Tetrahedron Lett.*, **47**, 1315; (f) Colarusso, S., Attenni, B., Avolio, S., Malancona, S., Harper, S., Altamura, S., Koch, U., and Narjes, F. (2006) *ARKIVOC*, **vii**, 479; (g) Koch, U., Attenni, B., Malancona, S., Colarusso, S., Conte, I., Filippo, M., Harper, S., Pacini, B., Giomini, C., Thomas, S., Incitti, I., Tomei, L., De Francesco, R., Altamura, S., Matassa, V.G., and Narjes, F. (2006) *J. Med. Chem.*, **49**, 1693; (h) Ferrara, M., Crescenzi, B., Donghi, M., Muraglia, E., Nizi, E., Pesci, S., Summa, V., and Gardelli, C. (2007) *Tetrahedron Lett.*, **48**, 8379; (i) Pye, P.J., Zhong, Y.-L., Jones, G.O., Reamer, R.A., Houk, K.N., and Askin, D. (2008) *Angew. Chem. Int. Ed.*, **47**, 4134; (j) Zhong, Y.-L., Pipik, B., Lee, J., Kohmura, Y., Okada, S., Igawa, K., Kadowaki, C., Takezawa, A., Kato, S., Conlon, D., Zhou, H., King, A.O., Reamer, R.A., Gauthier, D.R., Jr., and Askin, D. (2008) *Org. Process. Res. Dev.*, **12**, 1245–1252; (k) Naidu, B.N. (2008) *Synlett*, 547–550; (l) Pacini, B., Avolio, S., Ercolani, C., Koch, U., Migliaccio, G., Narjes, F., Pacini, L., Tomei, L., and Harper, S. (2009) *Bioorg. Med. Chem. Lett.*, **19**, 6245–6249.

10 Ogilvie, W., and Rank, W. (1987) *Can. J. Chem.*, **65**, 166.

11 (a) Newcomb, M. (1993) *Tetrahedron*, **49**, 1151; (b) Newcomb, M., Tanaka, N., Bouvier, A., Tronche, C., Horner, J.H., Musa, O.M., and Martinez, F.N. (1996) *J. Am. Chem. Soc.*, **118**, 8505; (c) Horner, J.H., Musa, O.M., Bouvier, A., and Newcomb, M. (1998) *J. Am. Chem. Soc.*, **120**, 7738.

7
Cyclopentane-Based NK1 Receptor Antagonist

Jeffrey T. Kuethe

The NK1 receptor antagonist **1** was discovered at Merck Research Laboratories in Rahway, NJ for the potential treatment of depression [1]. Located in specific areas of the central nervous system and primarily associated with sensory neurons, neurokinin-1 (NK-1) is a member of the G-protein-coupled receptor family. The natural ligand for NK-1 is the tachykinin peptide substance P which has been implicated in the pathophysiology of a wide range of conditions. The prevention of chemotherapy-induced emesis has been established with Merck's aprepitant Emend®, the only approved drug in this class [2–4]. Efforts to target other potent, orally active NK-1 antagonists led to the discovery of a series of cyclopentane-based compounds, such as **1**, which have significant binding affinity (subnanomolar) for the human NK-1 receptor.

Compound **1** was the first cyclopentane-based NK-1 receptor antagonist development candidate at Merck. It contains five stereocenters: a central core possessing three contiguous all-trans stereocenters, a pendent bis(trifluoromethyl)-benzylic ether, and a nipecotic acid moiety (Figure 7.1). Key to the successful preparation of **1** was construction of the *trans*, *trans*-cyclopentyl core and installation of the unsymmetrical secondary-secondary (*sec-sec*) ether. The preparation of **1** is the focus of this chapter.

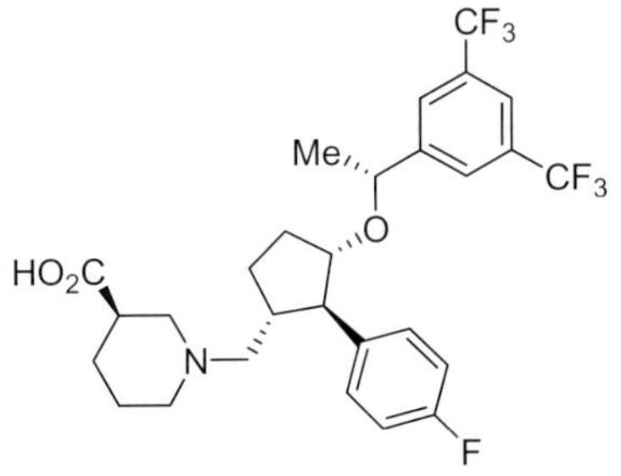

1

Figure 7.1 Structure of a cyclopentane-based NK-1 antagonist.

The Art of Process Chemistry. Edited by Nobuyoshi Yasuda

ISBN: 978-3-527-32470-5

In Section 7.1, process development for compound **1** is described in detail. In Section 7.2, further discussion of the key chemical transformations from this project will be addressed.

7.1
Project Development Compound 1

7.1.1
Medicinal Route

The medicinal chemistry route to compound **1** is depicted in Schemes 7.1–7.3. It involved 17 linear steps and proceeded in 1.9% overall yield. The original synthetic method for the core cyclopentanol **10** having three contiguous chiral centers is summarized in Scheme 7.1. The route began with the condensation of ethyl cyanoacetate (**3**) and 4-fluorobenzaldehyde (**4**) in the presence of catalytic piperidine, followed by a Michael addition with NaCN, alkylation with ethyl 3-chloropropionate, and subsequent hydrolysis to afford triacid **5** in 68% overall yield. Esterification of **5** was followed by Dieckman cyclization, decarboxylation, and re-esterification to give racemic cyclopentanone **6** in 63% overall yield. Reduction of **6** with $NaBH_4$ gave a 70% yield of a 3 : 1 mixture of desired cyclopentanol **7** and undesired **8**, which were separated by chromatography. Compound **7** was saponified to acid **9**, which was resolved by formation of the (*R*)-α-methylbenzyl amine salt. The optically pure salt was then converted to enantiopure cyclopentanol **10** in 35% yield from racemic **7**.

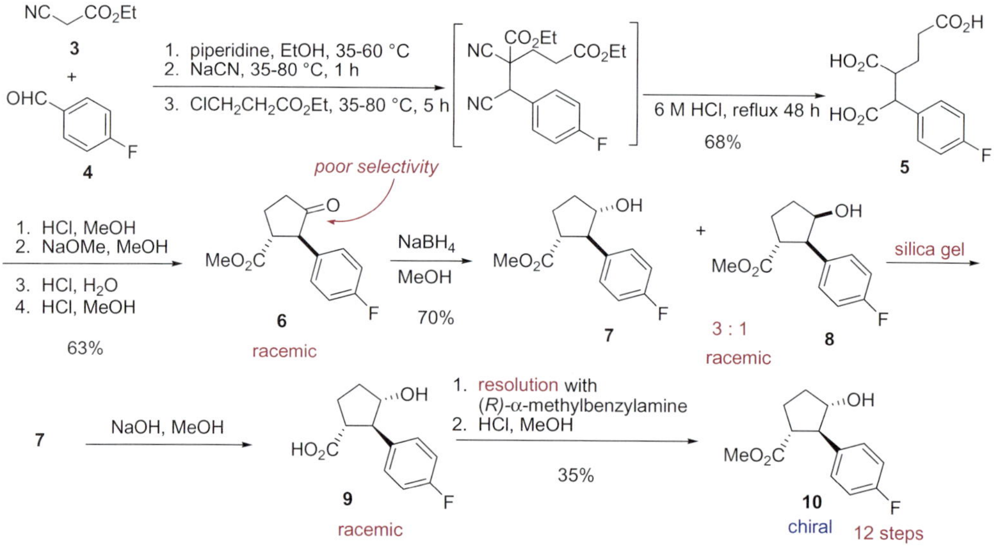

Scheme 7.1 Medicinal route for the preparation of the cyclopentane core **10**.

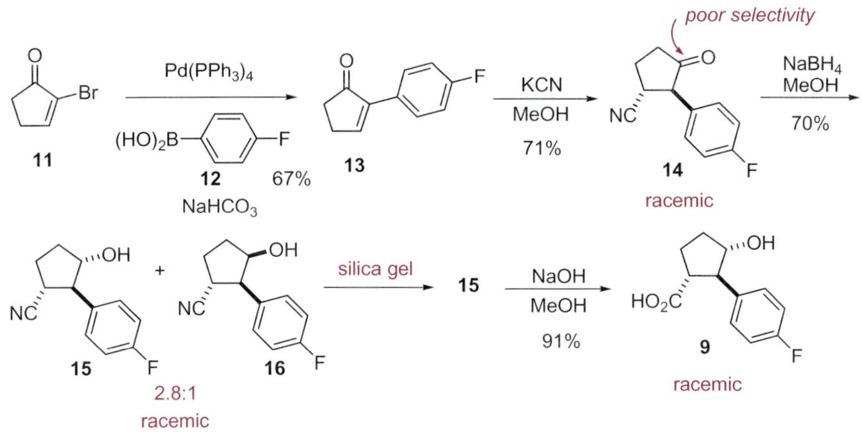

Scheme 7.2 Improved Medicinal route for the preparation of cyclopentane core **10**.

The medicinal chemists subsequently discovered an improved route to racemic acid **9** that started with 2-bromo-2-cyclopente-1-one **11** (Scheme 7.2) [5]. Suzuki–Miyaura cross-coupling of **11** with 4-fluorophenyl boronic acid **12** provided **13** in 67% yield. Conjugate addition of cyanide furnished ketone **14** in 71% yield. Reduction of **14** with $NaBH_4$ gave a 2.8 : 1 mixture of desired **15** and undesired **16** which were separated by silica gel chromatography. The observed diastereoselectivity with the cyano group was similar to ester **6**. Hydrolysis of **15** with 5 M NaOH in MeOH gave racemic acid **9** in 91% yield, which was resolved as outlined in Scheme 7.1.

The drug candidate **1** was prepared from chiral cyclopentanol **10** as shown in Scheme 7.3. Reaction of **10** with racemic imidate **17**, prepared from the corresponding racemic benzylic alcohol, in the presence of catalytic TfOH furnished a 1 : 1 mixture of diastereomers **18** and **19** which were only separated from one another by careful and tedious chromatography. Reduction of ester **18** with $LiBH_4$ and subsequent Swern oxidation gave aldehyde **20** in 68% yield. Reductive amination of **20** with (*R*)-ethyl nipecotate L-tartrate salt **21** and $NaBH(OAc)_3$ and subsequent saponification of the ester moiety yielded drug candidate **1**.

7.1.1.1 Problems of the Original Route

Numerous problems with the original route were identified at the beginning of the project and included:

1) Length of synthesis (17 linear steps, 20 total steps).
2) The synthesis of the cyclopentanol **8** was racemic and required chiral resolution.
3) Chiral ethyl nipecotate L-tartrate salt **21** was not available on a large scale.
4) Multiple protection–deprotections that were highly inefficient for scale-up.

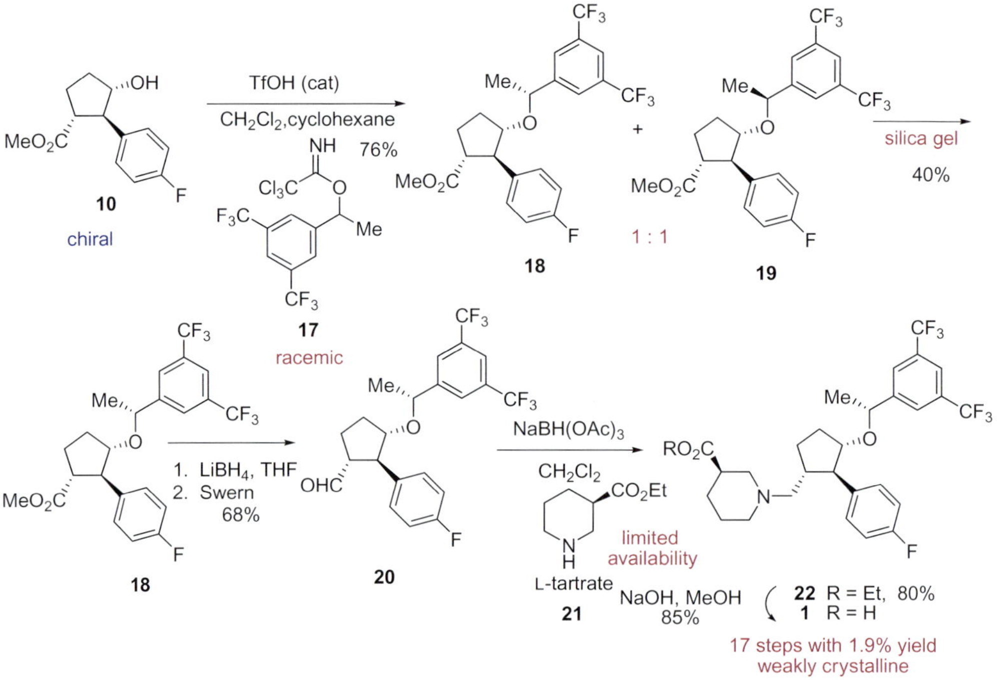

Scheme 7.3 Medicinal route for drug candidate **1**.

5) Etherification of chiral **10** with racemic imidate **17** provided a diastereomeric mixture of **18** and **19**, which had to be separated by silica gel column.
6) Drug candidate **1** was reported to be noncrystalline as the free base.

Due to these issues, it was decided to completely abandon the original route to compound **1**.

7.1.2
Process Development

In order to prepare multi-kilogram quantities of **1** our efforts were strictly focused on the development of an asymmetric route. Our retrosynthetic approach was centered on the preparation of cyclopentenone **27** which, we envisioned, could be elaborated to chiral hydroxy acid **26** through a series of asymmetric transformations (Scheme 7.4). Etherification of the hydroxy group of **26** with benzylic alcohol **25** followed by installation of (*R*)-nipecotate **23** at the acid position of **24**, would furnish the drug candidate **1**. This section will address the following:

1) Preparation of cyclopentenone **27**.
2) Conversion of cyclopentenone **27** to chiral hydroxy acid **26**.
3) Etherification of **26**.
4) Preparation of (*R*)-nipecotate **23** and completion of the synthesis of **1**.

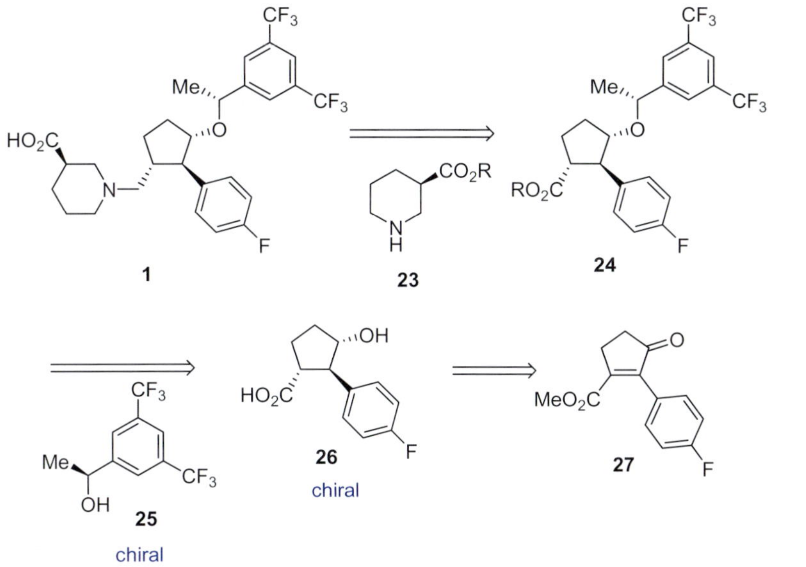

Scheme 7.4 Retrosynthetic strategy.

7.1.2.1 Preparation of Cyclopentanone 27

Selection of starting materials is always key for early development. Starting materials should be readily available within a reasonable lead time. Cyclopentenone **27** was prepared by four different routes [6].

Our first approach started with commercially available methyl cyclopentencarboxylate **28** (Scheme 7.5). Allylic oxidation of **28** with CrO_3 afforded **29** in yields ranging from 20–40%. Bromination of **29** by first treating **29** with Br_2 followed by the addition of NEt_3 gave the desired bromide **30** in variable yields ranging from 20–60% where the typical yield was 30%. Suzuki–Miyaura cross-coupling [7] of **30** with 4-fluorophenyl boronic acid (**12**) gave the desired cyclopentenone **27** in 89% yield. While in the early stages of our development program this route was utilized for the preparation of only gram quantities of **27** for downstream investigations, it was never considered for scale-up due to the use of CrO_3, as well as the extremely low yields associated with the use of these reagents.

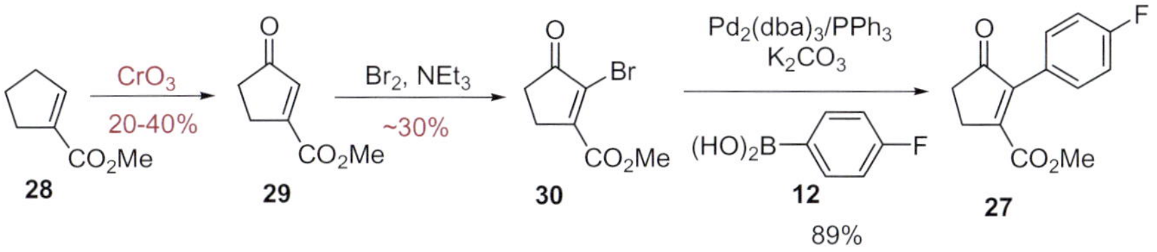

Scheme 7.5 First synthetic approach to cyclopentenone **27**.

The second route was based on installation of the ester group of **27** via carbonylation, which offered improved reaction efficiency and allowed the preparation of multi-hundred gram quantities of **27** (Scheme 7.6). Bromination of commercially available 3-methoxy-2-cyclopentenone (**31**) with NBS provided bromide **32** in quantitative yield [8]. Suzuki–Miyaura cross-coupling with **12** afforded the coupled product **33** in 89% isolated yield. It was discovered that quenching the cooled reaction mixture with aqueous K_3PO_4 significantly improved the purity of **33** by effectively removing the excess boronic acid from the organic layer prior to crystallization. Conversion of the methoxy group to the corresponding bromide was effected with PBr_3 in refluxing 1,2-dichloroethane, providing bromide **34** in 65% isolated yield after purification by silica gel plug [9]. The key palladium-catalyzed carbonylation (40 psi CO, 100 °C, MeOH, 3 mol% $Pd(PPh_3)_2Cl_2$) worked well and provided the target compound **27** in 90% isolated yield [10].

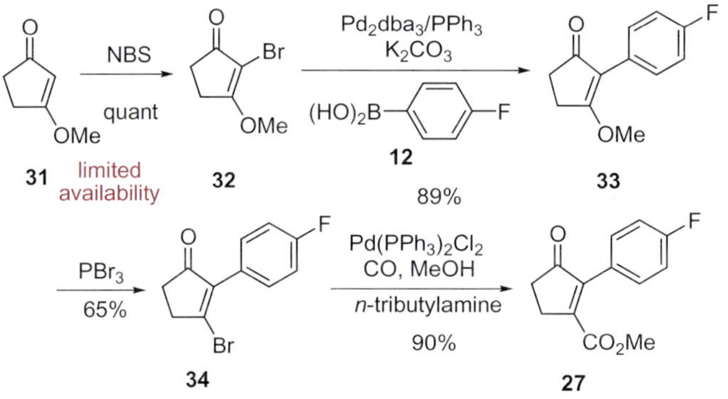

Scheme 7.6 Second synthetic approach to cyclopentenone **27**.

Our third approach to **27** addressed the unavailability of 3-methoxy-2-cyclopentenone (**31**) in bulk quantities which necessitated the discovery of an alternative route (Scheme 7.7). Fortunately, the precursor to **31**, 1,3-cyclopentandione (**35**), was available in the required quantities and our efforts shifted to the use of this reagent. Bromination of **35** with NBS, employing either $KHCO_3$ or KOH as the base, gave brominated dione **36** in 85% isolated yield. Unfortunately, direct cross-coupling of alkyl bromide **36** with boronic acid **12** under a variety of Suzuki–

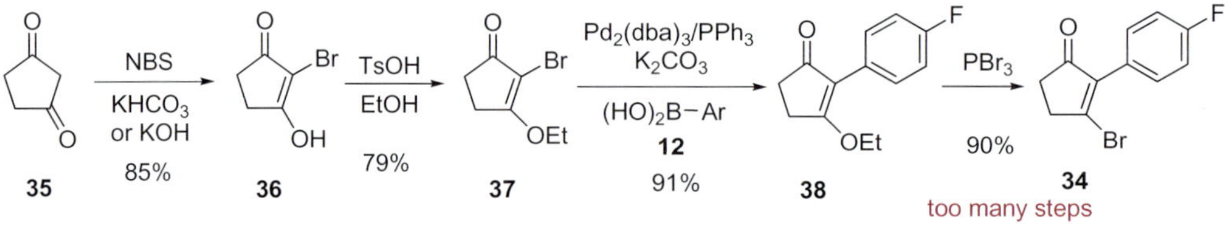

Scheme 7.7 Third synthetic approach to cyclopentenone **27**.

Miyaura reaction conditions did not give any of the desired coupled product. Therefore, conversion of **36** to the corresponding enol bromide **37** was performed using constant addition/distillation of ethanol in the presence of catalytic TsOH giving **37** in 79% isolated yield. After solvent switching from EtOH to toluene, the Suzuki–Miyaura cross coupling of **37** with boronic acid **12**, as previously described, afforded enone **38** in 91% yield. Bromination of **38** with PBr_3 gave vinyl bromide **34** in 90% yield. However, during the kilogram preparation of **27**, it was recognized that this approach would need to be slightly modified to shorten the length of the synthesis.

The final route to **27** was fully optimized as shown in Scheme 7.8 and successfully scaled to multi-hundred kilogram scale. The first step in streamlining this route involved the direct preparation of **41**. In an extension of the chemistry developed by Buchwald [11], heating **35** in the presence of 1-bromo-4-fluorobenzene (**39**), anhydrous K_3PO_4, $Pd(OAc)_2$ (1 mol%), and 2-(di-*tert*-butylphosphino)biphenyl (2 mol%) in refluxing 1,4-dioxane gave 4-fluorophenyl-1,3-cyclopentandione (**41**) in 85% yield. The reaction could also be carried out with the corresponding chloride **40** and was optimized in terms of catalyst/ligand loading with chloride **40**. Anhydrous, powdered K_3PO_4 was found to be the best base, giving the most consistent results, and the product **41** was isolated in 92% yield. Other bases that were examined either gave no product, or significant amounts of aldol adduct **42** [12].

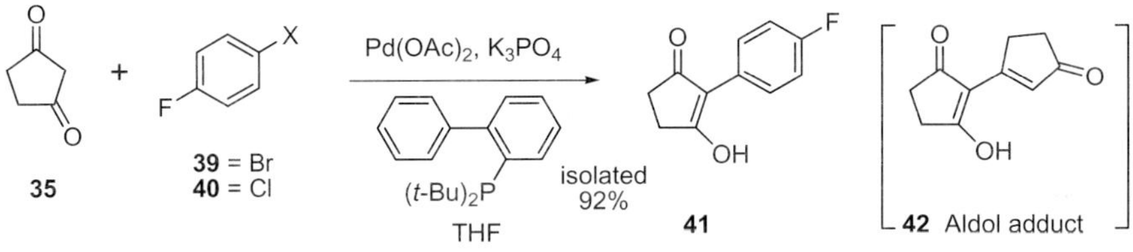

Scheme 7.8 Buchwald cross-coupling in the final route.

Due to the toxicity of 1,4-dioxane, a change in the reaction solvent was necessary prior to scale-up to pilot plant scale. After screening a number of solvents, THF was chosen as the optimal solvent. The reaction was sluggish in refluxing THF, resulting in low conversion to **41** (<35%). However, it was found that at higher reaction temperatures, and conducting the reaction in a pressurized vessel, gave better results (e.g., 73% of **41** and 12% of **35** at 100 °C [25 psi]; 83% of **41** and 3–5% of **35** at 105 °C [30 psi]). The best result was obtained by heating the reaction mixture in a pressurized vessel to 110 °C (36 psi) which afforded **41** in 90% yield. Each of these reactions was carefully monitored with the aid of a RC1 reaction calorimeter[1)] which not only allowed safe execution of this reaction in the pilot

1) The internal temperature and pressure were monitored by a Mettler-Toledo RC1 reaction calorimeter.

plant, but also enabled complete understanding of the reaction profile with respect to pressure in the head space of the reactor. In order to isolate **41** in acceptably high yields, it was discovered that the THF had to be removed prior to acidification of the crude reaction mixture with HCl. Upon dilution of the crude reaction mixture with water, the THF was removed by distillation. Addition of 6 M HCl at 50 °C, cooling to room temperature, and filtration gave **41** as a crystalline solid which was isolated in 86% yield.

The bromination of enol **41** was next examined (Scheme 7.9). Reaction of **41** with PBr_3 in refluxing 1,2-dichloroethane gave **34** in 78% yield where the mass balance of the reaction was the remaining starting material. Bromination with phosphorus oxybromide ($POBr_3$) gave better results and bromide **34** could be obtained in up to 90% yield. Initial conditions employed $POBr_3$ (1.0 equiv) and Na_2HPO_4 (0.5 equiv) in MeCN at 45 °C for 7 h. The use of Na_2HPO_4 was found to be crucial as it appeared to attenuate the acidity of the reaction mixture and allowed higher yields of **34** (80%) to be obtained after crystallization of the product from MTBE/heptane, Also observed in the crude reaction mixture was approximately 8–12% of tribromide impurities **43** and **44** (~6 : 1 by HPLC). Even though these impurities were easily removed in the crystallization step, this process needed to be further refined prior to pilot plant implementation. The optimal conditions that were finally employed involved treatment of **41** with 0.75 equiv of $POBr_3$ in MeCN in the presence of 0.5 equiv of Na_2HPO_4 at 65 °C for 1.5 h. Under these conditions, the formation of **43** and **44** was greatly reduced to <4%. Improved work-up conditions involved quenching the reaction with 1 M KOH, and separating the bottom aqueous layer. The resulting MeCN layer was then concentrated slightly and water was added which resulted in the crystallization of **34**. After filtration and drying the desired product **34** was isolated in 92% yield. During the course of the KOH quench, it was noted that trace amounts of **43** and **44** were converted to a new impurity with a molecular weight of 210, which was tentatively assigned as **45**. These impurities were removed to <5% combined amounts during the crystallization step.

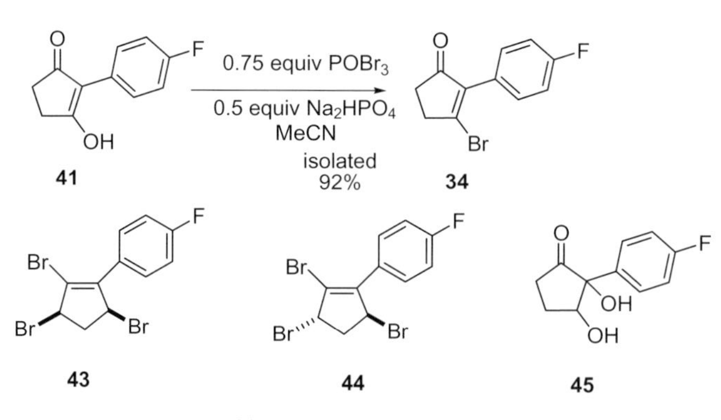

Scheme 7.9 Optimized bromination reaction in the final route.

The carbonylation of **34**, depicted in Scheme 7.6, needed further improvements prior to scale-up (Scheme 7.10). The initial conditions produced significant amounts (5–8%, HPLC A%) of methyl ether **33** which could only be removed by chromatography. In addition, the use of $Pd(PPh_3)_2Cl_2$ as a homogeneous catalyst source left the crude product contaminated with extremely high levels of palladium. As part of our experimental design, a heterogeneous catalyst was sought for this reaction to simplify product isolation and reduce the amount of palladium contamination. Our efforts were primarily focused on the use of Pd/C and the reaction was fully optimized in terms of solvent, catalyst, temperature, and reagent charges. The optimal conditions that were adopted for pilot plant manufacture involved running the carbonylation in *N,N*-dimethylacetamide (1 M in **34**) with 5 equiv of MeOH, 2 equiv of *n*-Bu_3N, 1 mol% of 5 wt% Pd/C at 60 °C at 10 psi CO for 10–12 h [13]. The assay yield of **27** under these reaction conditions was 98–99%. There were no detectable amounts of reduction, dimerization, or amide formation [14]. The product was isolated after distillation of the MeOH and the addition of 1 M HCl, and **27** was isolated as a crystalline solid in 90% yield. The product contained 200 ppm of Pd, as determined by ICP-AES. While bromide **34** was contaminated by small amounts of **43–45**, these impurities remained either unreactive **45** or decomposed under the reaction conditions (**43** or **44**). The highly effective three-step optimized synthetic route for cyclopentenone **27** is summarized in Scheme 7.10 and was successfully demonstrated on a multi-hundred kilogram scale.

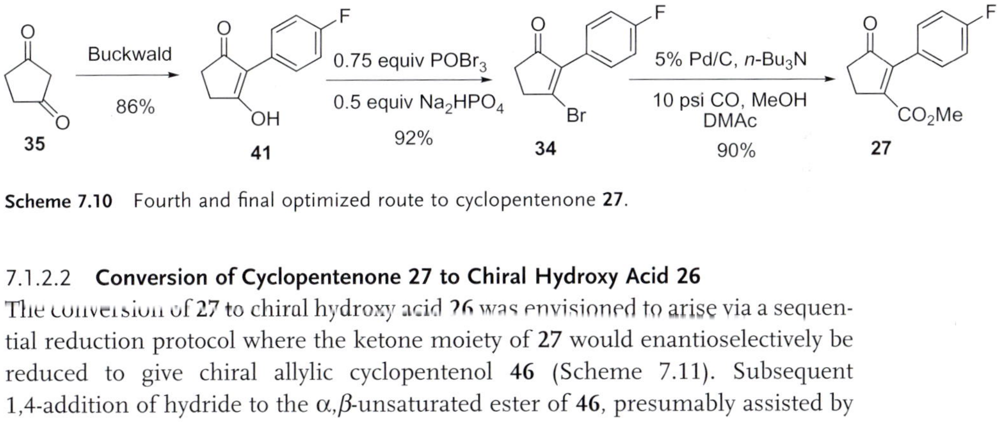

Scheme 7.10 Fourth and final optimized route to cyclopentenone **27**.

7.1.2.2 Conversion of Cyclopentenone 27 to Chiral Hydroxy Acid 26

The conversion of **27** to chiral hydroxy acid **26** was envisioned to arise via a sequential reduction protocol where the ketone moiety of **27** would enantioselectively be reduced to give chiral allylic cyclopentenol **46** (Scheme 7.11). Subsequent 1,4-addition of hydride to the α,β-unsaturated ester of **46**, presumably assisted by

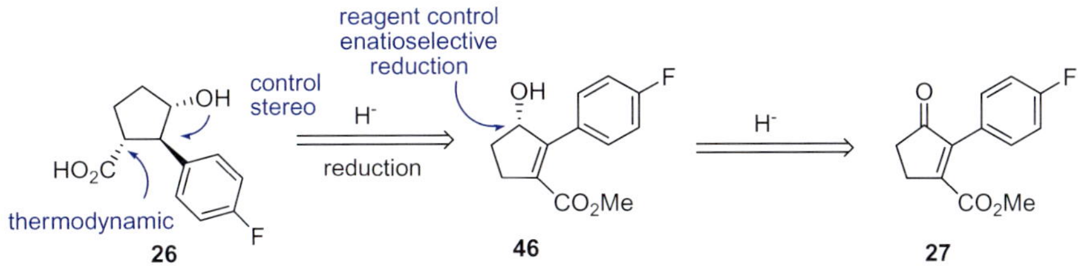

Scheme 7.11 Sequential reduction to chiral cyclopentane acid **26**.

the proximal hydroxy group would deliver **26**. For the second reduction, it was essential to establish the *trans*-orientation between the hydroxy group and the 4-fluorophenyl group, since the configuration of the carboxylic acid could be controlled through epimerization to the thermodynamically all-*trans* relationship in a subsequent step.

7.1.2.2.1 Ketone Reduction

Based on a wealth of in-house experience, the enantioselective reduction of ketone **27** was achieved by addition of **27** to a preformed mixture of (*R*)-2-methyloxazaborolidine (**47**) (10 mol%) and $BH_3 \cdot SMe_2$ (0.6 equiv) in toluene at −20 °C (Scheme 7.12) [15]. The chiral allylic alcohol **46** was formed in >95% HPLC assay yield and in 93% ee. Attempts to conduct the reaction either with lower amounts of catalyst or at elevated temperatures were successful, but at the expense of the enantiomeric excess (typically <90%). It was also discovered that the use of $BH_3 \cdot SMe_2$ was crucial since the use of $BH_3 \cdot THF$ resulted in significant erosion in ee (56–84%). When $BH_3 \cdot NHMe_2$ was used as the reductant, the starting material was recovered unchanged. Upon completion of the reaction, the reaction mixture was quenched with 6 equiv of MeOH and warmed to room temperature. After being washed with 1 M HCl, the toluene solution was azeotropically dried and the stream carried forward without further purification in the stereoselective reduction of the double bond.

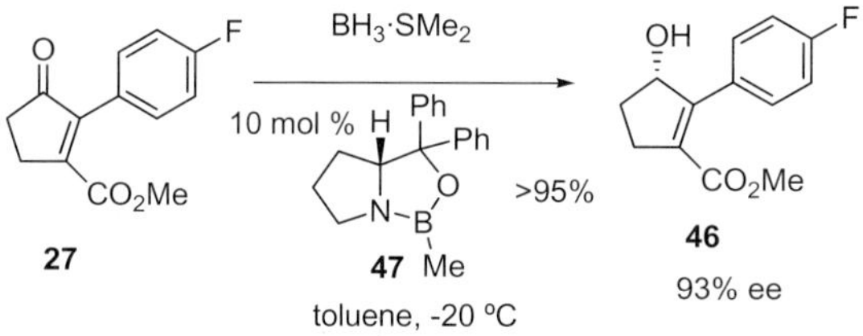

Scheme 7.12 Oxazaborolidine reduction to set chiral center.

7.1.2.2.2 *Anti*-Selective Reduction of Double Bond

With an efficient synthesis of allylic alcohol **46** in our hands, our attention turned to the selective reduction of the double bond. As stated above, we intended to use the hydroxy group in **46** to deliver hydride from the same face as the hydroxy group. Mainly there were two methods available: (i) transition metal-mediated hydrogenation and (ii) metal hydride reduction.

Transition metal-mediated hydrogenation A high degree of stereocontrol has been reported for transition metal-catalyzed hydrogenations directed by proximal hydroxy groups. Crabtree reduction [16] of **46** using 20 mol% $Ir(COD)(py)(PCy_3)PF_6$ in an atmosphere of 500 psi H_2 for 12 h gave the desired 1,2-*anti* product **48** as a single isomer (>99 : 1), where the stereochemistry was confirmed by nOe studies, but the yield was only 63% (Scheme 7.13). The use of rhodium catalysts

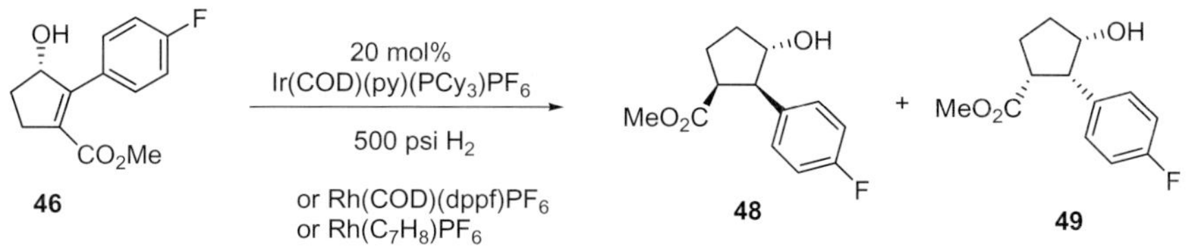

Scheme 7.13 Transition metal-mediated hydrogenation of **46**.

Rh(COD)(dppb)PF_6 and Rh(C_7H_8)PF_6 was also briefly examined [17]. Although these reactions proceed to completion under nearly identical conditions to those employed for the Crabtree reduction, they were generally nonselective and gave varying amounts of **48** and the undesired all-*cis* product **49**. When the more stable iridium-carbene complexes described by Nolan (chlorobenzene, 80 °C, 50 psi H_2) were explored, excellent stereoselectivity was observed (>99.9 : 1) in favor of **48**; however, the catalyst only survived about three turnovers and low conversions resulted in all attempts [18]. Due to incomplete reduction, even under rather forcing conditions, and the expense of Crabtree's catalyst that would make it use on a larger scale prohibitive, this approach was abandoned.

Metal hydride reduction Given the ability of hydride delivery through oxygen atom coordination, various metal hydrides were screened for the reduction of **46**. After examination of a variety of hydride sources, it was discovered that reaction of **46** with sodium bis(2-methoxyethoxy)aluminum hydride (Red-Al®) resulted in clean reduction and gave the desired 1,2-*anti* stereochemistry exclusively (Scheme 7.14) [19, 20]. The optimal conditions involved addition of 1.5 equiv of Red-Al® to a solution of **46** in 3 : 2 toluene/THF at −40 °C. Warming the reaction mixture to −25 °C drives the reaction to completion. The crude toluene stream of **46** from the previous step was diluted with THF and used in the reduction step as-is. It should also be pointed out that the addition of less than 1.5 equiv of Red-Al® resulted in incomplete conversion, while the use of more than 1.5 equiv of Red-Al® resulted in significantly more over-reduction to diol **50**. The use of THF as a cosolvent was important since it helped solubilize the Red-Al® at low temperatures. Once the reaction was complete, the crude mixture was inversely quenched into a 2 M

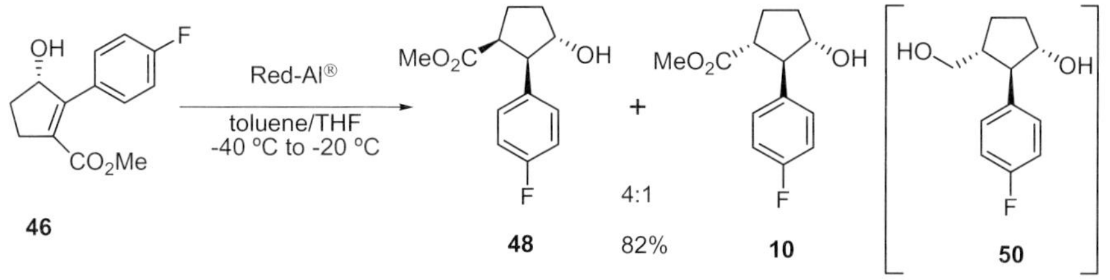

Scheme 7.14 Red-Al® reduction to set 1,2-*anti* stereochemistry.

solution of $NaHSO_4$ which afforded a 4 : 1 mixture of the desired 1,2-*anti* diastereomers **48** and **10** in 82% combined yield, with 5–8% of diol **50**.

The intriguing reaction mechanism of the Red-Al® reaction is discussed in detail in Section 7.2.

7.1.2.2.3 Epimerization to Set All-*trans* Configuration

Epimerization of the crude toluene solution containing *anti-syn* **48** and the desired *anti-anti* **10** to the thermodynamically more stable **10** was conducted by adding 0.4 equiv of NaOMe to the dry toluene solution from the Red-Al® reduction at 50 °C and then further heating the reaction mixture at 75 °C for 1 h when the diastereomeric ratio of **10** : **48** was >17 : 1 (Scheme 7.15). Saponification of the mixture by the direct addition of 6 M NaOH (3.5 equiv), water, and MeOH (2.6 equiv) to the toluene solution and vigorous stirring for 2 h at room temperature gave **26** in nearly quantitative conversion, together with small amounts (<3%) of the *anti-syn* isomer **51**. The reaction mixture was then diluted with IPAc and acidified by the addition of concentrated HCl. To the resulting IPAc stream containing **26** was added heptane, which resulted in the crystallization of **26**. The crystallization not only completely removed all traces of **51** but also increased the ee of **26** from 92–94% to >99.9% ee and the isolated yield was 94%. Acid **26** was converted to methyl ester **10** by treatment of **26** with a catalytic amount of HCl in MeOH in quantitative yield (Scheme 7.15).

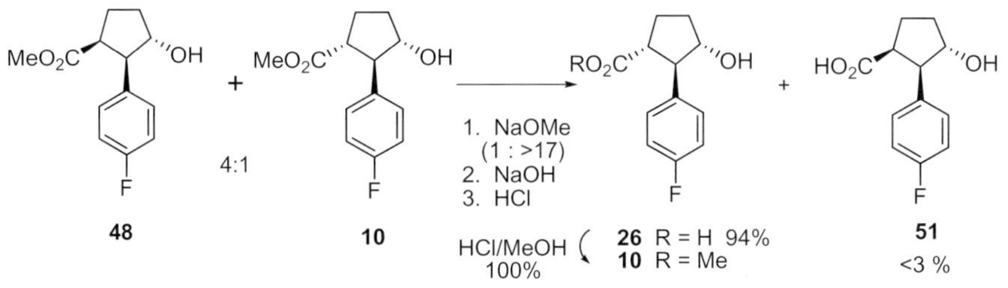

Scheme 7.15 Epimerization to complete preparation of cyclopentanol **26**.

7.1.2.3 Etherification of 10

Arguably the most challenging aspect for the preparation of **1** was construction of the unsymmetrically substituted *sec-sec* chiral bis(trifluoromethyl)benzylic ether functionality with careful control of the relative and absolute stereochemistry [21]. The original chemistry route to ether intermediate **18** involved an unselective etherification of chiral alcohol **10** with racemic imidate **17** and separation of a nearly 1 : 1 mixture of diastereomers, as shown in Scheme 7.3. Carbon–oxygen single bond forming reactions leading directly to chiral acyclic *sec-sec* ethers are particularly rare since known reactions are typically nonstereospecific. While notable exceptions have surfaced [22], each method provides ethers with particular substitution patterns which are not broadly applicable.

We envisioned three different approaches for this most challenging problem:

1) Application of the Emend® Process (Tebbe Reaction followed by hydrogenation)
2) *In situ* reductive etherification
3) S_N2 substitution reactions

7.1.2.3.1 **Application of the Emend® Process**

Substrate-controlled diastereoselective hydrogenation of vinyl ethers has been employed in the preparation of chiral secondary ethers and can often proceed with complete control of the absolute stereochemistry [23]. A diastereoselective hydrogenation approach was successfully utilized on a large scale for the preparation of kilogram quantities of Merck's Aprepitant (Emend®) where a similar bis(trifluoromethyl)benzylic ether was required [4]. With a great deal of in-house experience, we elected to explore the preparation and catalytic hydrogenation of vinyl ether **55** (Scheme 7.16). Reduction of **10** with $LiAlH_4$ in THF gave the intermediate diol **50** in quantitative yield. Selective protection of the primary alcohol was accomplished with TBDMSCl/imidazole in DMF and furnished the mono-protected alcohol **52** in an unoptimized 93% yield. Reaction of **52** with bis(trifluoromethyl)benzoyl chloride **53** in the presence of NEt_3 afforded ester **54** which was isolated in 97% yield. Conversion of **54** to vinyl ether **55** was conducted with 2.3 equiv of dimethyltitanocene [24] in toluene at 80 °C for 8 h to give **55** in 89% yield. Unlike the Emend® case, catalytic hydrogenation of **55** employing 5% Pd/C under an atmosphere of hydrogen (20 psi) for 8 h afforded a 1 : 4 mixture of **56** (desired) : **57** (undesired) in 93% combined yield. Desilyation of this mixture

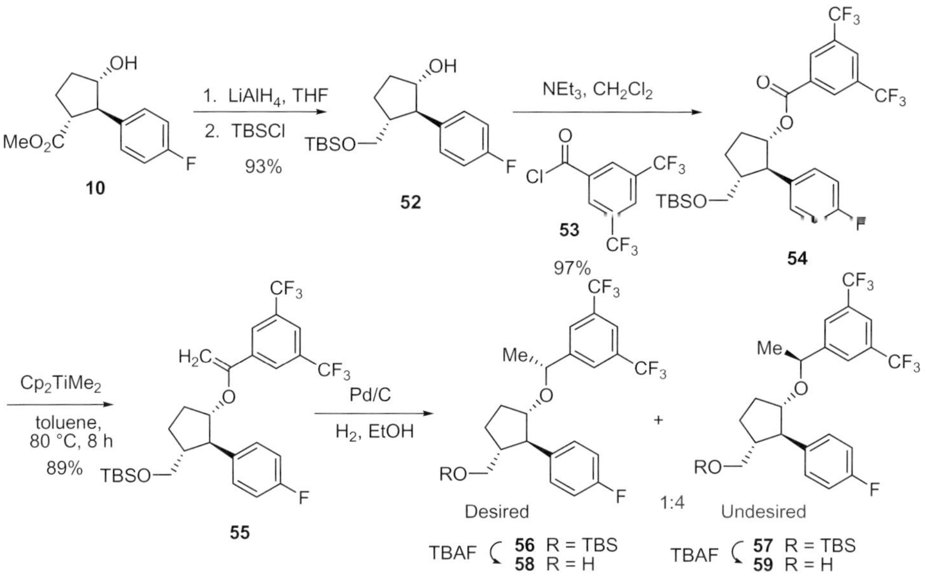

Scheme 7.16 Emend® approach.

with TBAF gave **58 : 59** which unequivocally established the absolute stereochemistry. This approach was abandoned.

7.1.2.3.2 ***In situ*** **Reductive Etherification Approach**

The reductive etherification of carbonyl compounds with alkoxytrimethylsilanes in the presence of TMSOTf and a trialkylsilane has been reported to be a method for the preparation of ethers under nonbasic conditions [25, 26]. Although the use of this methodology for the preparation of acyclic *sec-sec* ethers has not been reported, in certain cases acceptable levels of diastereoselectivity have been achieved. Trimethylsilyl ether **60** was prepared in 94% yield by reaction of **10** with TMSCl and NEt_3 in CH_2Cl_2 at 0 °C (Scheme 7.17). Treatment of a mixture of **60** and acetophenone **61** with sub-stoichiometric amounts (0.1–0.5 equiv) of TMSOTf and Et_3SiH (1.2 equiv) in CH_2Cl_2 at −78 °C and then allowing it to warm to −20 °C afforded a 2.3 : 1 mixture of desired **18** and undesired **19** in a combined 37% HPLC assay yield. The remaining mass balance was identified as **10**, **61**, and racemic benzylic alcohol arising from competitive reduction of **61**. Even after extensive optimization, the diastereomeric ratio was improved only to 3.2 : 1 (**18 : 19**) in 65% combined HPLC assay yield by reaction of **60** with **61** in the presence of 5 equiv of TMSOTf and 1.2 equiv of Et_3SiH for 18 h. Thus this route was abandoned, but remained mechanistically interesting. The mechanism will be discussed in greater detail in Section 7.2.

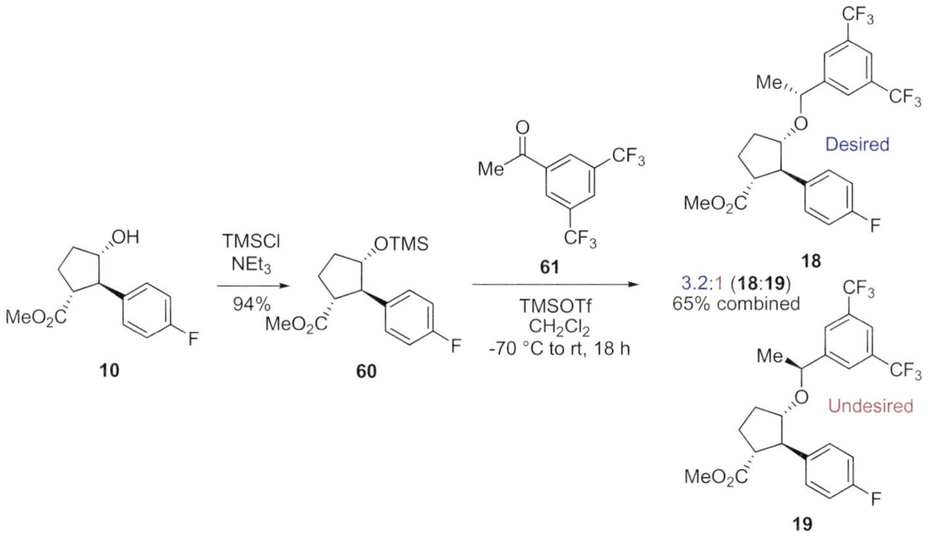

Scheme 7.17 Reductive etherification approach.

7.1.2.3.3 S_N2 Substitution Approaches

The formation of carbon–oxygen single bonds via the attack of an alkoxide on an alkyl halide, (Williamson ether synthesis) is an extremely important reaction that is still extensively used on an industrial scale. The preparation of intermediate **18** was envisioned to arise from two possible disconnections, as shown in Scheme 7.18. Disconnection A would involve formation of the ether bond by displacement of a suitable leaving group on the cyclopentane core **62** with chiral benzylic alcohol **63**. Disconnection B would involve formation of the ether bond by displacement of a suitable benzylic leaving group on the bis(trifluoromethyl)phenyl moiety **64** with alcohol **10**. Each of these disconnections posed concerns, including low substrate reactivity, stereocenter scrambling, and elimination. From a synthetic standpoint, only Disconnection B appeared to offer significant advantages since both **10** and **64** would be available in a limited number of steps. Disconnection A was never seriously considered since alcohol **63** suffers from poor reactivity and elimination of any potential leaving group leading to cyclopentene by-products would be problematic.

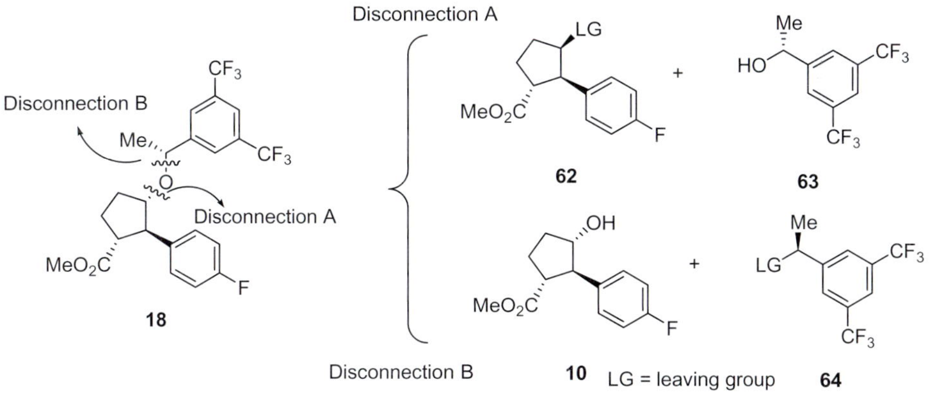

Scheme 7.18 S_N2 Substitution approach–two disconnections.

Our investigations began by examining the direct displacement of **64** bearing various leaving groups with alcohol **10** under strictly S_N2 reaction conditions (Scheme 7.19). Treatment of a mixture of **10** with 10 equiv of mesylate **65** in the

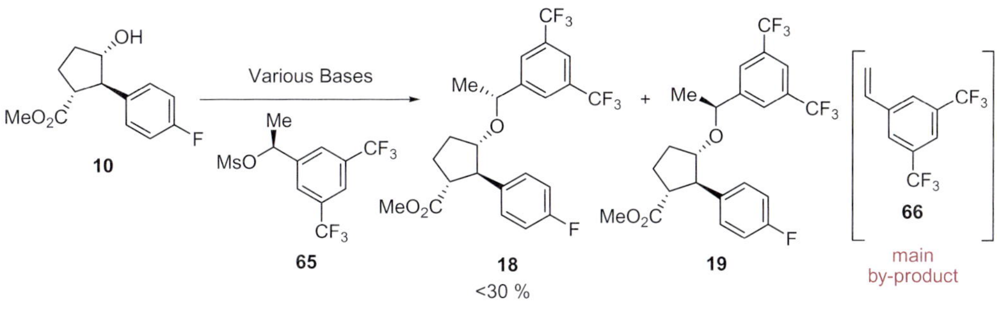

Scheme 7.19 Initial S_N2 substitution attempts.

presence of various bases (NaO*t*-Bu, KO*t*-Bu, BuLi, NEt_3, pyridine, NaH) led to extremely low conversions (<10%) to the desired ether **18**. However, the observed diastereoselectivity was >20 : 1 for **18** : **19**. In the presence of 30 equiv of **65** the yield of **18** improved to 22%. Under essentially all conditions examined, elimination of the mesylate to styrene **66** was the major reaction pathway. The use of other leaving groups such as OTs, *p*-methoxybenzenesulfonate, OTf, and Br did not improve the reaction profile and resulted in extensive elimination to **66** and led to only minor amounts (<30%) of **18**.

The original medicinal chemistry synthesis of ether **18** involved reaction of alcohol **10** with racemic imidate **17** in the presence of a catalytic amount of TfOH and furnished an approximately 1.1 : 1 mixture of **18** : **19** (Scheme 7.3) [1]. We thought it worthwhile to reinvestigate this reaction with chiral imidate **67** in an effort to explore the diastereoselectivity of the etherification.

The use of trichloroimidates for the preparation of ethers is an effective method for *O*-alkylation of alcohols [27]. This method has found widespread use in the protection of alcohols as benzyl ethers since the corresponding trichlorobenzylimidate is inexpensive and commercially available. The mechanism involves activation of the imidate with a catalytic amount of a strong acid (typically TfOH) which leads to ionization of the electrophile and the formation of carbocation which is rapidly trapped by an alcohol. For the preparation of *sec-sec* ethers, this protocol has been limited to glycosidation reactions, due to the S_N1 nature of the reaction which often leads to diastereomeric mixtures of products [26].

Chiral imidate **67** was initially prepared under standard conditions employing catalytic NaH and an excess of trichloroacetonitile (1.5 equiv). Although these conditions provided **67** in generally high yield (~90% yield, > 99% ee), the crude reaction mixture contained a number of unidentified by-products and required purification through a plug of silica gel to give **67** of acceptable quality. Furthermore, the use of NaH on a preparative scale, let alone full scale manufacture, was deemed unacceptable. A rapid screen of a number of bases and solvent combinations was performed. The optimal conditions involved treatment of **25** in a 4 : 1 mixture of cyclohexane/CH_2Cl_2 with 0.1 equiv of DBU in the presence of 1.05 equiv of trichloroacetonitrile (Scheme 7.20). Following an aqueous work-up to remove trace amounts of DBU, crude imidate **67** was obtained in 96% HPLC assay yield and 99.5% ee. Crude **67** was used without further purification in the following etherification as a stream in cyclohexane.

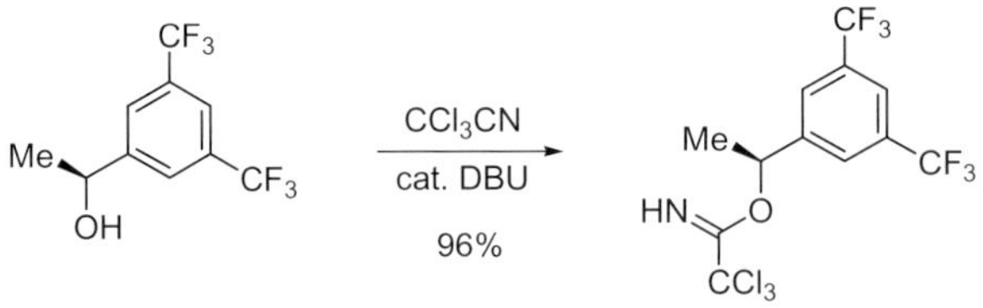

25 **67** 99.5 %ee

Scheme 7.20 Optimized imidate preparation.

Treatment of a mixture of alcohol **10** and chiral imidate **67** with catalytic TfOH only afforded a 1.2 to 1.3 : 1 mixture of **18** : **19** in a combined HPLC assay yield of 91%. Clearly, under these conditions, the reaction was proceeding under an S_N1 reaction pathway. The use of other acid catalysts (TMSOTf, HCl, H_2SO_4, TFA, MsOH) in a variety of solvent systems and under a number of reaction conditions did not improve the diastereomeric ratio of **18** : **19** (typically 1.2 : 1), or simply resulted in no reaction.

During the course of our investigations, interesting reactivity was observed when a mixture of **10** and 2 equiv of imidate **67** was treated with 20 mol% of $BF_3 \cdot OEt_2$ in CH_2Cl_2 at −15 °C. While HPLC analysis of the crude reaction mixture after 3 h indicated low conversion (<10%), the diastereomeric ratio of **18** : **19** was 8 : 1. Upon warming the reaction mixture to room temperature and stirring for an additional 15 h the ratio of **18** : **19** dropped to 1.6 : 1. This indicated that the initial reaction between **10** and **67** may be proceeding through an unprecedented S_N2 reaction pathway. Warming the reaction mixture appeared to divert the reaction between **10** and **67** into a conventional S_N1 pathway. Unfortunately, prolonged reaction at −15 °C did not improve the overall conversion to **18** and **19** (36% combined). Even after considerable experimentation with all of the possible reaction parameters, the reaction of **10** and **67** in the presence of $BF_3 \cdot OEt_2$ could never be improved.

It was during these investigations that the serendipitous discovery of HBF_4 as the superior catalyst was made. The optimal conditions involved reaction of a mixture of alcohol **10** and imidate **67** (1.5 equiv) with 10 mol% HBF_4[2] in a solvent system of 1,2-dichloroethane (DCE)/heptane (1 : 2) at −15 °C for 12 h and warming to room temperature, and gave **18** and **19** as a 17 : 1 mixture of diastereomers and in a combined 75% HPLC assay yield (Scheme 7.21). The remaining mass balance

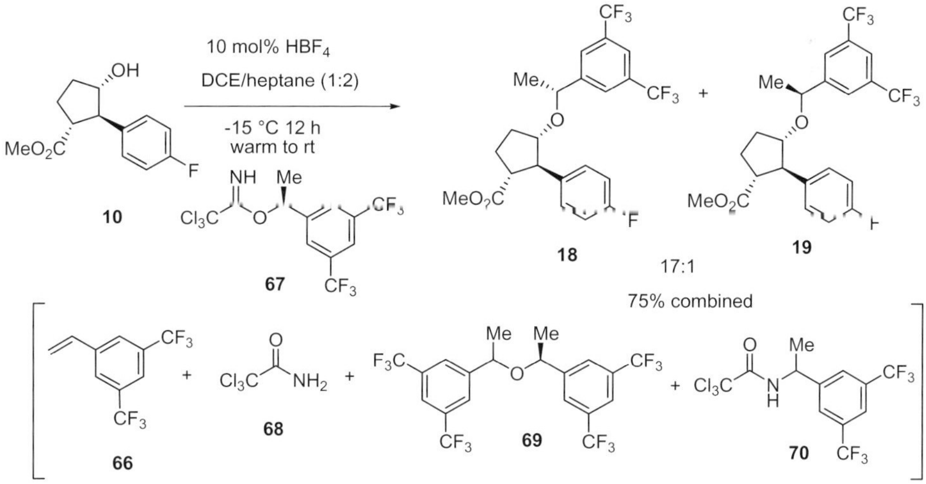

Scheme 7.21 Optimized etherification with chiral imidate.

2) Commercially available 54 wt% tetrafluoroboric acid solution in diethyl ether was used in all transformations.

of the reaction was predominately unreacted starting material **10** (10–13%). Also identified by NMR as minor reaction products in the crude reaction mixture were styrene **66**, trichloroacetamide **68**, bis-ether **69**, and the racemic amide **70** arising from rearrangement of imidate **67**. Since ether **18** was the major reaction product, the etherification reaction was occurring with nearly complete inversion of configuration of the imidate reaction center, most likely proceeding through an S_N2 reaction pathway.

This unprecedented S_N2 reaction prompted us to further investigate the mechanistic details of this reaction. The results will be discussed in detail in Section 7.2.

After identifying the optimal etherification conditions, our attention turned to isolation of **18** in diastereomerically pure form. Diastereomers **18** and **19** were not crystalline, but, fortunately, the corresponding carboxylic acid **71** was crystalline. Saponification of the crude etherification reaction mixture of **18** and **19** with NaOH in MeOH resulted in the quantitative formation of carboxylic acids **71** and **72** (17 : 1) (Scheme 7.22). Since the etherification reaction only proceeded to 75–80% conversion, there still remained starting alcohol **10**. Unfortunately, all attempts to fractionally crystallize the desired diastereomer **71** from the crude mixture proved unfruitful. It was reasoned that crystallization and purification of **71** would be possible via an appropriate salt. A screen of a variety of amines was then undertaken. During the screening process it was discovered that when NEt_3 was added

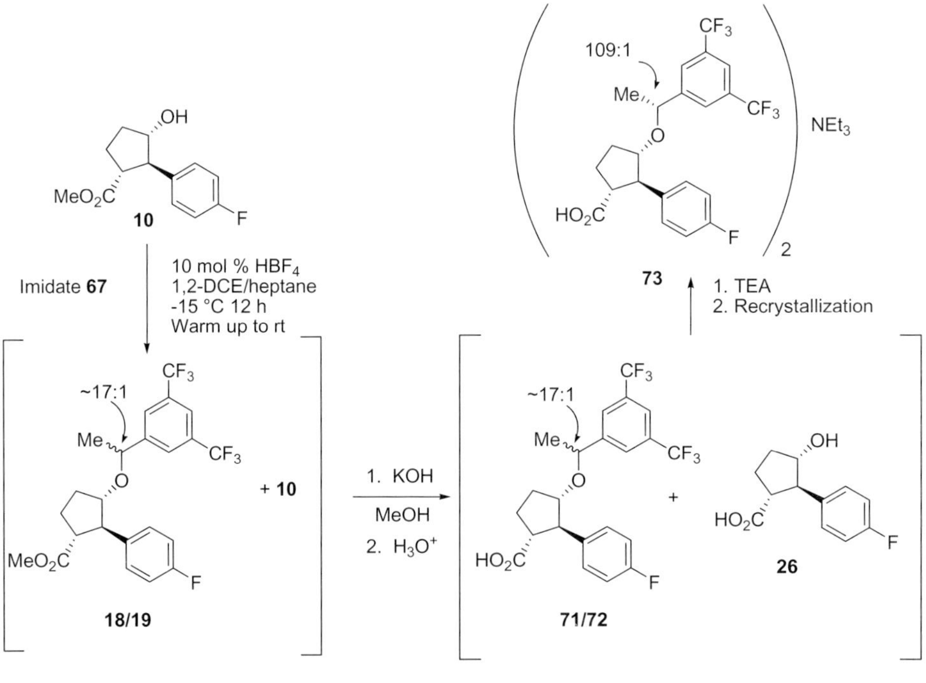

Scheme 7.22 Optimized etherification process.

to the crude mixture of carboxylic acids, the desired diastereomer was obtained as the triethylamine solvate **73** with a significant upgrade in diastereomeric excess (~ 40 : 1) without significant contamination of **72** or the corresponding carboxylic acid **26**. The absolute stereochemistry and confirmation of the structural assignment of **73** was unambiguously established via single-crystal X-ray analysis.

The isolation of **73** was then fully optimized. Upon completion of the etherification reaction, the insoluble trichloroacetamide **68** was filtered, leaving a 17 : 1 mixture of **18** and **19** as a DCE/heptane solution, together with starting material **10**. The solvent was switched to MeOH and the esters were saponified with KOH. The carboxylic acid was isolated after neutralization and the addition of NEt_3 which gave the highly crystalline triethylamine solvate **73** as a 40 : 1 mixture of diastereomers. Recrystallization from MTBE/heptane gave a 109 : 1 diastereomeric mixture of **73** in 54% overall yield from **10**. This final process was successfully implemented in the pilot plant without incident.

7.1.2.4 Preparation of (*R*)-Nipecotate 76 and Completion of the Synthesis of 1

The completion of the synthesis of **1** required installation of the (*R*)-nipecotic moiety. The original method used (*R*)-ethyl nipecotate L-tartrate **21**, which was commercially available, but the availability of this intermediate on multi-kilogram scale required long lead times and cost was a major factor. In addition, it was also discovered that saponification of the ethyl ester in the final stages of the synthesis, as shown in Scheme 7.3, was accompanied by small amounts of epimerization at the carboxylic acid center of **1**, resulting in diastereomeric contamination of the final product.

In order to circumvent this epimerization under strongly basic conditions in the final stages of the synthesis, a *tert*-butyl group was selected for its ease in removal under acidic conditions. *tert*-Butyl nipecotate **76** was prepared via two separate protocols (Scheme 7.23). For example, reaction of *N*-Cbz (*R*)-nipecotic acid **74** [28] with isobutylene in the presence of a catalytic amount of H_2SO_4 gave *tert*-butyl ester **75** in 97% yield. Catalytic hydrogenation over 10% Pd/C afforded **76** in quantitative yield. Since the preparation of **74** began with (*R*)-nipecotic acid **77**, a more streamlined approach was developed for large scale processing and involved the direct conversion of **77** to **76** via trans-esterification. The final process required treatment of **77** with $BF_3 \cdot OEt_2$ (6 equiv) in *tert*-BuOAc and furnished the desired product **76** in 70% yield for the one-pot procedure [29]. While **76** could be isolated

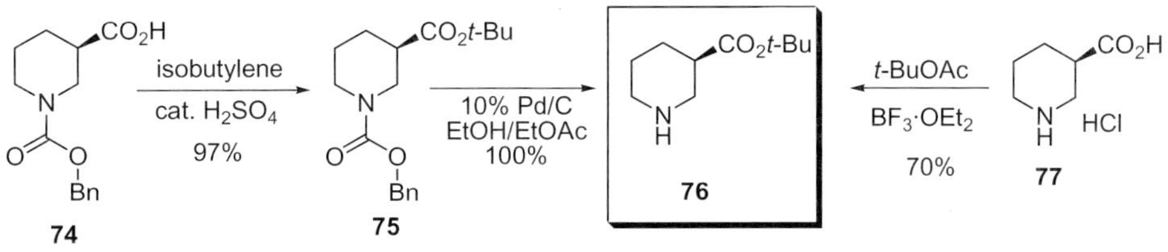

Scheme 7.23 Preparation of *tert*-butyl (R)-nipecotate.

in analytically pure form by distillation, the purity of the material from either synthetic route was sufficiently high to allow its use in subsequent transformations without the need for additional purification.

Conversion of triethylamine solvate **73** to the final product **1** involved a series of high-yielding transformations where **1** was the only isolated product (Scheme 7.24). We selected a S_N2 displacement approach instead of the original reductive amination. Reduction of **73** to alcohol **78** was conducted with 2 equiv of $BH_3 \cdot THF$ in toluene at 65 °C. After an aqueous work-up, alcohol **78** was obtained in 96% assay yield and used crude as a toluene stream in the next reaction. Treatment of crude **78** with 1.2 equiv of Ms_2O in the presence of *i*-Pr_2NEt gave the intermediate mesylate **79**. Direct addition of **76** to the reaction mixture, followed by additional *i*-Pr_2NEt and heating to reflux, afforded **80** in 90% assay yield for the two-step one-pot process. The use of MsCl was found to produce elevated levels of both chloride intermediate **81** and elimination by-product **82**.

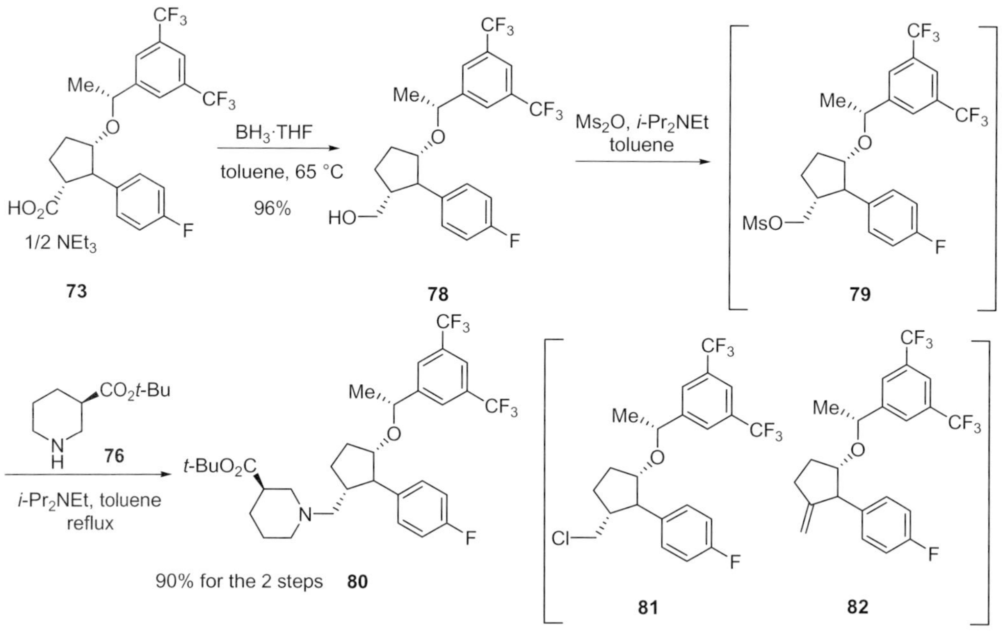

Scheme 7.24 Coupling of **79** with *tert*-butyl nicopetate.

The completion of the synthesis required removal of the *tert*-butyl group and crystallization of **1** as the HCl salt. Reaction of crude **80** with TFA in DCE at 75 °C gave the free base of **1** in near quantitative yield. The free base was not crystalline, but was converted to its crystalline HCl salt by treatment with 2 M HCl in MTBE. The product was isolated in 85% yield after filtration and drying. These conditions were also successfully transferred to the kilogram scale, providing **1** in the expected yields. Thus, the final overall optimized route to **1** is summarized in Scheme 7.25. The improved synthesis of **1** proceeded in 14 steps and in 22% overall yield from **35**.

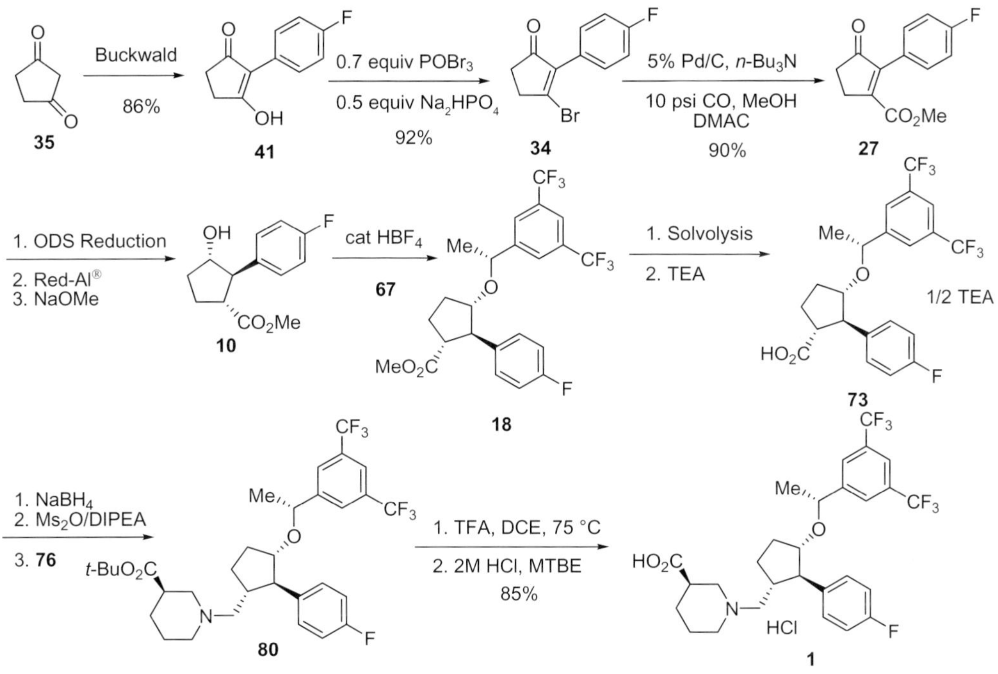

Scheme 7.25 Final optimized synthetic route.

7.2
Chemistry Development

During the development of the final process for the large scale preparation of drug candidate **1**, we encountered several interesting chemistry issues. Here we discuss three topics in greater detail:

1) Reduction of allylic alcohol **46** with Red-Al®.
2) Ether bond formation via reduction of the oxonium species with Et_3SiH.
3) Ether bond formation with chiral imidate **67**.

7.2.1
Reduction of the Allylic Alcohol 46 with Red-Al®

To gain mechanistic insight into the reduction of allylic alcohol **46**, the reaction was monitored by measuring the hydrogen evolution and ReactIR[3)]. The addition of 0.5 equiv of Red-Al® resulted in the evolution of 2 mol of H_2 and the formation of the dimeric intermediate **83** (Scheme 7.26) [19b, 30]. Figure 7.2 shows the cumulative evolution of hydrogen (mmol) and the cumulative amount of Red-Al®

3) ReactIR refers to the real time, *in situ* monitoring of the IR spectrum of the reaction mixture with a ReactIR 4000 Reaction Analysis System available from ASI Applied Systems.

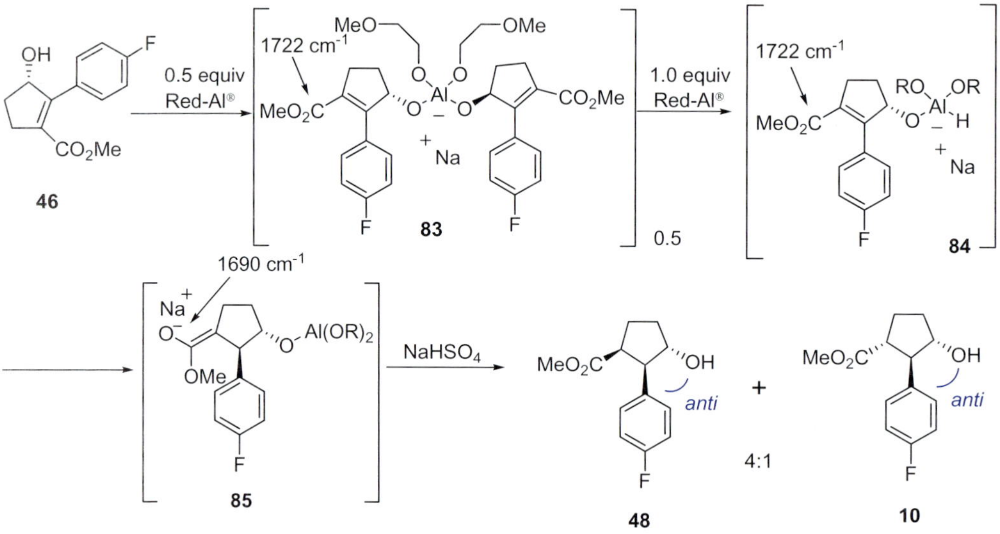

Scheme 7.26 Reaction intermediates for reduction with Red-Al.

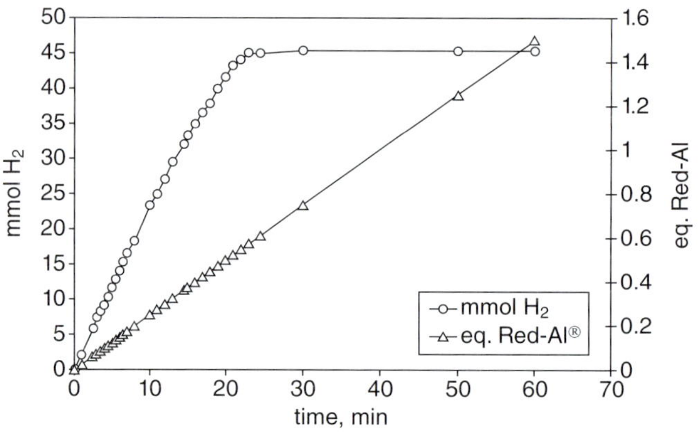

Figure 7.2 Amount of H_2 (mmol) evolved and amount of Red-Al® (equiv) added versus time.

added versus time. The hydrogen evolution abruptly stopped after 0.5 equiv of Red-Al® was added, and the molar rate of hydrogen release was approximately twice the molar rate of Red-Al® added.

The IR showed the complete disappearance of the OH stretch of **46** after 0.5 equiv of Red-Al® was added. Quenching the reaction at this stage resulted only in the recovery of starting material. However, the addition of an additional 1.0 equiv of Red-Al® to intermediate **83** led to the rapid formation of **48** and **10**. It was speculated that the addition of the excess Red-Al® led to the formation of **84** [31], where intramolecular hydride delivery occurred diastereoselectively from the α-face to give the observed 1,2-*anti* stereochemistry of **48** and **10**. The formation of intermediate **85** was monitored by IR, where the disappearance of the carbonyl absorption of the ester of **84** ($1722\,cm^{-1}$) and formation of the enolate ($1690\,cm^{-1}$) was observed. Upon quenching the reaction mixture, the kinetic product **48** was formed as the major product, and the thermodynamic product **10** was formed as the minor product. A complete understanding of the mechanistic details, including hydrogen generation, of this reaction allowed successful implementation on preparative and pilot plant scale.

7.2.2
Oxonium Reduction Configuration Issue

This chemistry was mechanistically interesting, and in order to rationalize the observed diastereoselectivities, molecular modeling calculations were conducted on the proposed oxonium ion intermediates.[4] There are *E*- and *Z*-oxonium ion conformers. Since the major *Z*-oxonium ion conformers (**86**, **87**) from these calculations were approximately $3\,kcal\,mol^{-1}$ higher in energy, these conformers most likely did not significantly contribute to the stereochemical outcome of the reaction (Figure 7.3).

The *E*-conformation had two prevalent conformers **88** and **89** where the steric bulk of the 4-fluorophenyl group blocks the β-face of the oxonium ions (Scheme 7.27). Reduction occurs from the less hindered α-face on each conformer leading to the diastereomeric mixture of products. Since there was only a $0.6\,kcal\,mol^{-1}$ energy difference in these lowest energy conformers **88** and **89**, this nicely supported the observed product distribution of approximately 3 : 1 of **18** : **19** under the best reaction conditions.

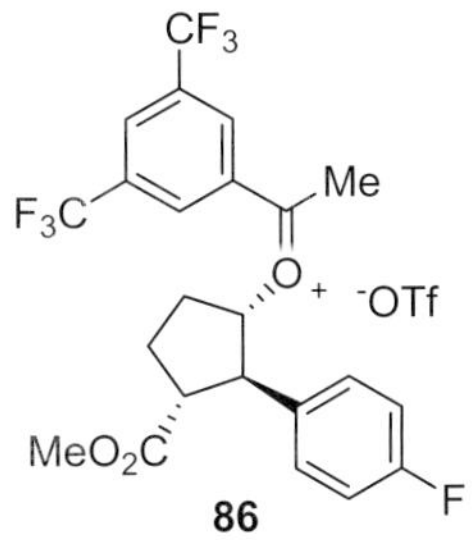

86

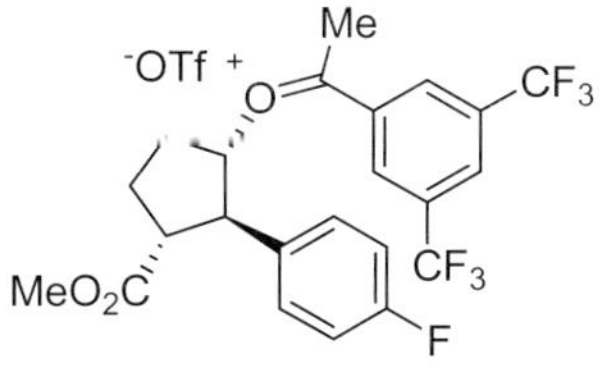

87

Figure 7.3 Higher energy *Z*-conformers (*Z*-conformations approximately $3\,kcal\,mol^{-1}$ higher in energy).

4) The Titan software package was used for analysis. The "Conformer Study" option was selected as a computation type, and PM3 was selected as the method, the results in conformations being generated using the included MMFF94 molecular mechanics force field and the resulting structures being minimized with the PM3 semiempirical method. Calculations were performed in the gas phase.

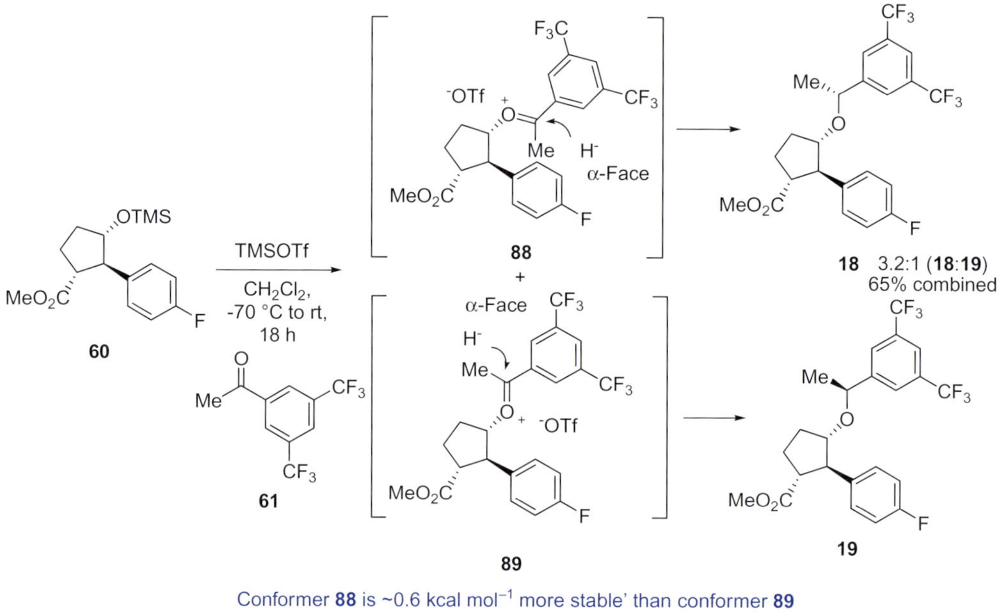

Scheme 7.27 Reduction outcome rationalization.

7.2.3
Ether Bond Formation with Chiral Imidate 67

The etherification between alcohol **10** and imidate **67** was one of the key transformations in the successful preparation of compound **1**. The use of HBF_4 as the catalyst for the etherification was crucial for obtaining high levels of diastereoselectivity and relatively high conversion to the desired product **18**. The fact that *sec-sec* ethers have rarely, if ever, been obtained with high levels of diastereocontrol in S_N2 fashion under typical S_N1 reaction conditions prompted us to investigate the complex mechanistic details of this exceptional reaction.

In order to fully confirm that the reaction was proceeding through an S_N2 mechanism, two sets of experiments were devised. Reaction of **10** with imidate enantiomer **90** with 10 mol% HBF_4 under identical reaction conditions as for imidate **67** afforded a 1:13 mixture of **18** and **19** in an unoptimized 59% assay yield (Scheme 7.28). The product distribution of this reaction was in perfect agreement with the expected result with diastereomer **19** being the major reaction product. The origin of the inversion of the imidate center was first established by kinetic isotope experiments (Scheme 7.29). For example, reaction of **10** with a mixture of 3 equiv of racemic imidate **17** and 3 equiv of deuterated racemic imidate **91** was conducted under the optimized conditions described above. Examination of the crude NMR at different time intervals indicated a negligible secondary kinetic isotope effect ($K_{H/D}$ ~1.0), consistent with a minimal hybridization change

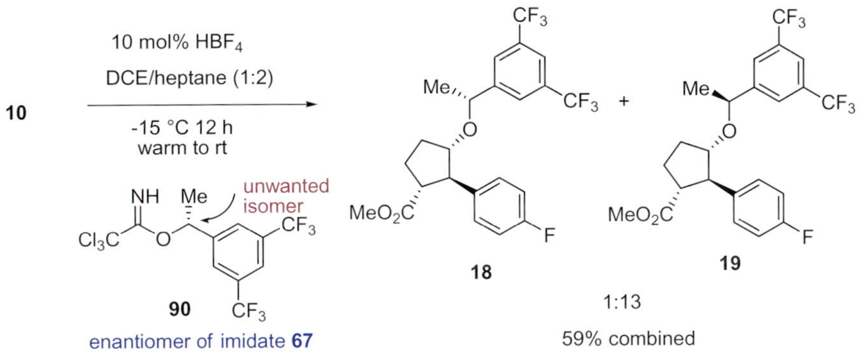

Scheme 7.28 Etherification with wrong imidate enantiomer **90**.

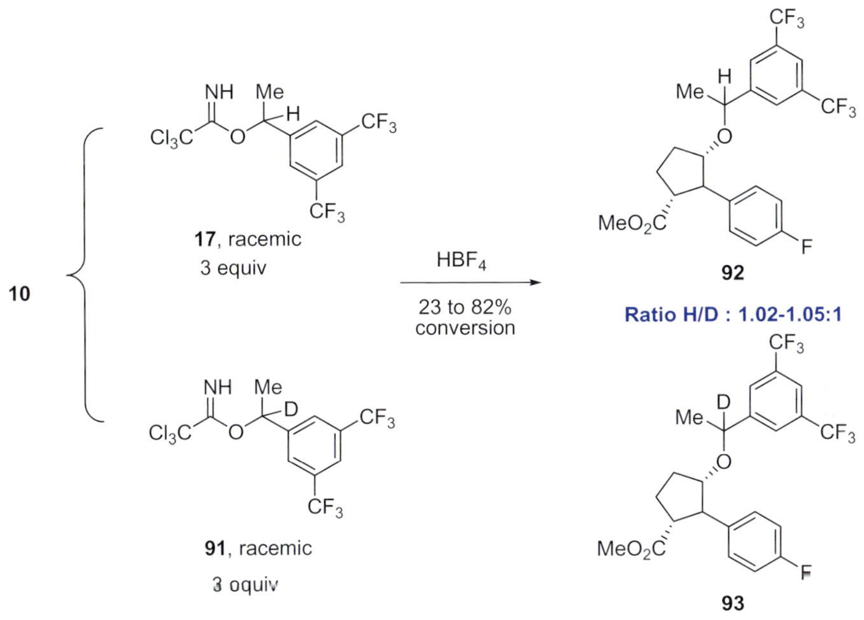

Scheme 7.29 Isotope effects on etherification.

at the reacting center [32]. From these results it was unambiguously established that under the reaction conditions, the reaction was proceeding almost exclusively via an S_N2 mechanism.

In an effort to identify the origin of the formation of the minor diastereomer **19** and understand whether its formation was a function of a breakdown in the S_N2 pathway leading to an S_N1 pathway, the activation of the chiral imidate **67** was next investigated. In the etherification reaction between **10** and **67**, the acid catalyst increases the electrophilicity of imidate **67** through coordination between the acid catalyst and the nitrogen atom in **67**, leading to activated intermediate **94** (Scheme 7.30). However, simultaneously, the acid catalyst can also reduce the nucleophilicity of **10** via coordination between the acid catalyst and an oxygen atom, leading to **96**. It is generally believed that etherification reactions employing imidates and strong acids other than HBF_4 proceed through a carbocation of type **95**, leading to a 1 : 1 mixture of diastereomers, as was seen when TfOH was utilized in the etherification reaction. However, in the etherification with HBF_4, excellent diastereoselectivity was observed that indicated that the equilibrium between activated imidate **94** and carbocation **95** is shifted far toward to **94**. All attempts to observe carbocation **95** by spectroscopic means were unsuccessful and suggested that its formation as a transient intermediate and rearrangement were rapid. However, if carbocation **95** formed, epimerization of imidate **67** should be observed.

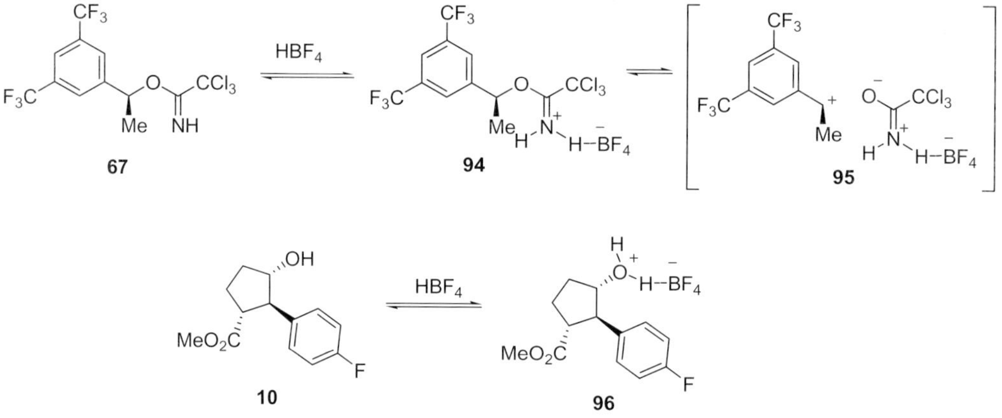

Scheme 7.30 Imidate reaction intermediates.

Indeed, epimerization of imidate **67** was observed in the presence of 10% HBF_4 in dichloroethane without any alcohol **10** present, as shown in Table 7.1. Under these conditions, approximately 5% epimerization was observed in the initial stages of the reaction and no further epimerization was observed afterward. The

Table 7.1 Etherification reaction profile under optimized conditions.

Entry	Time	Temperature (°C)	ee of 67 (%)	% conversion (18 : 19)
1	0	−15	99.5	0
2	5 min	−15	90	5
3	1 h	−15	90	20
4	6 h	−15	90	35
5	12 h	−15	90	72
6	15 h	20	90	75

Table 7.2 Epimerization of chiral imidate **67**.

Time	DCM (%ee)	DCE (%ee)
0	100.0	100.0
1	86.8	84.9
3	77.9	77.9
5	72.6	78.0
23	58.4	78.0

Table 7.3 Etherification impacts on aging.

Age time (h)	Lewis acid	Equiv	Assay yield **18**:**19** (%)
0	HBF_4	0.15	75
1	HBF_4	0.15	56
3	HBF_4	0.15	20
5	HBF_4	0.15	0
1	$BF_3 \cdot OEt_2$	0.15	<5
0	$BF_3 \cdot OEt_2$	1.00	16
0	HBF_4	1.00	42

rate of epimerization was found to be solvent dependent (Table 7.2). For example, treatment of **67** with HBF_4 in CH_2Cl_2 for 3 h at −5 °C led to a decrease in ee from 100% to 77% ee. After 23 h, the ee had been reduced to 58.4% ee. Therefore, imidate **67** was far more stable in dichloroethane than in dichloromethane. We speculate that the epimerization is likely due to a localized heat generation upon addition of HBF_4. To confirm this, the etherification reaction was repeated with **10** and imidate **67** at −78 °C. The diastereomeric ratio of **18**:**19** jumped to 55:1; however, the overall conversion after warming to room temperature was only ~55%. We concluded that HBF_4 is the correct acid catalyst to activate imidate **67** toward S_N2 reaction, but not too active to cause a breakdown in the S_N2 pathway leading to an S_N1 pathway. We hypothesized that the formation of the minor diastereomer **19** was a result of reaction of **10** with the epimerized imidate **90** under S_N2 conditions, although we never could conclusively rule out the involvement of **95** under S_N1 conditions.

Even after extensive optimization, complete conversion of the starting material **10** to products **18** and **19** was never achieved. This observation was only noticed when HBF_4 and $BF_3 \cdot OEt_2$ were employed as catalysts for the etherification. In addition, the amount of time cyclopentanol **10** was in contact with HBF_4 had a profound impact upon the conversion to products. For example, when HBF_4 was added in one portion to a mixture of **10** and imidate **67** under the optimized reaction conditions, typical conversion to products was observed (75%). On the other hand, the longer **10** was aged with HBF_4 prior to the addition of imidate **67**, the lower the conversion to products (Table 7.3). This phenomenon is a clear indication of deactivation of a nucleophile. In an effort to understand this, a series of NMR experiments was devised. The course of the deactivation of **10** was conducted by the addition of HBF_4 or $BF_3 \cdot OEt_2$ to a solution of **10** in CD_2Cl_2 and analysis of the reaction mixture by ^1H, ^{11}B, and ^{19}F NMR. Treatment of **10** with 1.0 equiv of $BF_3 \cdot OEt_2$ at −10 °C gave an approximately 1 : 1 mixture of **10** and its BF_3 complex **97**. Analysis of the ^{11}B NMR was consistent with coordination of BF_3 to the hydroxy group of **97**. In addition, there were no detectable amounts of borates of **10** [33]. Subsequent warming of the sample led to the exclusive formation of **97**, which was irreversible under the reaction conditions. Complex **97** does not participate in the etherification reaction with imidate **67**. The nearly complete coordination of BF_3 nicely accounted for the lack of reactivity of alcohol **10** in the etherification reaction and explained why the reaction stalled at ~80% conversion.

When compound **10** was treated with 1.0 equiv of HBF_4 at −10 °C for 1 h, <20% coordination was observed. The ^{11}B NMR spectrum indicated the presence of a different coordinated intermediate which we speculated was protonated **96**, although there were also some detectable amounts of **97** present (Scheme 7.31). It is hypothesized that a rapid equilibrium between **10** and complex **96** exists, unlike between **10** and **97**. Upon warming to room temperature, both the ^1H and ^{11}B NMR spectra of the crude reaction mixture became increasingly broad and complex, most likely due to exchange processes. As the etherification reaction progresses, increased amounts of BF_3 and HF are present in solution since HBF_4 can be considered as an equilibrium mixture of BF_3 and HF. As the concentration of BF_3 in the reaction medium increases, increased amounts of **97** are formed. Since HBF_4 is a more active catalyst than BF_3, activation of **67**, and subsequent etherification, is the preferred reaction pathway in the early stages of the reaction.

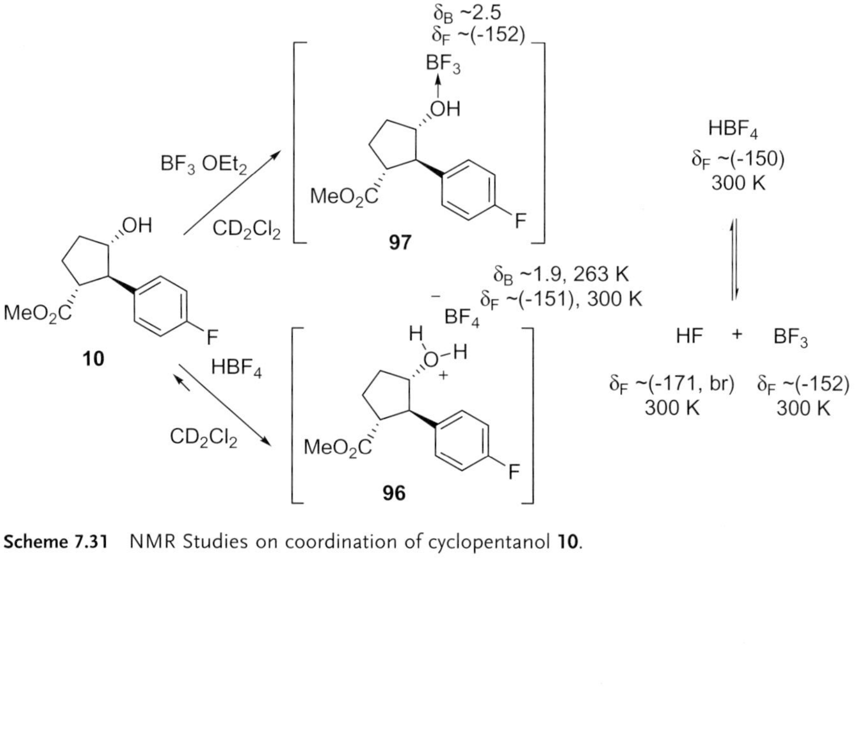

Scheme 7.31 NMR Studies on coordination of cyclopentanol **10**.

When the concentration of BF_3 increased, competitive deactivation leading to the formation of **97** resulted in unreacted starting material at the end of the etherification reaction. Efforts to break up this coordination and increase the conversion by the addition of certain additives, such as water, $NaPF_6$, KPF_6, $LiPF_6$, or Na_sSiF_6, either led to no improvement in conversion or to a complete shut down in the S_N2 pathway and significant erosion in the diastereoselectivity resulted.

In conclusion, the kilogram preparation of the structurally intriguing NK-1 receptor antagonist 1 required the development of novel chemistry for both the preparation of the chiral hydroxy acid **26** and formation of the challenging *sec-sec* ether bond. Key developments for the preparation of **26** involved asymmetric reduction of the ketone **27** to give cyclic allylic alcohol **46**, which, in conjunction with Red-Al® and epimerization/saponification, gave a highly stereoselective synthesis of *anti-anti*-1,2,3-trisubstituted cyclopentanol **26**. Key improvements in the preparation of **1** also involved the displacement of alcohol **10** with chiral imidate **67** in the presence of 10 mol% HBF_4 giving the desired ether product **18** in a 17 : 1 ratio. Upgrade of the diastereomeric ratio was realized by crystallization of the triethylamine solvate **73** to greater than 109 : 1.

A complete understanding of the mechanistic details of each reaction allowed the successful implementation of the chemistry on a multi-hundred kilogram scale.

Acknowledgments

The author would like to thank Ian Davies, Audrey Wong, Jimmy Wu, Jean-François Marcoux, Peter Dormer and Michael Hillier for their valuable contributions to the chemistry highlighted in this chapter. In addition, the author would like to acknowledge Nobuyoshi Yasuda for the opportunity to contribute this chapter and for his tireless attention to detail.

References

1 For the discovery of **1**, see: (a) Meurer, L.C., Finke, P.E., Owens, K.A., Tsou, N.N., Ball, R.G., Mills, S.G., MacCoss, M., Sadowski, S., Cascieri, M.A., Tsao, K.-L., Chicchi, G.C., Egger, L.A., Luell, S., Metzger, J.M., MacIntyre, D.E., Rupniak, N.M.J., Williams, A.R., and Hargreaves, R.J. (2006) *Bioorg. Med. Chem. Lett.*, **16**, 4504; (b) Finke, P.E., Meurer, L.C., Levorse, D.A., Mills, S.G., MacCoss, M., Sodowski, S., Cascieri, M.A., Tsao, K.-L., Chicchi, G.C., Metzger, J.M., and MacIntyre, D.E. (2006) *Bioorg. Med. Chem. Lett.*, **16**, 4497.

2 Houn, F. (2003) FDA approval letter is available at the website. http://www.access.fda.gov/drugsatfolatda_docs/appletter/2003/21549ltr.pdf (accessed August 2010).

3 Hale, J.J., Mills, S.G., MacCoss, M., Finke, P.E., Cascieri, M.A., Sadowski, S., Ber, E., Chicchi, G.G., Kurtz, M., Metzger, J., Eiermann, G., Tsou, N.N., Tattersall, F.D., Rupniak, N.M.J., Williams, A.R., Rycroft, W., Hargreaves,

R., and MacIntyre, D.E. (1998) *J. Med. Chem.*, **41**, 4607.

4 Pendergrass, K., Hargreaves, R., Petty, K.J., Carides, A.D., Evans, J.K., and Horgan, K.J. (2004) *Drugs Today*, **40**, 853.

5 Desai, R.C., Cicala, R., Meurer, L.M., and Finke, P.E. (2002) *Tetrahedron Lett.*, **43**, 4569.

6 We previously published only two of these routes, see: Kuethe, J.T., Wong, A., Wu, J., Davies, I.W., Dormer, P.G., Welch, C.J., Hillier, M.C., Hughes, D.L., and Reider, P.J. (2002) *J. Org. Chem.*, **67**, 5993.

7 Suzuki, A. (1985) *Pure Appl. Chem.*, **57**, 1749.

8 Belmont, D.T., and Paquett, L.A. (1985) *J. Org. Chem.*, **50**, 4102.

9 Kress, M.H., and Kishi, Y. (1995) *Tetrahedron Lett.*, **36**, 4583.

10 Schoenberg, A., Bartoletti, I., and Heck, R.F. (1974) *J. Org. Chem.*, **39**, 3318.

11 Fox, J.M., Huang, X., Chieffi, A., and Buchwald, S.L. (2000) *J. Am. Chem. Soc.*, **122**, 1360.

12 Eskola, S. (1957) *Suom. Kemistil.*, **30B**, 34; Chem. Abstr. 1957, 53, 16014f.

13 Davies, I.W., Matty, L., Hughes, D.L., and Reider, P.J. (2001) *J. Am. Chem. Soc.*, **123**, 10139.

14 Schnyder, A., Beller, M., Mehltretter, G., Nsenda, T., Struder, M., and Indolese, A.F. (2001) *J. Org. Chem.*, **66**, 4311.

15 Corey, E.J., Bakshi, R.K., Shibata, S., Chem, C.P., and Sinch, V.K. (1987) *J. Am. Chem. Soc.*, **109**, 7925; (b) Mathre, D.J., Jones, T.K., Xavier, L.C., Blacklock, T.J., Reamer, R.A., Mohan, J.J., Jones, T.T., Hoogsteen, K., Baum, M.W., and Grabowski, E.J.J. (1991) *J. Org. Chem.*, **56**, 751.

16 (a) Stork, G., and Kahne, D.E. (1983) *J. Am. Chem. Soc.*, **105**, 1072; (b) Crabtree, R.H., and Davies, M.W. (1983) *Organometallics*, **2**, 681.

17 Evans, D.A., and Morrissey, M.M. (1984) *J. Am. Chem. Soc.*, **106**, 3866.

18 Lee, H.M., Jiang, T., Stevens, E.D., and Nolan, S.P. (2001) *Organometallics*, **20**, 1255.

19 For leading references describing the reduction of α,β-unsaturated esters using Red-Al, see: (a) SarKar, A., Rao, B.R., and Konar, M.M. (1989) *Synth. Commun.*, **19**, 2313; (b) Malek, J. (1988) *Org. React.*, **36**, 249.

20 For leading references describing the reduction of α,β-unsaturated esters using Red-Al and copper (I) bromide, see: (a) Semmelhack, M.F., Stauffer, R.D., and Yamashita, A. (1977) *J. Org. Chem.*, **42**, 3180; (b) Semmelhack, M.F., and Stauffer, R.D. (1975) *J. Org. Chem.*, **40**, 3619.

21 For a full account of our etherifications studies documented in this section, see: Kuethe, J.T., Marcoux, J.-F., Wong, A., Wu, J., Hillier, M.C., Dormer, P.G., Davies, I.W., and Hughes, D.L. (2006) *J. Org. Chem.*, **71**, 7378.

22 For leading references using palladium-iridium, and zinc-catalyzed allylic etherifications, see: (a) Kim, H., and Lee, C. (2002) *Org. Lett.*, **4**, 4369; (b) Roberts, J.P., and Lee, C. (2005) *Org. Lett.*, **7**, 2679; for asymmetric aldol additions, see: (c) Crimmins, M.T., and Tabet, E.A. (2000) *J. Am. Chem. Soc.*, **122**, 5473; (d) Crimmins, M.T., and She, J. (2004) *Synlett*, 1371; (e) Crimmins, M.T., and Ellis, J.M. (2005) *J. Am. Chem. Soc.*, **5**, 17200; for diastereoselective additions to α-acetoxy ethers using α-(trimethylsilyl) benzyl auxiliaries, see: (f) Rychnovsky, S.D., and Cossrow, J. (2003) *Org. Lett.*, **5**, 2367; for oxa-Michael additions of alkoxides to Michael acceptors, see: (g) Enders, D., Hartwig, A., Raabe, G., and Runsink, J. (1998) *Eur. J. Org. Chem.*, **9**, 1771 and references cited therein; for addition of silyl enol ethers to 1,3-butadienes in the presence of SO_2, see: (h) Narkevitch, V., Schenk, K., and Vogel, P. (2000) *Angew. Chem. Int. Ed.*, **39**, 1806; (i) Narkevitch, V., Megevand, S., Schenk, K., and Vogel, P. (2001) *J. Org. Chem.*, **66**, 5080.

23 For leading references on diastereoselective hydrogenations, see: (a) Hoveyda, A.H., Evans, D.A., and Fu, G.C. (1993) *Chem. Rev.*, **93**, 1307; (b) Bouzide, A. (2002) *Org. Lett.*, **4**, 1347; (c) Ikemoto, N., Tellers, D.M., Dreher, S.D., Liu, J., Huang, A., Rivera, N.R., Njolito, E., Hsiao, Y., McWilliams, J.C., Williams, J.M., Armstrong, J.D., III, Sun, Y., Mathre, D.J., Grabowski, E.J.J., and

Tillyer, R.D. (2004) *J. Am. Chem. Soc.*, **126**, 3048 and references cited therein.

24 Payack, J.F., Huffman, M.A., Cai, D., Hughes, D.L., Collins, P.C., Johnson, B.K., Cottrell, I.F., and Tuma, L.D. (2004) *Org. Process Res. Dev.*, **8**, 256 and references cited therein.

25 (a) Hatakeyama, S., Mori, H., Kitano, K., Yamada, H., and Nishizawa, M. (1994) *Tetrahedron Lett.*, **35**, 4367; (b) Sassaman, M.B., Kotian, K.D., Prakash, G.K.S., and Olah, G.A. (1987) *J. Org. Chem.*, **52**, 4314; (c) Komatsu, N., Ishida, J., and Suzuki, H. (1997) *Tetrahedron Lett.*, **38**, 7219; (d) Kato, J., Iwasawa, N., and Mukaiyama, T. (1985) *Chem. Lett.*, 743.

26 (a) Tsunoda, T., Suzuki, M., and Noyori, R. (1979) *Tetrahedron Lett.*, **20**, 4679; (b) Tsunoda, T., Suzuki, M., and Noyori, R. (1980) *Tetrahedron Lett.*, **21**, 1357.

27 For leading references, see: (a) Iversen, T., and Bundle, D.R. (1981) *J. Chem. Soc. Chem. Commun.*, 1240; (b) Schmidt, R.R., and Michel, J. (1980) *Angew Chem. Int. Ed. Engl.*, **19**, 731; (c) Schmidt, R.R., and Hoffmann, M. (1983) *Angew Chem. Int. Ed. Engl.*, **22**, 406; (d) Wessel, H.P., Iversen, T., and Bundle, D.R. (1985) *J. Chem. Soc., Perkin Trans. 1*, 2247; (e) Nakajima, N., Horita, K., Abe, R., and Yonemitsu, O. (1988) *Tetrahedron Lett.*, **29**, 4139; (f) Nakajima, N., Saito, M., and Ubukata, M. (1998) *Tetrahedron Lett.*, **39**, 5565; (g) Eichler, E., Yan, F., Sealy, J., and Whitfield, D.M. (2001) *Tetrahedron*, **57**, 6679; (h) Kusumoto, T., Hanamoto, T., Sato, K., Hiyama, T., Takehara, T., Shoji, T., Osawa, M., Kuiyama, T., Nakamura, K., and Fujisawa, T. (1990) *Tetrahedron Lett.*, **31**, 5343.

28 Das, J., Kimball, S.D., Reid, J.A., Wang, T.C., Lau, W.F., Roberts, D.G.M., Seiler, S.M., Schumacher, W.A., and Ogletree, M.L. (2002) *Bioorg. Med. Chem. Lett.*, **12**, 41.

29 Grigan, N., Musel, D., Veinberg, G.A., and Lukevics, E. (1996) *Synth. Commun.*, **26**, 1183.

30 Malek, J. (1985) *Org. React.*, **34**, 1.

31 Casensky, B., Machacek, J., and Abraham, K. (1971) *Collect. Czech. Chem. Commun.*, **36**, 2648.

32 Carpenter, B.K. (1984) *Determination of Organic Reaction Mechanisms*, John Wiley & Sons, Inc., New York.

33 Nöth, H., and Wrackmeyer, B. (1978) *Nuclear Magnetic Resonance Spectroscopy of Boron Compounds*, Springer-Verlag, New York.

8
Glucokinase Activator

Artis Klapars

This chapter will detail early process development and a kilogram delivery of a Merck glucokinase activator **1**, shown in Figure 8.1. Glucokinase, an enzyme that catalyzes phosphorylation of D-glucose, is a target for the treatment of type 2 diabetes. Glucokinase controls the conversion of glucose to glycogen and regulates glucose production in the liver. The expression of glucokinase is controlled by insulin in the liver and D-glucose in the pancreatic β-cells [1].

The key synthetic challenges in the target molecule **1** were the chiral 2-arylpyrrolidine fragment, the densely functionalized benzimidazole ring, and the hindered biaryl ether linkage.

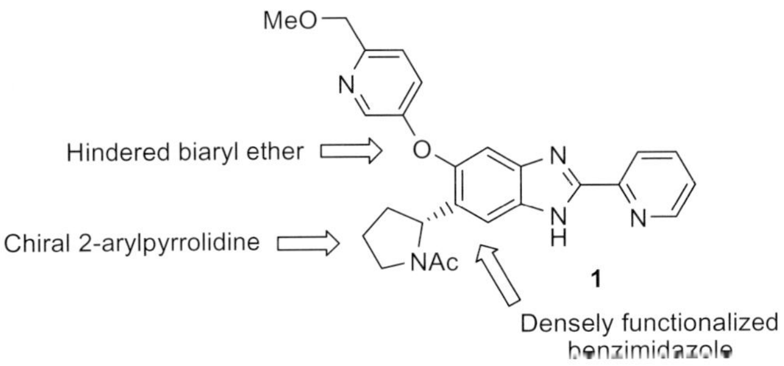

Figure 8.1 Structure of the glucokinase activator.

8.1
Project Development

8.1.1
Medicinal Route

The drug discovery route to compound **1** started out with the expensive and poorly available boronic acid **2**, which was coupled with aryl bromide **3** (Scheme 8.1). Hydrogenation of the resulting pyrrole **4** provided the racemic pyrrolidine **5**. At

The Art of Process Chemistry. Edited by Nobuyoshi Yasuda

ISBN: 978-3-527-32470-5

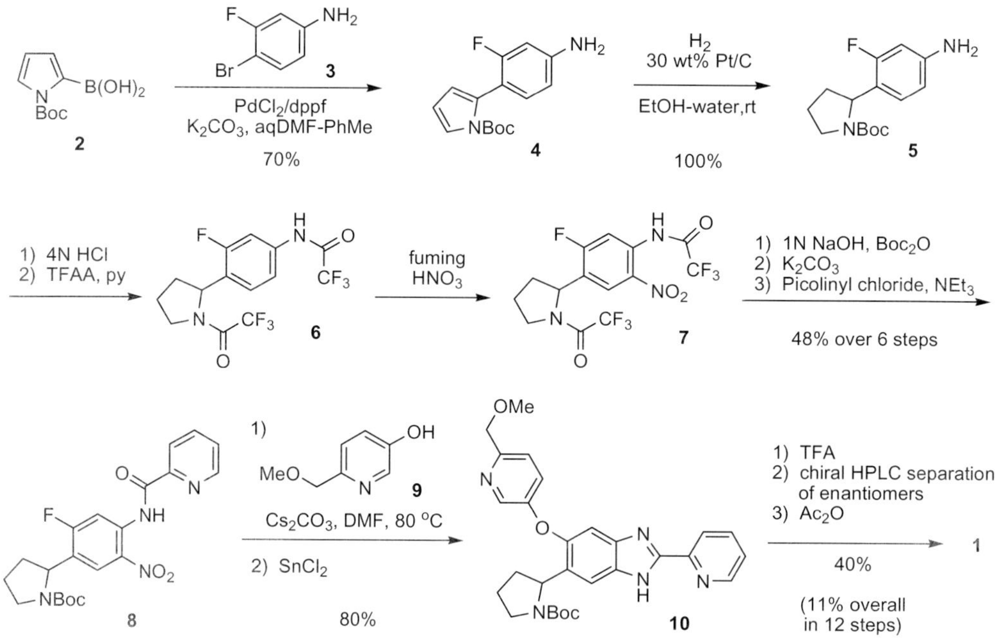

Scheme 8.1 Medicinal Chemistry route to **1**.

this point, protecting group manipulation was required to ensure compatibility of the substrate with the strongly acidic conditions of the nitration reaction. To this end, the *N*-Boc protecting group was removed with 4 M aqueous HCl and the resulting amine was bis-protected with the trifluoroacetyl group. This allowed selective nitration of **6**; however, another exchange of protecting groups was required at this point to introduce the picolinyl group on the aniline nitrogen. Upon treatment with 1 M aqueous NaOH and Boc_2O, the protecting group on the pyrrolidine nitrogen was exchanged to *N*-Boc and the other triflouroacetyl group was partially cleaved (an additional treatment with K_2CO_3 was required to fully reveal the aniline NH_2 group). The resulting aniline was acylated with picolinyl chloride to provide nitroarene **8**, which was set up for nucleophilic aromatic substitution of the fluoride with hydroxypyridine **9**. Notably, the nitro group in compound **8** served a dual role of activating the aryl fluoride and providing a building block for the benzimidazole ring in **10** after reduction with $SnCl_2$ and cyclization. The synthesis was completed by deprotection of the *N*-Boc group in **10**, HPLC separation of enantiomers, and acetylation to provide **1** in 12 steps and 11% overall yield [2].

8.1.1.1 Problems of the Original Route

The Medicinal Chemistry route suffered from several shortcomings that prohibited implementation on a kilogram scale while meeting the project timeline requirements:

1) Boronic acid **2** was expensive and not available on scale. It would have to be prepared, thus adding extra step(s) to the route.
2) Multiple protecting group manipulations drastically reduced the efficiency of the route.
3) The use of $SnCl_2$ as the reducing agent for the nitro group was not desirable because of environmental issues and difficult work-up (formation of emulsions).
4) Most importantly, the inefficiency of the chiral HPLC separation of the enantiomers in the penultimate step was detrimental to the speed and throughput required for the project.

8.1.1.2 Advantages of the Original Route

Despite the shortcomings listed above, two key features of the Medicinal Chemistry route were highly attractive:

1) The highly selective nitration of **6** set the stage for the installation of the benzimidazole as well as solving the problem of the biaryl ether formation, which would have been a challenge for existing transition metal-catalyzed technologies.
2) Coupling of aryl fluoride **8** and hydroxypyridine **9** provided a highly convergent strategy that allowed parallel preparation of the two fragments.

8.1.2 Process Development

With consideration of the shortcomings and advantages of the Medicinal Chemistry route discussed above, a retrosynthesis of **1** was designed to incorporate the convergent coupling of hydroxypyridine **9** and fluoroarene **11** as the key step (Scheme 8.2). Enantioselective preparation of the α-arylpyrrolidine **12** was identified as the key challenge.

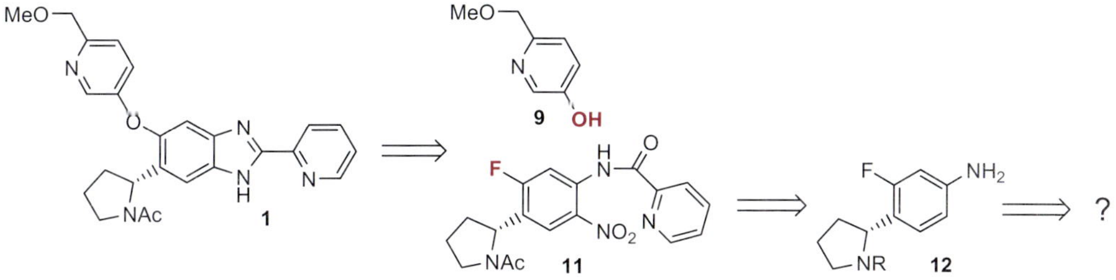

Scheme 8.2 Retrosynthetic analysis of **1**.

The discussion of process development will be outlined as follows:

1) Preparation of hydroxypyridine fragment **9**
2) Enantioselective preparation of the α-arylpyrrolidine **12**
3) Elaboration of **12** to the final product **1**.

8.1.2.1 Preparation of Hydroxypyridine Fragment 9

Hydroxypyridine **9** presented only moderate synthetic challenges. Utilizing slightly modified literature procedures [3], *O*-benzylation of **13** was followed by oxidation with *m*-CPBA to provide the *N*-oxide **14** (Scheme 8.3). A classical reaction in pyridine chemistry, the treatment of pyridine *N*-oxide **14** with a hot mixture of acetic anhydride and acetic acid, provided the 2-hydroxymethylpyridine **16**. This interesting transformation presumably proceeds via electrocyclic rearrangement of intermediate **15**, followed by hydrolytic cleavage of the acetate upon work-up [4]. *O*-Methylation of the resulting alcohol **16** and deprotection of the benzyl ether via catalytic hydrogenation gave the desired hydroxypyridine **9**.

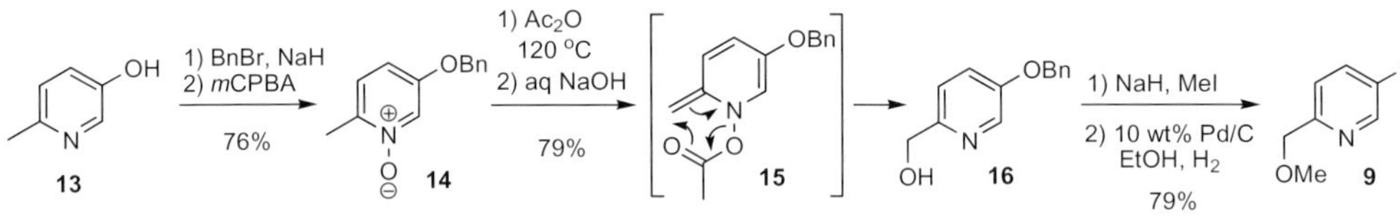

Scheme 8.3 Preparation of hydroxypyridine **9**.

Clearly, the sequence in Scheme 8.3 did not excel in terms of elegance. In particular, the use of NaH was not desirable due to the potentially dangerous properties of the reagent and the hydrogen gas produced in the reaction. Metal alkoxide bases such as NaO*t*-Bu are generally much preferred to NaH in process development. Nevertheless, the route depicted in Scheme 8.3 provided a workable access to hydroxypyridine **9**. A decision was made to temporarily withhold from improving the preparation of **9** and to outsource the existing route. This strategy allowed us to dedicate all resources to the more challenging problems in the synthesis of **1**.

8.1.2.2 Enantioselective Preparation of the α-Arylpyrrolidine 12

With the synthesis of the hydroxypyridine **9** progressing in parallel, most of the effort was focused on the investigation of the far more challenging preparation of the chiral α-arylpyrrolidine **12**. At the outset of our studies, very few practical methods for enantioselective preparation of α-arylpyrrolidines were known [5]. The four most attractive approaches are summarized in Scheme 8.4.

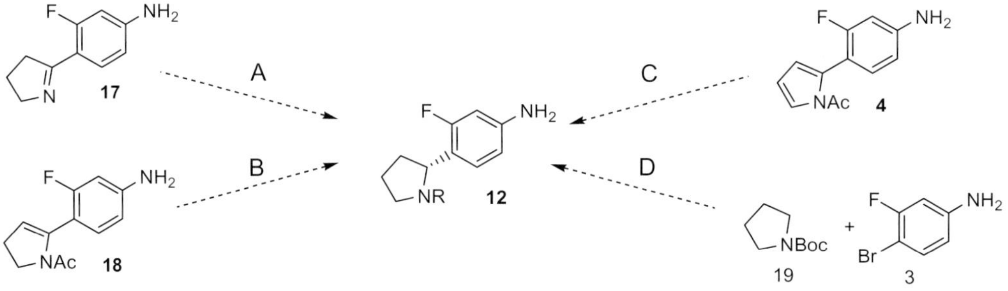

Scheme 8.4 Enantioselective approaches to α-arylpyrrolidine **12**.

Asymmetric hydrosilylation of the cyclic imine **17** (Approach A) was precedented on simpler substrates by Buchwald but the method requires an expensive and highly air-sensitive chiral titanocene catalyst (Scheme 8.5) [6].

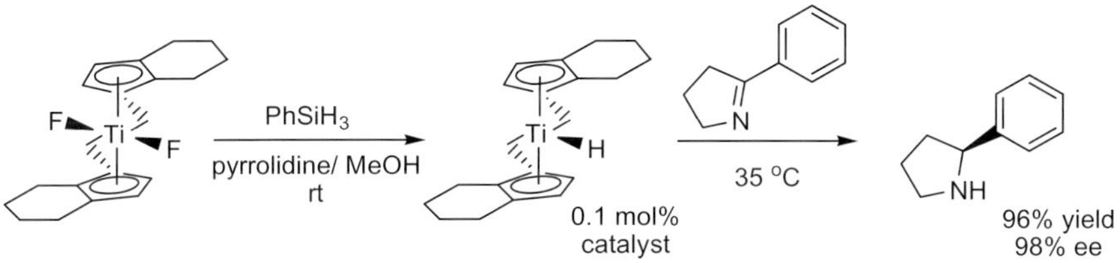

Scheme 8.5 Buchwald's example of asymmetric hydrogenation of cyclic imine.

Asymmetric hydrogenation of a cyclic enamide (Approach B) had very sparse literature precedents [7]. It should also be noted that preparation of these cyclic imines and enamides is not straightforward. The best method for the synthesis of cyclic imines involves *C*-acylation of the inexpensive *N*-vinylpyrrolidin-2-one followed by a relatively harsh treatment with refluxing 6 M aqueous HCl, which accomplishes deprotection of the vinyl group, hydrolysis of the amide, and decarboxylation (Scheme 8.6) [8].

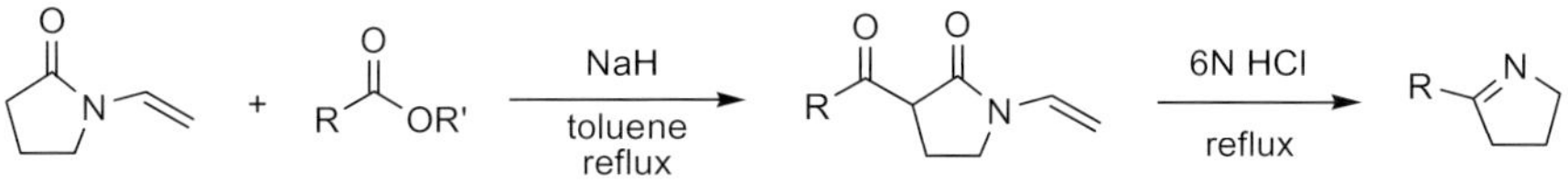

Scheme 8.6 Known preparation method for cyclic imines.

Approach C, direct asymmetric hydrogenation of pyrrole **4**, had no precedent at the time, and was deemed to require extensive development. Only recently, the first successful cases of asymmetric hydrogenation of certain pyrroles have been reported by Kuwano (Scheme 8.7) [9].

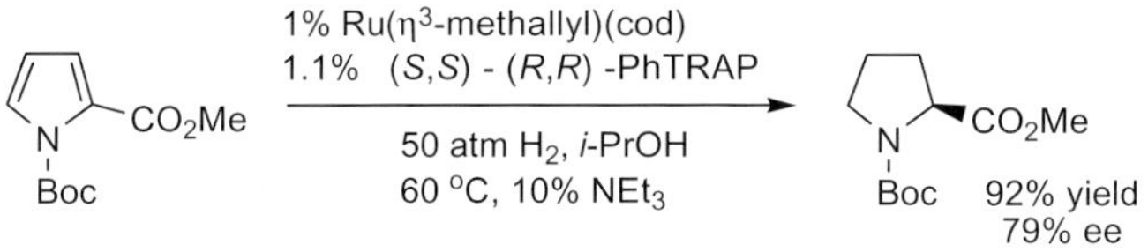

Scheme 8.7 Kuwano's asymmetric hydrogenation of pyrroles.

We became interested in a disconnection between the pyrrolidine and the aryl group (Approach D) as the most convergent method for enantioselective construction of **12** [10]. Although (–)-sparteine mediated enantioselective lithiation of *N*-Boc pyrrolidine **19** is well established by Beak [11], arylation of the resulting chiral

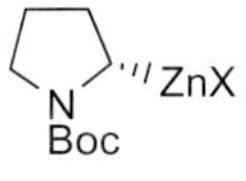

21

Figure 8.2 Chiral organozinc reagent.

2-pyrrolidinolithium was only known to provide racemic product in the presence of a Pd/Cu catalyst system [12]. Indeed, coupling reactions of secondary alkylmetal reagents typically encounter a multitude of challenges, such as configurational lability, propensity toward β-hydride elimination, and slow rate of transmetalation. Nevertheless, we decided to investigate the related organozinc reagent **21** (Figure 8.2) based on the following precedents:

1) Configurational integrity of secondary alkylzinc reagents had been demonstrated by Knochel although the preparation of the zinc reagent was not practical (Scheme 8.8) [13].

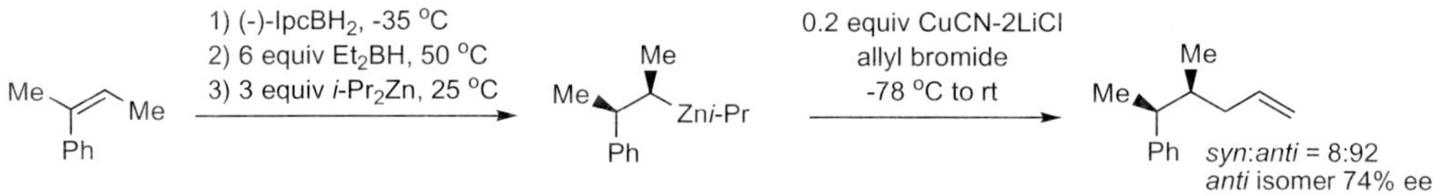

Scheme 8.8 Knochel's precedent for configurational integrity of organozinc reagents.

2) Pd-catalyzed coupling of racemic secondary alkylzincs with aryl halides had been reported by Hayashi (Scheme 8.9) [14].

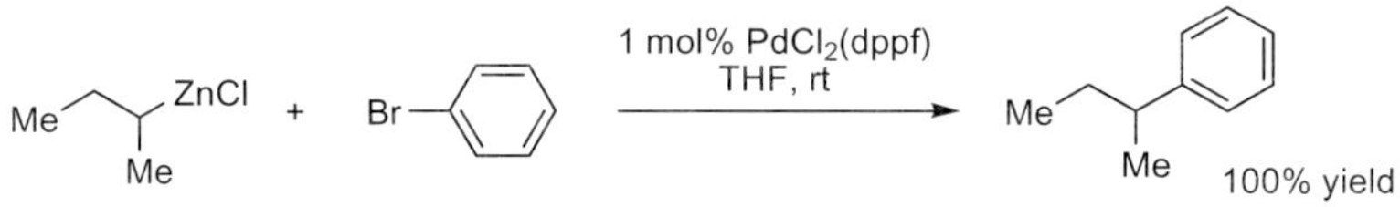

Scheme 8.9 Hayashi's cross coupling of *sec*-alkylzincs with aryl bromides.

3) Transmetalation with Zn on a chiral lithiated alkyl carbamate had been reported by Taylor via addition of a solution of $ZnCl_2$ to an *in situ* generated chiral organolithium (Scheme 8.10) [15].

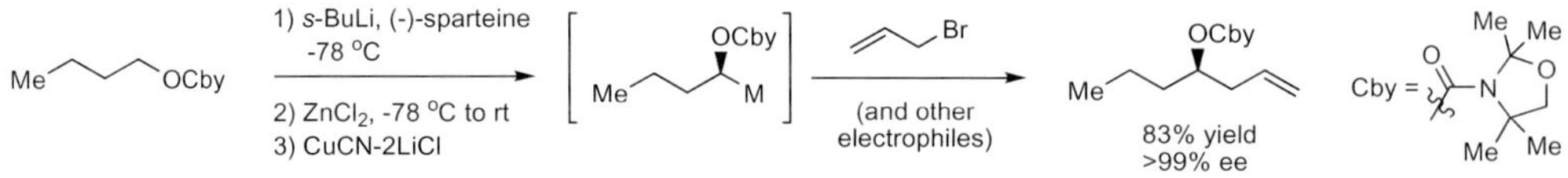

Scheme 8.10 Taylor's transformation of chiral organolithium to chiral organozinc.

To test this hypothesis, *N*-Boc pyrrolidine **19** was lithiated with *s*-BuLi and (–)-sparteine **20** and transmetalated *in situ* with $ZnCl_2$. We were very delighted to find that the resulting chiral organozinc reagent **21** could be coupled with aryl bromides using Pd catalysts comprising hindered electron-rich phosphine ligands such as *t*-Bu_3P, to provide 2-arylpyrrolidines in 92% ee with retention of stereointegrity throughout the process (Scheme 8.11) [16]. The development and scope of this new reaction will be discussed in more detail in Section 8.2.

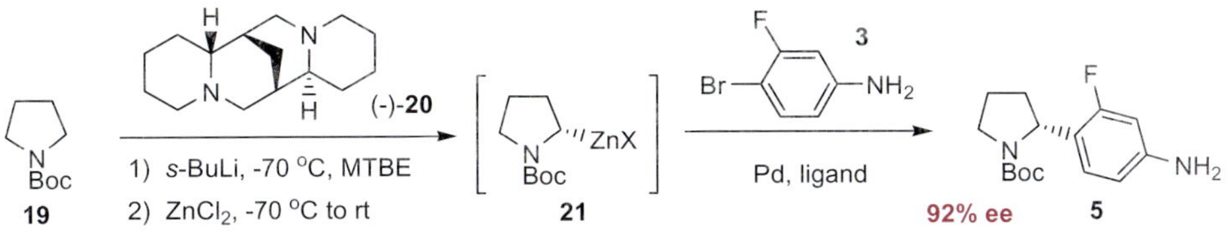

Scheme 8.11 Enantioselective coupling of *N*-Boc pyrrolidine with aryl bromide **3**.

Aryl halide **3** presented an additional challenge for the cross-coupling due to the presence of the acidic NH_2 group. Initial experiments using 2.5 mol% $Pd_2(dba)_3$ and 5 mol% *t*-Bu_3P-HBF_4 provided incomplete conversion of aryl halide **3** and a significantly lower yield of the product **5** (48%) than was observed in the case of simpler aryl halides such as bromobenzene (typically 82% yield) [16]. It was determined that the NH_2 group in aryl bromide **3** or the product **5** underwent slow yet competitive deprotonation during the Negishi coupling step. In addition to consuming the valuable chiral organozinc reagent, the deprotonated bromoaniline **3** underwent a much slower coupling reaction.

A more active catalyst was sought in order to increase the rate of Negishi coupling with respect to the rate of deprotonation of aryl bromide **3**. After screening several catalysts, it was found that $Pd(OAc)_2$ provided a significantly faster coupling rate and improved the yield of the coupled product from 48% to 74%, compared to $Pd_2(dba)_3$. It is possible that the acetate ion stabilizes the active catalyst (or a catalyst resting state) thereby preventing formation of inactive Pd black (formation of a palladacycle as a resting state from $Pd(OAc)_2$ and *t*-Bu_3P has been observed in a Negishi-type coupling reaction by Hartwig) [17]. With the more stable and reactive catalyst, the loading could be reduced from 5 mol% to 3 mol% without a significant decrease in the product yield. The use of the HBF_4 salt of *t*-Bu_3P, pioneered by Fu [18], provided a catalyst that was more practical and easier to handle than the highly air-sensitive *t*-Bu_3P free base.

The stoichiometry of $ZnCl_2$ was found to have a pronounced effect on the coupling reaction. Depending on the amount of $ZnCl_2$ added, either RZnCl, R_2Zn or R_3ZnLi species or their aggregates could theoretically be produced in the reaction mixture. It was found that the stoichiometry favoring the more basic "R_3ZnLi" species gave a poor yield of the coupled product **5**, presumably due to competitive deprotonation of the NH_2 group, while those favoring the less basic "R_2Zn" and "RZnCl" species provided optimal results.

Several stress tests were performed before scale-up. If the asymmetric deprotonation of **19** was carried out at −55 to −45 °C instead of −70 to −60 °C, product **5** from the coupling reaction was obtained in a significantly lower 85% ee although the yield was not affected (83%). An overcharge of (−)-sparteine provided only a marginal improvement in the ee of product **5**; for example, 1.1 equiv of sparteine provided a 93% ee while 0.90 equiv of sparteine gave a slightly eroded 89% ee.

With the safe operating range established, the coupling reaction was demonstrated in two 6.0 mol batches. A 1.3 M solution of *s*-BuLi was added to a solution of *N*-Boc pyrrolidine and (−)-sparteine in 15 litter of MTBE at −65 °C over 4 h. A 1 M solution of $ZnCl_2$ in diethyl ether was added over 2 h at the same temperature. After warming to +15 °C, solid aryl bromide **3** was added, followed by a mixture of solid $Pd(OAc)_2$ and the ligand. The reaction mixture was stirred at 20 °C for 15 h, quenched, filtered, extracted, and crystallized to provide the total of 2.13 kg of the coupled product **5** in 78–79% assay yield, 61–64% isolated yield, 91–93% ee and 99.1 wt% purity [19].

8.1.2.3 Elaboration of 12 to the Final Product 1

The protecting group issues in the synthesis of **1** were addressed next. The Medicinal Chemistry route (Scheme 8.1) resorted to multiple protecting group switches, primarily in order to ensure compatibility of the substrate with the relatively harsh nitration step, which was performed in fuming HNO_3. We speculated that the *N*-picolinyl group, containing an extremely electron-poor heteroaromatic ring (further deactivated by protonation), as well as the robust *N*-acetyl group should be compatible with the nitration reaction conditions. Thus, the use of protecting groups would be avoided altogether by strategically employing the *N*-picolinyl and *N*-acetyl groups, not only as building blocks for the final structure **1** but also as protecting group equivalents.

To test this hypothesis, the picolinamide **22** was prepared using *in situ* activated picolinic acid (Scheme 8.12). The *in situ* activation of picolinic acid was used because picolinyl chloride (available commercially as the HCl salt) is relatively expensive. The coupling reaction was not straightforward, and the best results were obtained by adding 1.4 equiv of thionyl chloride to a solution of 1.4 equiv of picolinic acid in acetonitrile, followed by addition of triethylamine. As soon as the addition of triethylamine was complete, aniline **5** was introduced immediately because the activated picolinic acid was unstable in the presence of triethylamine.

Picolinamide **22** could be isolated in 87% yield by crystallization from aqueous *i*-PrOH, which also resulted in an ee upgrade from 92% to 99.3%. Deprotection of the *N*-Boc group was performed by dissolving **22** in 5 M aqueous HCl. It was found that the subsequent *N*-acetylation of the revealed amine could be performed in the same pot under Schotten–Baumann conditions by simply adding 10 M aqueous NaOH and Ac_2O. After extraction with CH_2Cl_2, the organic phase was concentrated and used in the nitration step without any further purification.

The nitration step was the ultimate test of our decision to pursue a protecting group-free strategy. We were very pleased to find that the desired nitroarene

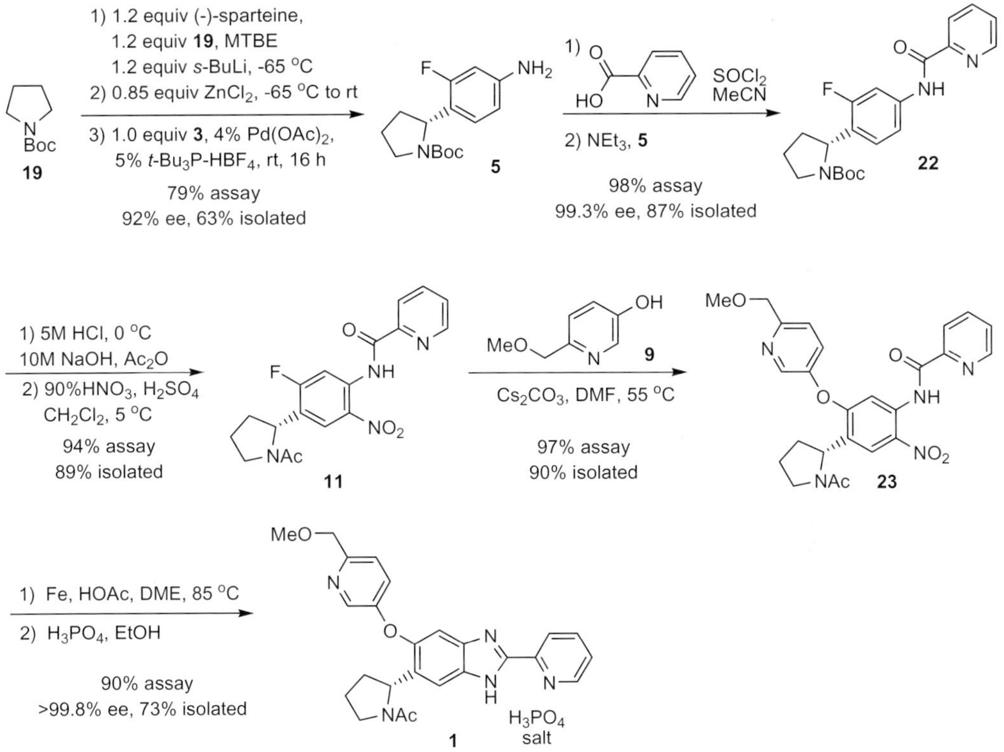

Scheme 8.12 Kilogram scale synthesis of **1**.

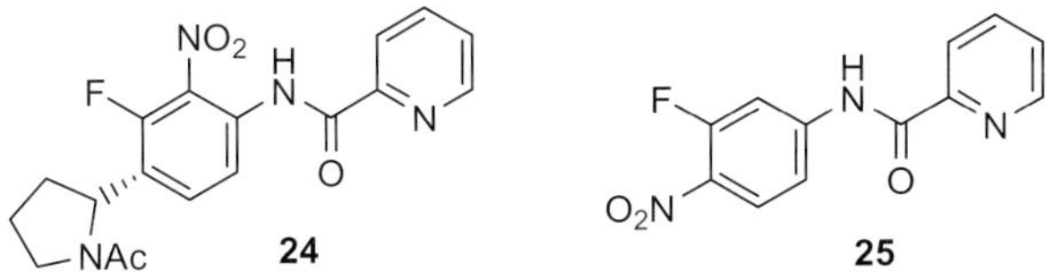

Figure 8.3 Identified impurities at nitration.

11 was formed in 94% assay yield using a mixture of 90% HNO_3 and concentrated H_2SO_4 as the nitrating reagent. Only two impurities were detected in the reaction mixture. The nitro-isomer **24** was formed in a 4% yield and could be partially rejected during crystallization of **11**. More interestingly, the product of an apparent *ipso*-nitration [20] **25** was also observed at a 1% level (Figure 8.3).

With the nitro group successfully introduced, the aromatic fluoride substituent in **11** was ready to undergo the nucleophilic aromatic substitution with the hydroxypyridine **9**. The reaction proceeded smoothly in DMF at 55 °C using an equimolar amount of cesium carbonate as the base and provided a 90% isolated yield of **23** after crystallization. With compound **23** in hand, only the reduction of the nitro group and cyclization to form the benzimidazole ring remained unsolved. To avoid the use of the toxic $SnCl_2$ reagent and the associated filtration and emulsion problems caused by the tin by-products, several alternatives for the reduction of the nitro group were explored. The preferred long-term solution, catalytic hydrogenation of **23**, suffered from catalyst poisoning, which resulted in incomplete conversion. Due to the time constraints, a temporary solution was found using iron powder in a mixture of acetic acid and DME. These conditions provided clean reduction of the nitro group as well as *in situ* acid-catalyzed cyclization directly to the desired product **1** in a 90% assay yield. Compound **1** was isolated as a phosphate salt (1 : 1 molar ratio) via slow addition of H_3PO_4 to a solution of the free base of **1** in a 73% yield, 99.4% purity and >99.8% ee [16].

8.1.2.4 **Summary of Process Development**

1) A short and practical synthesis of glucokinase activator **1** was developed utilizing a convergent strategy involving an S_NAr coupling of the activated aryl fluoride **11** with hydroxypyridine **9**.
2) The key to the success of the synthesis was the development of a novel method for enantioselective formation of α-arylpyrrolidines. In this method, (–)-sparteine-mediated, enantioselective lithiation of *N*-Boc-pyrrolidine **19** was followed by an *in situ* transmetallation to zinc and Pd-catalyzed coupling reaction with aryl bromide **3**, which afforded 2-arylpyrrolidine in 63% isolated yield and 92% ee. Notably, the acidic aniline NH_2 group was tolerated under the coupling reaction conditions.
3) The use of protecting groups in the synthesis was minimized by utilizing the *N*-picolinyl and *N*-acetyl groups not only as structural components of **1** but also to tune the reactivity of the intermediates.
4) Overall, 1.4 kg of compound **1** was prepared in excellent purity (>99% ee) and 31% yield over six steps.

8.2 Chemistry Development

8.2.1 Development of Enantioselective α-Arylation of *N*-Boc Pyrrolidines

The novel enantioselective α-arylation of *N*-Boc pyrrolidine **19** was initially optimized using bromobenzene as the arylating agent. The (–)-sparteine-mediated enantioselective lithiation of **19** was performed according to literature precedent by Beak [11a], and the resulting anion was treated with 1 equiv of $ZnCl_2$. The presumed organozinc reagent **21** was warmed to room temperature, to provide a homogeneous solution, which was readily subdivided for screening with a variety of palladium catalysts in the arylation with bromobenzene (Table 8.1) [16]. Despite the failure of $PdCl_2$(dppf) (entry 1), historically the catalyst of choice for Negishi

Table 8.1 Optimization of enantioselective α-arylation of *N*-Boc pyrrolidine.

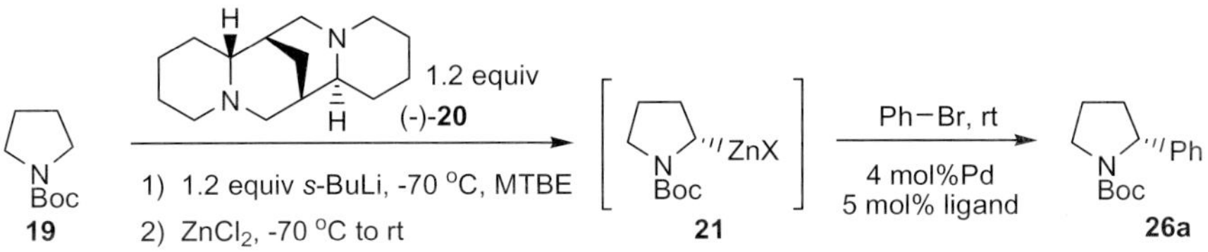

Entry	Ligand	Pd source	$ZnCl_2$ Equiv	% Yield (% ee)
1	–	$PdCl_2(dppf)$	1.0	<5 (nd)
2	1,1′-di-(t-Bu_2P)ferrocene	$Pd(OAc)_2$	1.0	<5 (nd)
3	Cy_3P-HBF_4	$Pd(OAc)_2$	1.0	12 (92)
4	t-Bu_2PMe-HBF_4	$Pd(OAc)_2$	1.0	<5 (nd)
5	t-Bu_3P-HBF_4	$Pd(OAc)_2$	1.0	83 (92)
6	Ru-phos	$Pd(OAc)_2$	1.0	80 (92)
7	Q-phos	$Pd(OAc)_2$	1.0	80 (92)
8	t-Bu_3P-HBF_4	Pd_2dba_3	1.0	82 (91)
9	t-Bu_3P-HBF_4	$PdCl_2$	1.0	70 (92)
10	–	$Pd(t\text{-}Bu_3P)_2$	1.0	70 (92)
11	–	$[PdBr(t\text{-}Bu_3P)]_2$	1.0	78 (92)
12	t-Bu_3P-HBF_4	$Pd(OAc)_2$	0.1	<5 (nd)
13	t-Bu_3P-HBF_4	$Pd(OAc)_2$	0.3	79 (92)
14	t-Bu_3P-HBF_4	$Pd(OAc)_2$	0.6	80 (92)

couplings [14], Pd-catalysts derived from Buchwald's Ru-phos [21], Hartwig's Q-phos [22], and Fu's t-Bu_3P-HBF_4 [18] delivered arylated product **26a** in good yield, and with the high enantioselectivity established in the deprotonation (entries 5–7, enantiomeric excess was determined by chiral HPLC on Chiralcel AD-H). The absolute configuration of **26a** was assigned by deprotection to 2-phenylpyrrolidine and comparison of the optical rotation with those reported in the literature, confirming that the transmetallation/Negishi coupling occurred with retention of configuration. For reasons of cost, availability and practicality, t-Bu_3P-HBF_4 was selected for further development.

The nature of the palladium source was found to have a profound effect on the rate of the coupling reaction. In particular, $Pd(OAc)_2$ provided a significantly faster reaction rate than all other palladium sources [17]. It is interesting to note that either a 1 : 1 or 2 : 1 ratio of ligand to Pd provided competent *in situ* generated catalysts; however the preformed catalyst $Pd[(Pt\text{-}Bu_3)_2]$ [23] afforded ~80% conversion whereas with $[PdBr(Pt\text{-}Bu_3)]_2$ [24], the reaction went to completion. These observations indicate that the acetate plays an important role in the catalytic system.

Because of the potential to form different organozinc species (RZnCl, R_2Zn, or R_3ZnLi), an examination of the dependence of yield and selectivity on the stoichiometry of $ZnCl_2$ was undertaken (entries 12–14). Interestingly, in the case of PhBr, the amount of $ZnCl_2$ could be reduced to as low as 0.33 equiv (with respect to **19**) without significant change in yield or enantioselectivity.

8.2.2
Scope of Enantioselective α-Arylation of *N*-Boc Pyrrolidines

2-Arylpyrrolidines constitute an important structural motif in biologically active compounds and effective chiral controllers in asymmetric synthesis [25]. Few methods exist for the synthesis of these privileged structures, and all suffer from either long synthetic sequences, low yields, lack of generality, or modest enantioselectivity [5, 26]. We were delighted to find that the enantioselective deprotonation/transmetallation/Negishi coupling was quite general, affording a diverse array of 2-arylpyrrolidines in good yield and 92% ee (Table 8.2). Not only were a variety of aryl bromides tolerated, but preliminary results suggested that aryl chlorides were also suitable coupling partners (entry 2, 60 °C). Phenyl triflate and phenyl tosylate were unreactive under the standard conditions.

Although 0.33 equiv of $ZnCl_2$ could be used in the coupling with bromobenzene, substrates containing acidic functionalities were incompatible with these reaction conditions, presumably due to proton transfer. Surprisingly, this could be circumvented simply by changing the amount of $ZnCl_2$. For example, the coupling of 4-bromoaniline with 0.33 equiv of $ZnCl_2$ provided only 18% of the desired product; however employing 0.6 equiv of $ZnCl_2$ under the same reaction conditions delivered arylated product **26g** in 70% yield (entry 10). Even unprotected indoles were tolerated with this protocol, providing adducts such as **26l** in good yield (entry 15).

The consistent observation of the arylated products with 92% ee confirms that the enantioselectivity of the asymmetric deprotonation was preserved during the transmetalation with $ZnCl_2$ and retained during the Pd-catalyzed coupling. In fact, the Negishi coupling with 3-bromopyridine (entry 16) was performed at 60 °C, and still provided **26m** in 92% ee, which constitutes a formal total synthesis of (*R*)-nicotine [27].

More recently, Coldham and O'Brien have extended this methodology to the arylation of *N*-Boc piperidine [28].

Having demonstrated a practical and reliable method to access 2-arylpyrrolidines in high enantioselectivity, we felt that a noteworthy extension of this methodology would lie in its application to bis-arylated products **27**, providing a rapid and efficient approach to enantiopure C_2-symmetric 2,5-diarylpyrrolidines, which have been identified as valuable chiral auxiliaries and chiral ligand manifolds [29]. Towards this end, substrate **26a** was subjected to the standard arylation conditions, which produced 2,5-diphenyl-*N*-Boc-pyrrolidine **27** in a 96 : 4 diastereomeric ratio, and 57% isolated yield (*s*-BuLi/TMEDA produced **27** in lower d.r. (66 : 34) and yield (42%)), as depicted in Scheme 8.13.

We have been delighted to see that this methodology, developed at Merck, has found its way into academic research. The α-arylation methodology has been

Table 8.2 Scope of Enantioselective α-Arylation of *N*-Boc Pyrrolidine.

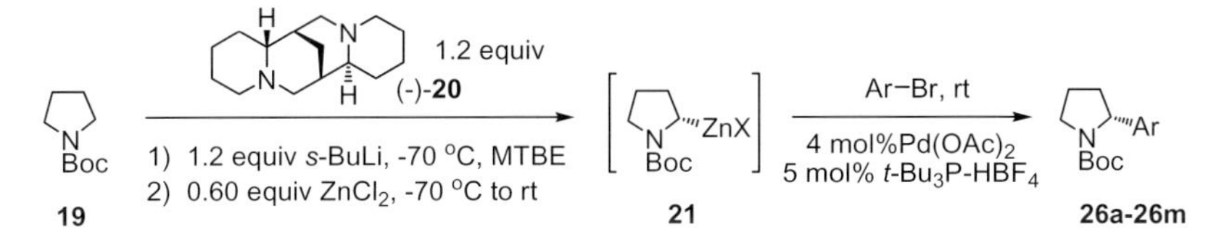

Entry		Ar	Product	Yield (ee), %
1	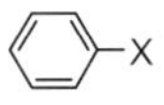	X = Br	**26a**	72 (92)
2		Cl	**26a**	48 (92)
3		OTf	**26a**	<5%
4		OTs	**26a**	<5%
5	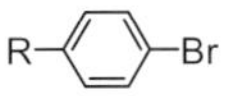	R = F	**26b**	75 (92)
6		NMe_2	**26c**	78 (92)
7		CO_2Me	**26d**	81 (92)
8		SO_2Me	**26e**	87 (92)
9		CN	**26f**	80 (92)
10		NH_2	**26g**	70 (92)
11	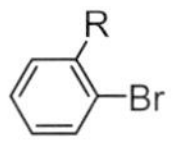	R = Me	**26h**	71 (92)
12		OMe	**26i**	72 (92)
13	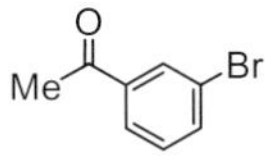		**26j**	78 (92)
14	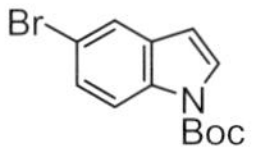		**26k**	81 (92)
15	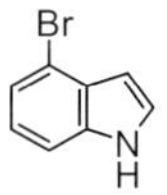		**26l**	77 (92)
16	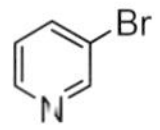		**26m**	60 (92)

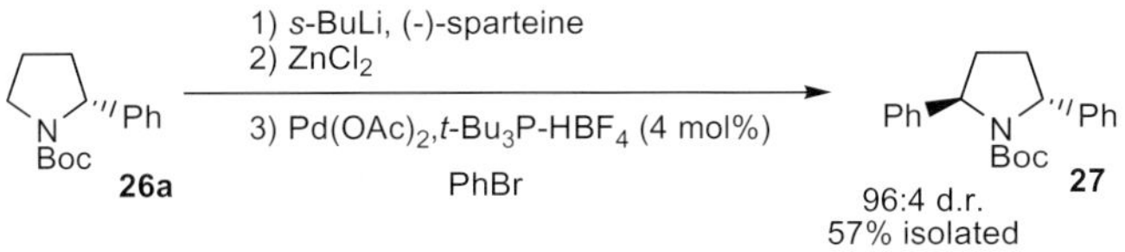

Scheme 8.13 Preparation of C_2-symmetric 2,5-diphenylpyrrolidine.

applied by Jacobsen in the synthesis of a chiral thiourea catalyst for enantioselective additions to oxocarbenium ions (Figure 8.4) [30].

Collaboration has been established with O'Brien's group to develop a catalytic asymmetric version of the reaction [31]. O'Brien has also used compound **26a** to

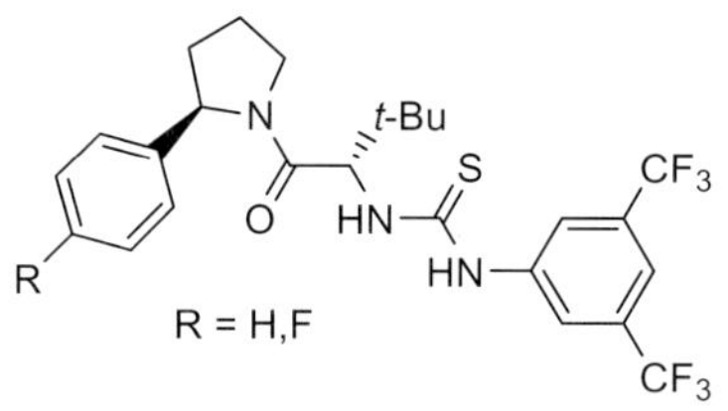

Figure 8.4 Application by Jacobsen.

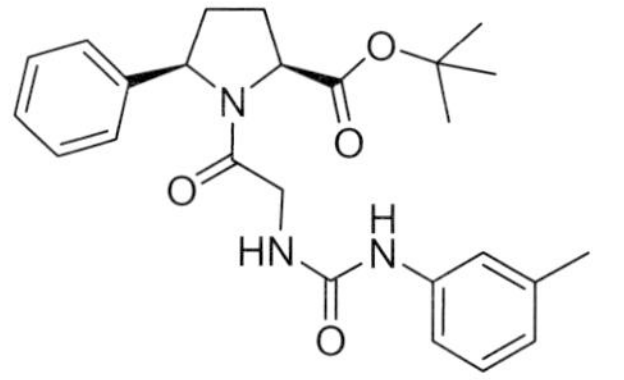

Figure 8.5 (+)-RP 66803.

prepare a key intermediate in the synthesis of the CCK antagonist (+)-RP 66803 (Figure 8.5) [32].

8.2.3 Detailed Examination of the Coupling Reaction

Although impurities generated in the coupling of *N*-Boc pyrrolidine and aryl bromide **3** were cleanly rejected during crystallization of **5** and, therefore, did not affect the quality of the product, we decided to isolate and identify these impurities for better understanding of the coupling reaction (Scheme 8.14). The debrominated compound **28** was observed at a very low level (<1%). The enamide impurity **29** (3–4%) seemed to arise from β-hydride elimination, and the unusual impurity **30** (3–4%) probably resulted from a subsequent reductive Heck reaction of enamide

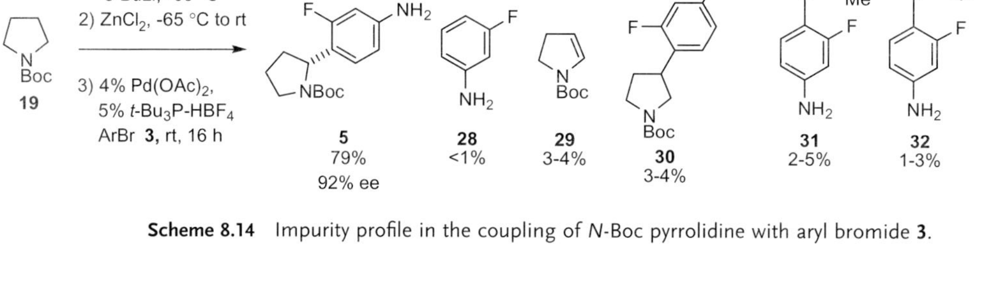

Scheme 8.14 Impurity profile in the coupling of *N*-Boc pyrrolidine with aryl bromide **3**.

29 with aryl bromide 3. Since a slight excess of *s*-BuLi was utilized in the lithiation step, some of the *s*-Bu coupled product 31 (2–5%) was observed.

Unexpectedly, the *n*-Bu impurity 32 (1–3%) was also isolated, which prompted a further investigation into the coupling of *s*-BuLi with aryl bromide 3 under the same conditions that were optimized for the preparation of 5 (Scheme 8.15). The arylation of *s*-BuLi turned out to be significantly less efficient than the coupling of metalated *N*-Boc-pyrrolidine 19, which suggests that the *N*-Boc group plays an important role in the Pd-catalyzed coupling reaction. Furthermore, a high level of *n*-Bu/*s*-Bu scrambling was observed, suggesting extensive β-hydride elimination/hydropalladation [33], not seen with metalated *N*-Boc-pyrrolidine 19. If the coupling of *s*-BuLi was performed in the absence of (–)-sparteine, even higher levels of the *n*-Bu product 32 were obtained and the Pd catalyst rapidly decomposed. These observations indicate that both sparteine and the *N*-Boc group in the substrate exert a beneficial effect on the Pd-catalyzed coupling reaction.

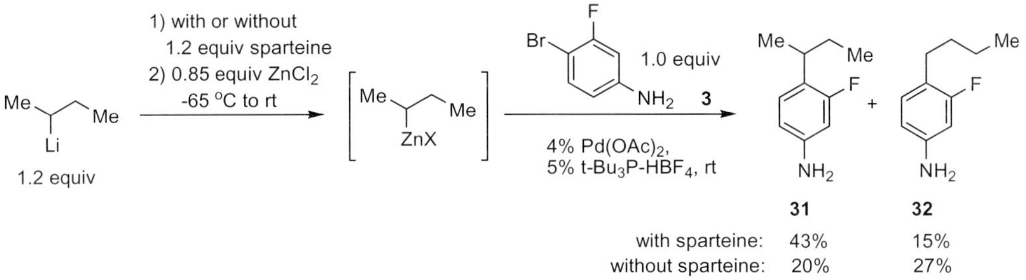

Scheme 8.15 Arylation of *s*-BuLi.

8.3 Conclusion

A short and practical synthesis of glucokinase activator 1 was demonstrated on a kilogram scale utilizing a convergent strategy involving an S_NAr coupling of the activated aryl fluoride 11 with hydroxypyridine 9. The key to the success was the development of an unprecedented asymmetric arylation of *N*-Boc-pyrrolidine, that relies on a (–)-sparteine-mediated asymmetric deprotonation, followed by transmetalation with $ZnCl_2$, and subsequent Pd-catalyzed Negishi coupling with an aryl bromide. In our opinion, the synthesis of 1 highlights the benefits of the problem-driven approach towards the development of new synthetic methodology.

The new method for asymmetric arylation was applicable to a host of aryl halides, to provide a diverse array of 2-arylpyrrolidines in good yield and 92% ee, regardless of the nature of the aryl bromide component. This sequence offers a number of advantages over existing methods, and represents the most convenient and practical synthesis of enantiopure 2-aryl pyrrolidines and 2,5-diarylpyrrolidines.

Acknowledgments

We thank Kevin Campos, Jacob Waldman, Daniel Zewge, Peter Dormer and Cheng-yi Chen for the key contributions to the project and for providing assistance with this manuscript.

References

1 Matschinsky, F. (2009) *Nat. Rev. Drug Discov.*, **8**, 399.

2 Nonoshita, K., Ogino, Y., Ishikawa, M., Sakai, F., Nakashima, H., Nagae, Y., Tsukahara, D., Arakawa, K., Nishimura, T., and Eiki, J. (2005) Patent Application WO 2005063738.

3 (a) Takeda, Y., Uoto, K., Chiba, J., Horiuchi, T., Iwahana, M., Atsumi, R., Ono, C., Terasawa, H., and Soga, T. (2003) *Bioorg. Med. Chem.*, **11**, 4431; (b) Akita, H., Takano, Y., Nedu, K., and Kato, K. (2006) *Tetrahedron Asym.*, **17**, 1705.

4 Koenig, T. (1966) *J. Am. Chem. Soc.*, **88**, 4045.

5 Brinner, K.M., and Ellman, J.A. (2005) *Org. Biomol. Chem.*, **3**, 2109 and references therein.

6 (a) Verdaguer, X., Lange, U.E.W., Reding, M.T., and Buchwald, S.L. (1996) *J. Am. Chem. Soc.*, **118**, 6784; (b) Willoughby, C.A., and Buchwald, S.L. (1992) *J. Am. Chem. Soc.*, **114**, 7562.

7 (a) Brunner, H., Kuerzinger, A., Mahboobi, S., and Wiegriebe, W. (1988) *Arch. Pharm.*, **321**, 73; (b) Kuwano, R., Karube, D., and Ito, Y. (1999) *Tetrahedron Lett.*, **40**, 9045.

8 Sorgi, K.L., Maryanoff, C.A., McComsey, D.F., and Maryanoff, B.E. (1998) *Org. Synth.*, **75**, 215.

9 Kuwano, R., Kashiwabara, M., Ohsumi, M., and Kusano, H. (2008) *J. Am. Chem. Soc.*, **130**, 808.

10 Campos, K.R. (2007) *Chem. Soc. Rev.*, **36**, 1069.

11 (a) Kerrick, S.T., and Beak, P. (1991) *J. Am. Chem. Soc.*, **113**, 9708; (b) O'Brien, P., and McGrath, M.J. (2005) *J. Am. Chem. Soc.*, **127**, 16378; (c) Coldham, I., Dufour, S., Haxell, T.F.N., Patel, J.J., and Sanchez-Jimenez, G. (2006) *J. Am. Chem. Soc.*, **128**, 10943; for an alternative approach to the undesired enantiomer of 2-aryl-*N*-Boc-pyrrolidines involving asymmetric deprotonation/ intramolecular alkylation of *N*-(arylmethyl)-*N*-(3-chloropropyl)- *N*-Boc-amines, see: (d) Wu, S., Lee, S., and Beak, P. (1996) *J. Am. Chem. Soc.*, **118**, 715.

12 Dieter, R.K., and Li, S. (1997) *J. Org. Chem.*, **62**, 7726.

13 Boudier, A., Flachsmann, F., and Knochel, P. (1998) *Synlett*, 1438.

14 Hayashi, T., Konishi, M., Kobori, Y., Kumada, M., Higuchi, T., and Hirotsu, K. (1984) *J. Am. Chem. Soc.*, **106**, 158.

15 Papillon, J.P.N., and Taylor, J.K. (2002) *Org. Lett.*, **4**, 119.

16 Campos, K.R., Klapars, A., Waldman, J.H., Dormer, P.G., and Chen, C.-y. (2006) *J. Am. Chem. Soc.*, **128**, 3538.

17 Wu, L., and Hartwig, J.F. (2005) *J. Am. Chem. Soc.*, **127**, 15824.

18 Netherton, M.R., and Fu, G.C. (2001) *Org. Lett.*, **3**, 4292.

19 Klapars, A., Campos, K.R., Waldman, J.H., Zewge, D., Dormer, P.G., and Chen, C.-y. (2008) *J. Org. Chem.*, **73**, 4986.

20 For precedents of ipso-nitration, see: (a) Moodie, R.B., and Schofield, K. (1976) *Acc. Chem. Res.*, **9**, 287; (b) Malecki, N., Carato, P., Houssin, P.C., and Hénichart, J.-P. (2005) *Monatsh. Chem.*, **136**, 1601.

21 Milne, J., and Buchwald, S.L. (2004) *J. Am. Chem. Soc.*, **126**, 13028.

22 Hama, T., Liu, X., Culkin, D.A., and Hartwig, J.F. (2003) *J. Am. Chem. Soc.*, **125**, 11176.

23 Dai, C., and Fu, G.C. (2001) *J. Am. Chem. Soc.*, **123**, 2719.

24 Stambuli, J.P., Kuwano, R., and Hartwig, J.F. (2002) *Angew. Chem. Int. Ed.*, **41**, 4746.

25 (a) Lewis, J.R. (2001) *Nat. Prod. Rep.*, **18**, 95; (b) Elliot, R.L., Kopeka, H., Lin, N.-H., He, Y., and Garvey, D.S. (1995) *Synthesis*, 772; (c) Lin, N.-H., Carrera, G.M., Jr., and Anderson, D.J. (1994) *J. Med. Chem.*, **37**, 3542; (d) Higashiyama, K., Inoue, H., and Takahashi, H. (1994) *Tetrahedron*, **50**, 1083 and references cited therein.

26 Wu, S., Lee, S., and Beak, P. (1996) *J. Am. Chem. Soc.*, **118**, 715 and references cited therein.

27 Girard, S., Robins, R.J., Villiéras, J., and Lebreton, J. (2000) *Tetrahedron Lett.*, **41**, 9245.

28 (a) Coldham, I., and Leonori, D. (2008) *Org. Lett.*, **10**, 3923. (b) Stead, D., Carbone, G., O'Brien, P., Campos, K.R., Coldham, I., and Sanderson, A. (2010) *J. Am. Chem. Soc.*, **132**, 7260.

29 (a) Kozmin, S.A., and Rawal, V.H. (1997) *J. Am. Chem. Soc.*, **119**, 7165; (b) He, S., Kozmin, S.A., and Rawal, V.H. (2000) *J. Am. Chem. Soc.*, **122**, 190; (c) Choi, Y.H., Choi, J.Y., Yang, H.Y., and Yong, H. (2002) *Tetrahedron Asym.*, **13**, 801.

30 Reisman, S.E., Doyle, A.G., and Jacobsen, E.N. (2008) *J. Am. Chem. Soc.*, **130**, 7198.

31 McGrath, M.J., and O'Brien, P. (2005) *J. Am. Chem. Soc.*, **127**, 16378.

32 Stead, D., O'Brien, P., and Sanderson, A. (2008) *Org. Lett.*, **10**, 1409.

33 Luo, X., Zhang, H., Duan, H., Liu, Q., Zhu, L., Zhang, T., and Lei, A. (2007) *Org. Lett.*, **9**, 4571.

9
CB1R Inverse Agonist–Taranabant

Debra Wallace

Sedentary lifestyles and easy access to high energy foods have led to a sharp increase in obesity throughout the developed world during the past 20 years; currently over 65% of adults in the US are overweight, with 30% classified as obese [1]. Obesity can also lead to a number of comorbidities including diabetes, hypertension, cardiovascular disease, cancer, and arthritis. The cannabinoid receptor system has been implicated in the regulation of feeding behavior [2], and hence selective cannabinoid-1 receptor (CB1R) inverse agonists are expected to be efficacious for suppression of food intake and thus weight reduction. Discovery efforts at Merck Research Laboratories identified taranabant **1** (Figure 9.1), as a potential selective CB1R inverse agonist [3], for the treatment of obesity, necessitating development of a synthesis of **1** to support both pre-clinical and clinical trials.

In this chapter the development of the synthesis of taranabant will be presented. In Section 9.1, we focus on evaluation and optimization of the Medicinal Chemistry route, and development of an asymmetric version of this initial route. In Section 9.2 the discovery and implementation of a new asymmetric route will be discussed, and extensions of the utility of this chemistry will also be discussed. Finally, the factors involved in selecting a route as a potential manufacturing approach for taranabant are presented.

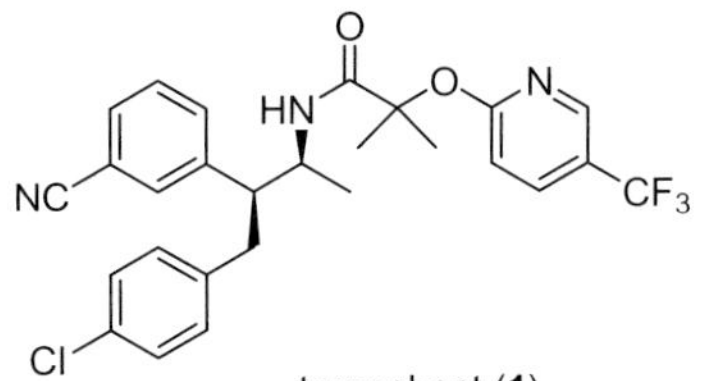

taranabant (**1**)

Figure 9.1 Structure of taranabant (**1**).

The Art of Process Chemistry. Edited by Nobuyoshi Yasuda

ISBN: 978-3-527-32470-5

9.1
Project Development

9.1.1
Introduction

Taranabant (**1**) is a hindered chiral secondary amide bearing two contiguous stereocenters, requiring any long-term route to address control of both absolute and relative stereochemistry. The molecule contains a number of sites that could be labile under acid or basic hydrolytic conditions, such as the aryl nitrile, the amide connection and the activated 2-position on the pyridine ring. Additionally, some functionalities were also anticipated to be reactive under hydrogenation or chemical reduction conditions, including the aryl nitrile and chloride groups. Given these concerns, developing a synthesis of taranabant that would be suitable for large scale implementation was anticipated to be a significant challenge for the process research group.

9.1.2
Medicinal Chemistry Route

The initial route to taranabant relied on a late stage amide bond coupling between racemic amine *rac*-**2** and pyridine acid **3** mediated by (benzotriazol-1-yloxy)tripyrrolidinophosphonium hexafluorophosphate (Py-Bop), followed by chiral HPLC separation of the product to afford a single enantiomer (Scheme 9.1).

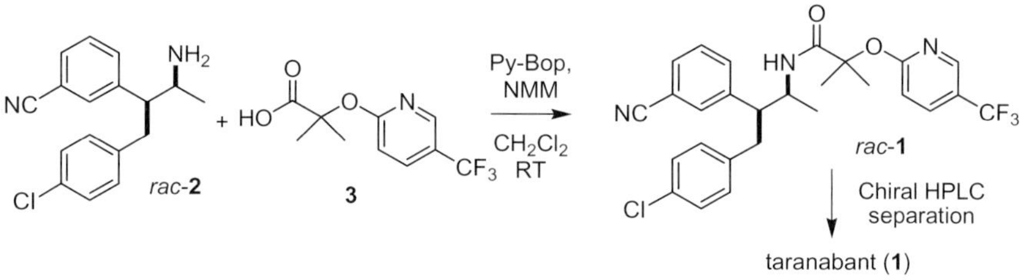

Scheme 9.1 Final steps in the Medicinal Chemistry synthesis of taranabant **1**.

The synthesis of the amine *rac*-**2** started with a tin-mediated coupling between 3-bromobenzonitrile and isopropenyl acetate to afford the phenyl acetone **4** [4]. Alkylation of phenylacetone **4** with 4-chlorobenzyl chloride under phase transfer conditions afforded the substituted compound *rac*-**5** along with quantities of bis-alkylated products and residual starting material. Reduction of the carbonyl group in *rac*-**5** using L-Selectride® followed by a standard oxidative work-up using basic hydrogen peroxide proceeded to give the desired *anti*-alcohol *rac*-**6** as the major disastereoisomer, in around 95% selectivity. Subsequent activation of the alcohol as its mesylate and S_N2 displacement with sodium azide afforded the *syn*-azide *rac*-**7**. Platinum oxide-mediated hydrogenation of the azide with concomitant *in situ* protection of the resulting amine with di-*tert*-butyl dicarbonate afforded the protected amine which was deprotected under acidic conditions to give the desired penultimate **2** as a racemate (Scheme 9.2).

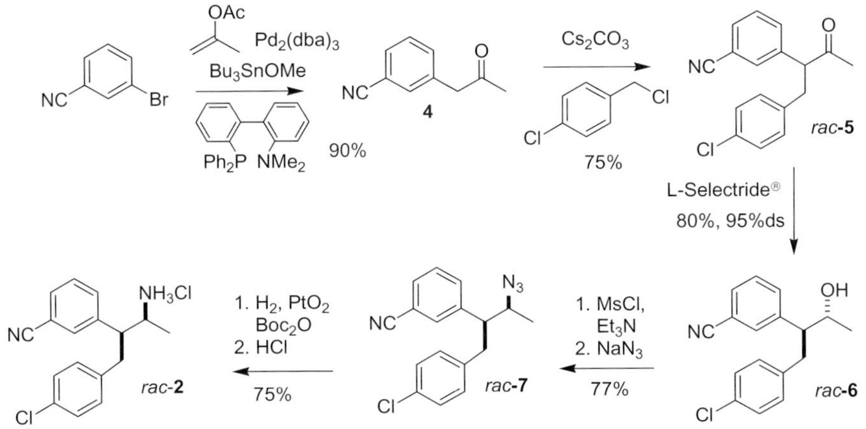

Scheme 9.2 Racemic synthesis of amine rac-**2**.

The synthesis of the pyridine acid **3** is outlined in Scheme 9.3. The 2-chloro-5-trifluoromethylpyridine was coupled with hydroxybutyric acid in the presence of an excess of sodium hydride to afford acid **3** in around 35% yield after necessary purification.

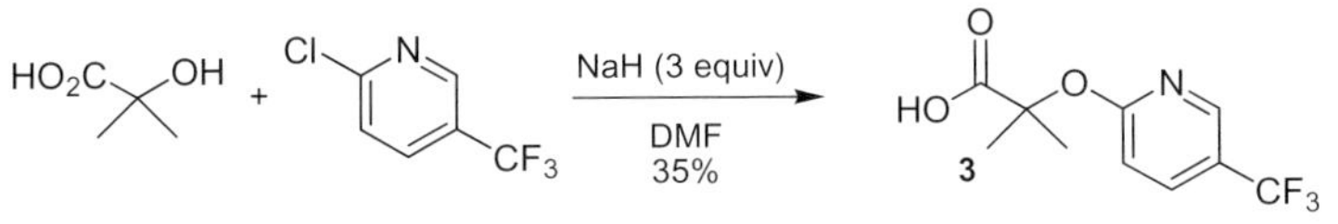

Scheme 9.3 Medicinal Chemistry approach to acid **3**.

9.1.3
Initial Strategy–Amide Bond Formation as the Final Step

The route described proved suitable for generation of multi-gram quantities of taranabant and allowed generation of sufficient data to confirm the molecule as a viable development compound. Formation of the amide bond as the final step in the synthesis allowed a convergent approach and was felt to be the best initial disconnection. Additionally, it was felt that synthesis of the acid **3** should be relatively straightforward subject to removal of the hazardous sodium hydride, and optimization was not deemed to be a major issue. However, for longer term implementation, a number of issues would have to be addressed surrounding the synthesis of amine **2** and the final steps of the synthesis. In evaluating the synthetic route the following were noted as short or longer term issues (Table 9.1).

For initial scale-up work the team recognized that addressing all the above issues would not be time efficient and our initial focus was on the two major issues, that

Table 9.1 Issues identified with the Medicinal Chemistry approach to **1**.

Development issue	Impact
Volume inefficient chiral chromatography required to deliver a single enantiomer	Separation at final API necessitated processing excess material through the synthesis to give the desired amount
Stoichiometric use of a tin reagent to prepare the phenylacetone **4**	Toxicity of tin compounds, and operational inefficiency based on high molecular weight of reagent and excessive chromatography required to reject tin residues
Chromatographic purification required for each intermediate, the majority of which were oils	Purification options would be limited
Cryogenic conditions (−78 °C) and an expensive reagent (L-Selectride®) to obtain high selectivity for carbonyl reduction in **5**	Additional cost for cryogenic step
Incompatibility of the nitrile group with the oxidative work-up of the L-Selectride® reduction using basic hydrogen peroxide	Significant hydrolysis to the primary amide with concomitant exothermic activity and loss of product
Use of sodium azide to introduce the nitrogen atom	Safety issues surrounding sodium azide and hydrazoic acid
Boc-protection needed during reductive hydrogenation of the azide group to prevent reaction medium becoming basic	Competitive de-chlorination of the aryl chloride group occurs under basic conditions
Use of sodium hydride in DMF to prepare the pyridine acid **3**	Safety concerns with use of sodium hydride, and a low yield of acid
Use of an expensive coupling reagent to promote the amide bond formation	A long-term cost implication

is, the removal of the tin-mediated reaction and improving or replacing the chiral chromatography. Further or more significant synthetic improvements would be initiated as the compound progressed through development.

9.1.3.1 Amide Bond Formation as the Final Step–Classical Resolution Approach

9.1.3.1.1 Choice of Starting Material

In considering an appropriate starting material for the synthesis of taranabant the commercial availability of a number of 3-substituted benzonitriles was evaluated, however, with the exception of the previously employed 3-bromobenzonitrile, cost effective options from recognized large-scale suppliers were limited (Figure 9.2).

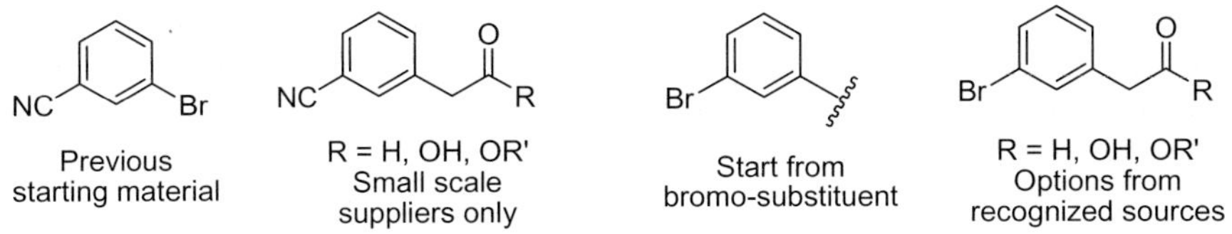

Figure 9.2 Potential starting materials for synthesis of amine **2**.

In contrast, a number of 3-substituted bromobenzenes were available on scale and it was felt that conversion of the bromide to the desired aryl nitrile group should be possible later in the synthesis, and might even offer some advantages. Indeed, in an alternate approach from the Medicinal Chemistry group such a strategy had also been employed, starting from methyl 3-bromophenylacetate (**8**), however, issues with the late stage conversion of the bromo group in amine **9**, to the cyano group had been encountered and the previously describe route used in preference (Scheme 9.4). We felt that 3-bromophenylacetic acid (**10**) should also be a suitable starting material and that the initially formed substituted acid **11** was a candidate for a classical resolution. Additionally, a number of intermediates are potential substrates for installation of the aryl nitrile group and any issues with reproducibility of this reaction were anticipated to be surmountable.

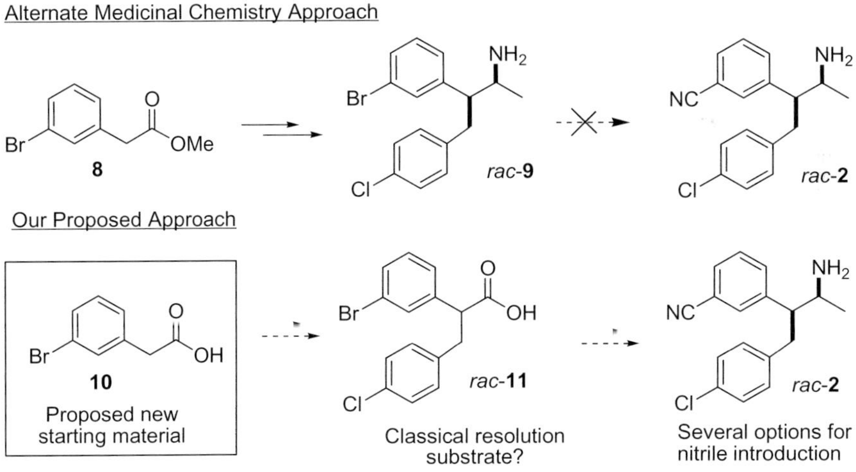

Scheme 9.4 Proposed route to amine **2** starting from 3-bromophenylacetic acid.

9.1.3.1.2 **Synthesis and Resolution of Acid 11**

Alkylation of the dianion of 3-bromophenylacetic acid (**10**) with *p*-chlorobenzyl chloride was accomplished using LiHMDS in THF to afford the substituted acid **11**. The only significant impurity was the bis-alkylated product, which could be minimized by control of the reaction temperature to below −20 °C to give an 88%

assay yield of the desired acid. A salt screen to explore potential resolutions led to identification of an (*S*)-α-methylbenzylamine salt **12** as the most promising lead. Diastereoselectivities of 96 : 4 could be obtained in EtOH, though with concomitant high loss of material. Further evaluation identified MeOH as a better solvent for recovery, albeit with a reduced selectivity requiring a second re-crystallization. After a second crystallization from MeOH a 30% yield of upgraded salt (97% ds in **12**, corresponding to 94% ee acid **11**) was obtained (Scheme 9.5).

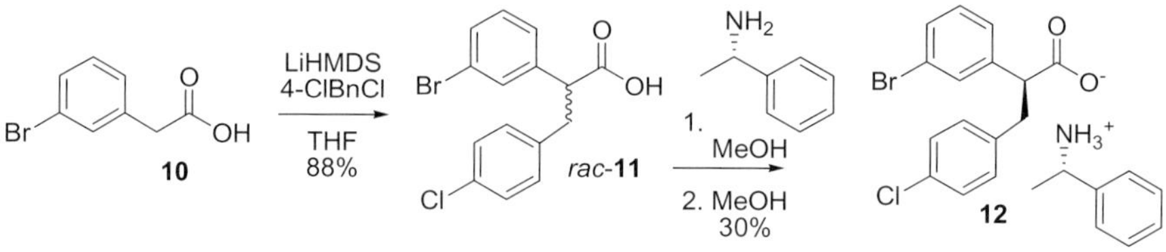

Scheme 9.5 Synthesis and resolution of acid **11**.

9.1.3.1.3 **Introduction of Nitrile Group**

A number of intermediates were screened as potential substrates for conversion of the aryl bromide to a nitrile group (Figure 9.3). Conversion of ketone **13** was high yielding, however, the basic reaction conditions typically employed for these cyanations led to some loss of enantiomeric purity. Attempted cyanation of the mesylate **14** led to partial elimination under the basic reaction conditions to give an alkene and, hence, only modest yield of the desired compound. In order to avoid excessive handling of potentially hazardous azide-containing intermediates, the cyanation of azide **15** was not considered. Data from Medicinal Chemistry had indicated difficulty converting the amine **9** and as such, conversion of the bromo-alcohol **16** to the corresponding cyano-alcohol **6** was found to be optimum.

After salt break chiral acid **11** was converted to the methyl ketone **13** in essentially quantitative yield via the intermediacy of the Weinreb amide and processed as an oil without further purification (Scheme 9.6). The carbonyl group in **13** was then reduced to the secondary alcohol **16** using L-Selectride® as previously described in 97% assay yield. We were pleased to find that an attainable reaction temperature of −50 °C was sufficient to obtain high selectivity of >98 : 2, and that, in the absence of the reactive nitrile group, no issues were observed or during

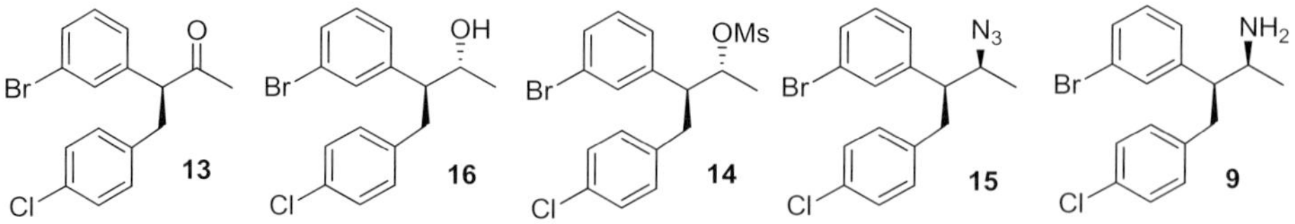

Figure 9.3 Potential substrates for introduction of the nitrile group.

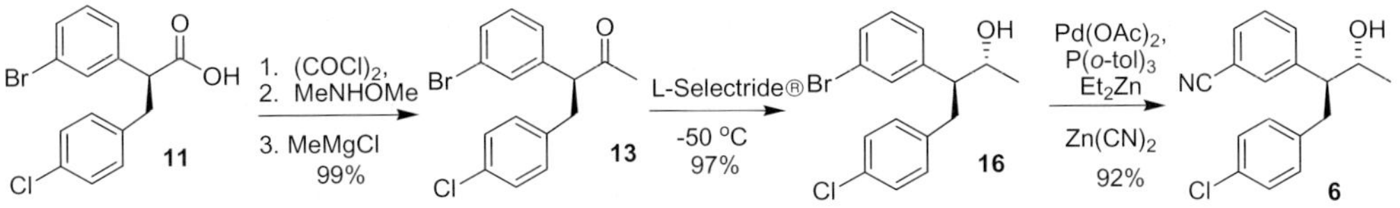

Scheme 9.6 Synthesis of alcohol **6** from resolved acid **11**.

oxidative work-up. The alcohol–an oil–was also processed through subsequent steps without any chromatographic purification. The cyanation was initially carried out using previously published conditions involving $Pd_2(dba)_3$ as catalyst, dppf as ligand, in the presence of zinc cyanide in DMF at 115 °C [5]. Unfortunately, the dppf ligand was difficult to remove in the downstream process without resorting to chromatography, and some reproducibility issues were seen, hence, alternatives were explored. This led to the discovery that $Pd[P(o\text{-tol})_3]_4$ formed *in situ* by addition of Et_2Zn to a mixture of $Pd(OAc)_2$ and $P(o\text{-tol})_3$ in DMF was the most successful catalyst for this transformation [6] allowing smooth cyanation under mild conditions with the advantage that the phosphine residues did not persist through the subsequent steps. On scale the reaction was typically complete within 12 h at 55 °C and afforded cyano alcohol **6** as an oil in 92% yield and sufficient purity for further processing without the need for chromatography.

9.1.3.1.4 **Synthesis of Amine 2**

With a common intermediate from the Medicinal Chemistry synthesis now in hand in enantiomerically upgraded form, optimization of the conversion to the amine was addressed, with particular emphasis on safety evaluation of the azide displacement step (Scheme 9.7). Hence, alcohol **6** was reacted with methanesulfonyl chloride in the presence of triethylamine to afford a 95% yield of the desired mesylate as an oil. Displacement of the mesylate using sodium azide in DMF afforded azide **7** in around 85% assay yield. However, a major by-product of the reaction was found to be alkene **17**, formed from an elimination pathway with concomitant formation of the hazardous hydrazoic acid. To evaluate this potential safety hazard for process scale-up, online FTIR was used to monitor the presence of hydrazoic acid in the head-space, confirming that this was indeed formed during the reaction [7]. It was also observed that the amount of hydrazoic acid in the headspace could be completely suppressed by the addition of an organic base such as diisopropylethylamine to the reaction, with the use of inorganic bases such as

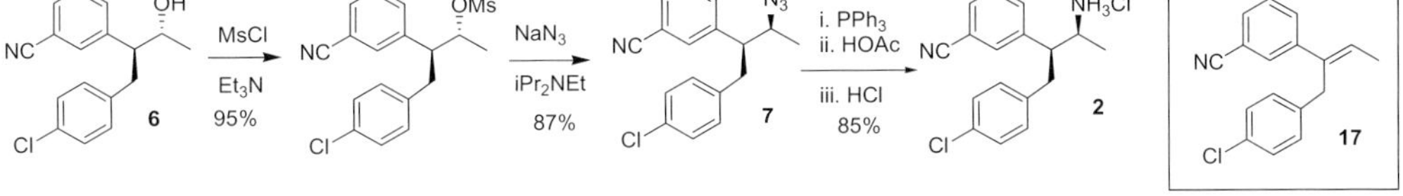

Scheme 9.7 Optimization of azide installation and reduction.

K_2CO_3 being less effective. Based on these observations, the mesylate displacement was carried out using a full equivalent of diisopropylethylamine with respect to sodium azide and proceeded smoothly to give the desired azide **7** in 87% assay yield along with 11% of alkene **17**.

Conversion of the azide **7** to amine **2** was then explored. The original hydrogenation conditions (H_2, PtO_2, EtOAc) were shown to lead to competing dechlorination as the reaction medium became basic during generation of the amine, which was circumvented by *in situ* Boc-protection of the amine. As the subsequent operation was removal of the protecting group prior to coupling, avoidance of this intermediate protection would be desirable for more efficient processing. A number of hydrogenation and transfer hydrogenation conditions for reduction of the azide **7** to amine **2** were screened, but all led to other impurities and/or suffered from lack of reproducibility and batch to batch variation. The variability was probably due to the multiple steps without any purification that had been used to generate the azide, which itself was also an oil and, hence, resistant to any purification procedures other than chromatography which we desired to avoid on scale. Success was finally found using Staudinger conditions [8]. Hence, treatment of azide **7** with PPh_3 (1.1 equiv) in toluene–water cleanly afforded amine **2**, with portion-wise addition of the reagent being necessary to control exothermic activity. For large scale operations an efficient removal of triphenylphosphine oxide had to be devised. Direct salt formation from the organic extract typically led to co-crystallization of the triphenylphosphine oxide with the amine salt, however, we were pleased to find that the amine could be extracted into 10% aqueous acetic acid leaving the triphenylphosphine oxide and the elimination product **17** (from the preceding displacement reaction) in the organic layer. Interestingly, other acids (aqueous HCl, citric acid, or phosphoric acid) resulted mainly in oiling and ineffective partition of the amine to the aqueous layer. After neutralization and re-extraction into the organic layer, isolation of the amine **2** as the hydrochloride salt in 85% yield allowed further purification and also an upgrade in diastereomeric purity.

To summarize the above section, an efficient synthesis of the desired amine HCl salt **2** as a single diastereomer and enantiomer was developed. In particular, this route avoided the need for tin chemistry, avoided the issues of nitrile hydrolysis during the reduction work-up procedure, and obviated the need for an interim protection of the amine group during the azide reduction step. Although all intermediates subsequent to acid salt **12** were found to be oils, reaction optimization allowed the generation of sufficiently high purity material for through processing without resorting to chromatography. Final formation of the amine hydrochloride salt allowed upgrade of purity, in preparation for the final coupling.

9.1.3.1.5 Synthesis of Acid 3

Although the synthesis of acid **3** developed by our Medicinal Chemistry colleagues proceeded in a single step from commercial starting materials the yield was modest, and there were significant safety concerns about the use of sodium hydride in DMF on a larger scale. After some experimentation it was found that use of the ester in conjunction with just 1 equiv of KHMDS as base gave a clean coupling; straightforward hydrolysis then afforded the acid **3** in 88% yield as a crystalline solid in high purity (Scheme 9.8).

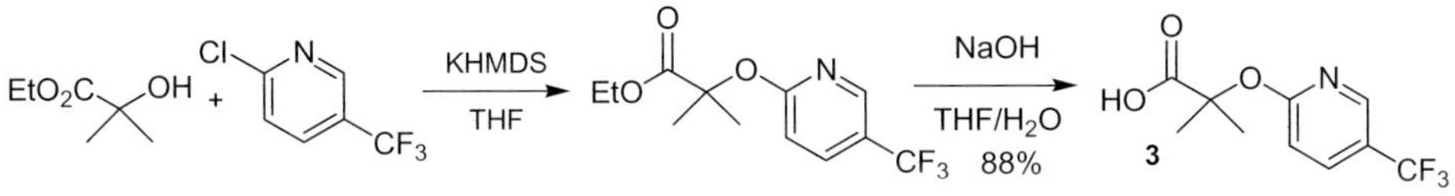

Scheme 9.8 Improved synthesis of acid **3**.

9.1.3.1.6 **Final Coupling and API Delivery**

As previously described, the Medicinal Chemistry approach to taranabant involved coupling of amine HCl salt **2** with the acid **3** mediated by Py-Bop. To avoid this expensive coupling reagent a number of alternatives were explored. Coupling of the amine with the acid chloride derived from acid **3** afforded product with a good impurity profile but led to the generation of high levels of color which was not readily removed, even after carbon or resin treatment. The coupling could also be achieved by reaction with EDC and pyridine, leading to less colored reaction mixtures, though cost and handling issues made this process less attractive for longer term use. The optimum conditions for the coupling were found to be use of the inexpensive and readily available cyanuric chloride as activating reagent, in conjunction with *N*-methylmorpholine as base in acetonitrile (Scheme 9.9). Less than 1 equiv of cyanuric chloride (0.6 equiv) was found to be sufficient for complete reaction and the amide was isolated as a white solid in 86% yield and 95%ee. Repeated attempts to upgrade the enantiopurity by direct crystallization failed, however, during these investigations it was noticed that at higher solvent volumes the residual solid showed a *downgrade* in ee, with upgraded material being left in the filtrate and, hence, that ee upgrade by initial removal of racemate might be possible. A number of solvent combinations were explored, leading to identification of 2:1 EtOH : water as optimum to upgrade to approximately 99% ee in the supernant. The low ee solid (typically around 30%ee and accounting for 5% of the material) was removed by filtration and after solvent adjustment, crystallization of the product from the filtrate afforded **1** in 90% recovery and excellent quality.

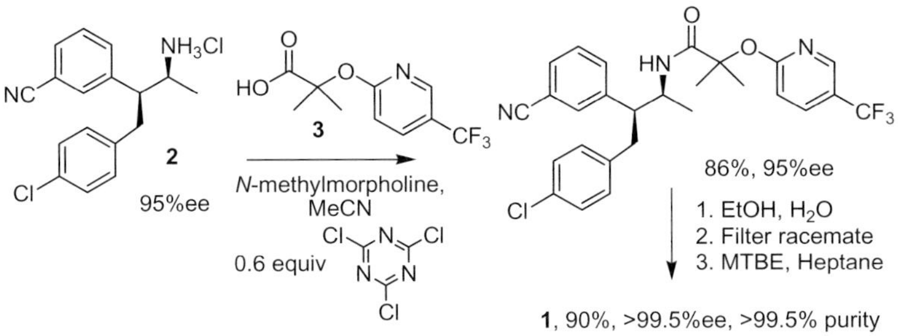

Scheme 9.9 Final coupling and upgrade for taranabant.

Table 9.2 Preliminary improvements to the taranabant synthesis.

Development issue	Outcome
Volume inefficient chiral chromatography required to deliver a single enantiomer	Classical resolution of early intermediate, acid **11**, implemented
Stoichiometric use of a tin reagent to prepare the starting phenylacetone	A different starting material obviates the need for the tin-promoted step
Chromatographic purification required for each intermediate, the majority of which were oils	High yielding reactions allowed for through processing from salt **12** to amine salt **2** The only three solid compounds (**12**, **2**, **1**) allowed for purity upgrades
Cryogenic conditions (−78 °C) and an expensive reagent to obtain high selectivity for reduction of the carbonyl group in **5**	L-Selectride® remains the reagent of choice, but an attainable temperature of −50 °C gave good selectivity
Incompatibility of the nitrile group with the oxidative work-up of the selectride reduction	Optimal introduction of nitrile group was identified as subsequent to reduction
Use of sodium azide to introduce the nitrogen atom	Sodium azide remains the source of nitrogen, however a thorough evaluation of the safety issues was conducted and safe operating conditions defined.
Boc-protection needed during reductive hydrogenation of the azide group	Staudinger reaction conditions obviate the need for interim protection
Use of sodium hydride in DMF to prepare the pyridine acid **3**	A high yielding procedure using KHMDS was developed
Use of an expensive coupling reagent to promote the amide bond formation	A number of alternatives were shown to be viable with cyanuric chloride chosen

Evaluation of the above route against our initial target objectives for the synthesis of taranabant indicated a high level of success, not just for the primary objectives of removing the tin chemistry and chiral chromatography, but for a number of other process improvements (Table 9.2). Of particular note was that the three crystalline intermediates were key for purification, first the phenethylamine salt **12** for the classical resolution, secondly the HCl salt of amine **2** allowed for upgrade of diastereomeric purity, and finally the API allowed for upgrade of enantiomeric purity via initial removal of racemic material.

Based on these improvements the route was found to be suitable for generation of up to a kilogram of API and allowed the rapid advancement of the program into early toxicology and clinical studies with relatively minimal time or manpower investments.

9.1.3.2 **Amide Bond Formation as the Final Step – Dynamic Kinetic Resolution**

Despite the improvements outlined above, the modest yield through the resolution process (around 30% yield of 97% ds salt) and the need for multiple crystallizations would become a limitation moving forward to prepare multi-kilo quantities of taranabant and an asymmetric version of the above synthesis was sought.

A potential asymmetric route to the chiral alcohol **16** could be envisioned by taking advantage of the inherent high diastereoselectivity obtained in reductions of the precursor ketone **13**, combined with the relative ease of epimerization of the α-chiral center in said ketone. As such, if a chiral reducing agent could show significant rate differences for reaction with each of the enantiomeric starting ketones (*R*-**13** vs. *S*-**13**), and epimerization of the starting material could be achieved under the same reaction conditions a dynamic kinetic resolution (DKR) should be possible [9, 10]. This was an attractive possibility for an efficient synthesis, potentially allowing both stereocenters to be set in a single step starting from a racemic starting material *rac*-**13**, thus addressing the key shortcoming of the resolution route without involving significant changes to the established chemistry (Scheme 9.10).

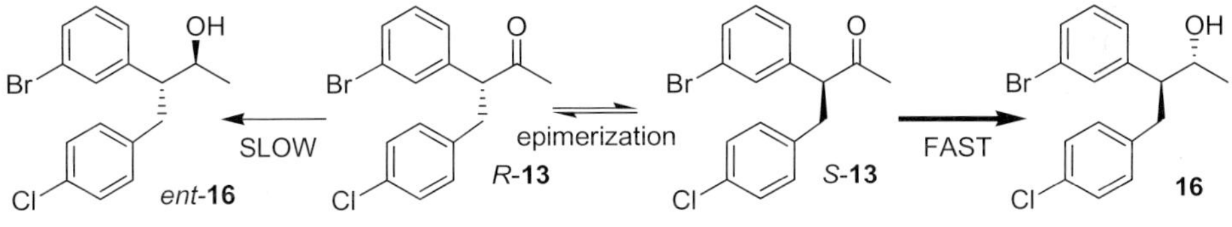

Scheme 9.10 Proposed dynamic kinetic resolution (DKR) of ketone *rac*-**13**.

Our work in this area started with an evaluation of the racemization rate of ketone **13**, leading to the observation that it can be readily epimerized upon treatment with 20 mol% of KO*t*-Bu in THF over a range of temperatures. A range of chiral catalysts and conditions could be envisaged to effect this transformation, however, based on a number of recent results within our department, reduction under hydrogenation conditions mediated by a chiral ruthenium catalyst was investigated. Indeed, ketone *rac*-**13** was hydrogenated at room temperature under basic conditions in the presence of the (xyl-BINAP)(DAIPEN)$RuCl_2$ catalyst to give the desired diastereoisomer **16** in 89% ee and 83% ds. This exciting lead validated the proposal and provided the impetus for further development. After many rounds of screening and optimization for both enantio- and diastereo- selectivity the use of *iso*-propanol with a KF of <500 ppm was found to be the best solvent with an optimal reaction temperature of 0 °C. The equivalents of base had minimal effect with 20 mol% being sufficient to maintain an appropriate racemization rate for the starting ketone **13**. Hydrogen pressure (15–90 psi) was found to affect the rate of the reaction but not the selectivity, unless racemization was very slow when a lower ee was observed at higher pressures. Under these optimized conditions the reaction proceeded in essentially quantitiative yield, 94%ee for the major desired diastereoisomer and around 88% diastereoselectivity (Scheme 9.11).

With this reaction in place the only change required to the synthesis was development of an isolation of the racemic acid **11** as a neutral compound rather than

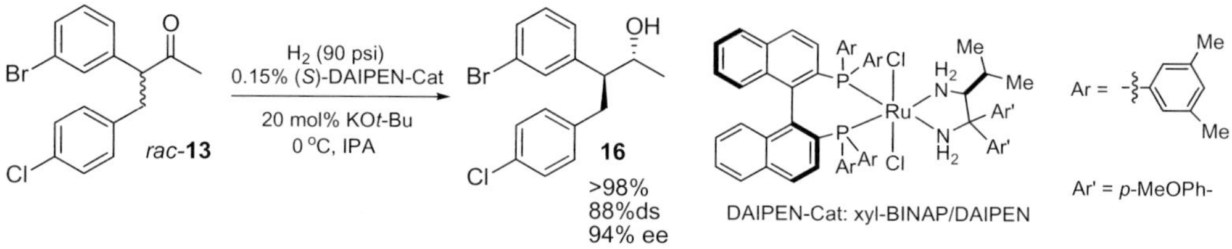

Scheme 9.11 Successful dynamic kinetic resolution of ketone rac-**13**.

a salt (as used in the resolution). A crystallization of the racemic acid **11** was rapidly developed, and conversion to racemic methyl ketone *rac*-**13** proceeded in the same manner as demonstrated in the resolution route. The racemic ketone **13** was an oil, however, as a testament to the robustness of the asymmetric hydrogenation, no purification was required prior to this reaction. After completion of the hydrogenation an aqueous work-up was employed and the chiral alcohol **16** processed without further purification. The remaining steps of the synthesis performed as expected with diastereomeric purity being upgraded at the amine salt **2**, and enantiomeric purity at API via removal as racemate, as previously described (Scheme 9.12).

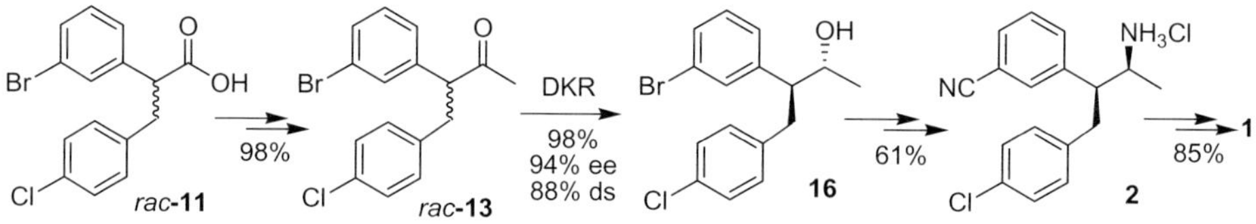

Scheme 9.12 Completion of the amine synthesis using the DKR.

As such, an asymmetric synthesis of tarantabant had been demonstrated centered around a DKR as the key step to set both stereocenters in a single catalytic step. The synthesis proceeded in six steps and 40% yield from the ketone *rac*-**13**. This route was found to be robust and reliable and generated in excess of 70 kg of API in various campaigns from both preparative laboratories and pilot plant facilities, and provided support for the project through longer term toxicology and clinical studies [13].

Box 9.1 Utility of the ruthenium-catalyzed dynamic kinetic resolution

The above described DKR allows generation of a single enantiomer and diastereoisomer of a compound from a racemic starting material, and, as such, is a very powerful transformation. There are a number successful implementations documented, and here we outline two recently published examples also using a ruthenium catalyst under hydrogenation conditions. In an example from Merck and Co. the racemic ketone **18** is converted to chiral alcohol **19** in excellent de and ee, which can be further upgraded by recrystallization (Scheme 9.13). The optimum catalyst and conditions are closely related to those employed for ketone **13** with xyl-SEGPHOS replacing the xyl-BINAP ligand [11]. In a second example, a method for the preparation of anti-β-hydroxy-α-amino acids such as **20** from racemic β-ketoesters **21** was developed (Scheme 9.14) [12]. In this case the ruthenium catalyst lacked the DIAPEN ligand and the greater acidity of the keto-ester chiral center obviated the need for additional base to promote the racemization.

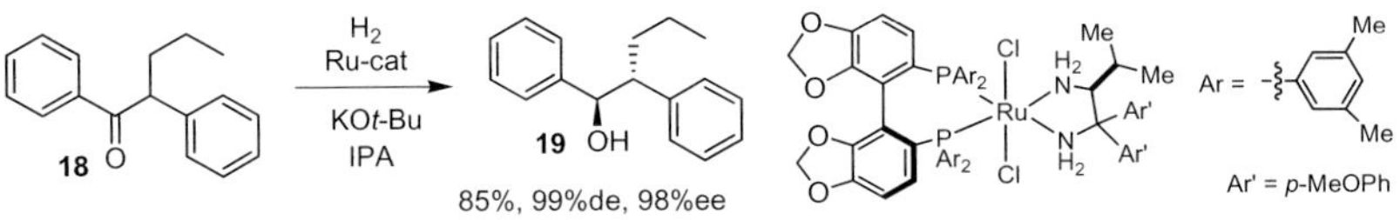

Scheme 9.13 DKR of α-aryl ketone **18**.

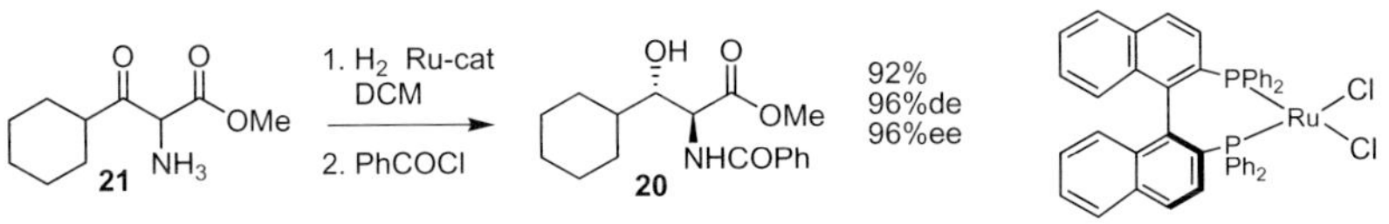

Scheme 9.14 DKR of α-amino-β-ketoester **21**.

9.2 Further Project Development

9.2.1 Introduction

As the program continued to move forward the team once again faced the need to evaluate the synthetic approach to taranabant, this time with regard to implementation on scales in excess of 100 kg and for potential manufacturing purposes. Despite the significant advancements outlined in Section 9.1, a route analysis indicated a number of issues and shortcomings still to be addressed, as outlined below (Table 9.3).

While it was felt that some of the individual issues above could be addressed using the same synthetic sequence (e.g., alternate catalysts for the reduction step) it seemed unlikely that all the above would be solvable, especially as efforts to replace sodium azide with other nucleophiles had failed. Based on this assessment the team felt it would be necessary to evaluate a fundamentally new approach to taranabant and, in particular, to look for a method for installation of the chiral centers without the intermediacy of an alcohol.

Table 9.3 Long term development issues still requiring resolution.

Long-term development issue	Impact
Use of sodium azide to introduce the nitrogen functionality	Despite safety evaluation, concerns on manufacturing scale remained
Use of an expensive and limited availability catalyst for the dynamic kinetic resolution	Long term cost/supply concerns
Lack of any solid intermediate throughout the synthesis other than acid **11** and amine salt **2**	Lack of purification options and opportunities to use outside vendors for intermediate supplies
Cumbersome method for ee upgrade of API involving three isolations	Additional processing required which translates to cost on large scale

9.2.2
New Synthetic Approach

An attractive new approach to achieving our goal was to explore an asymmetric reduction of a stereodefined enamide such as **22**, thereby setting both stereocenters of the target molecule in a single step starting from an achiral intermediate (Scheme 9.15). We recognized this as a very ambitious approach requiring significant resources and development. In particular, changes to a synthetic route at a late stage in the drug development process had implications with regard to ensuring comparable purity of the API produced via the new process. Notably, any new impurities generated would have to be controlled to a very low level (as recommended by ICH guidelines) or qualified in additional toxicology studies prior to use in any clinical program. However, the opportunity to address the previous shortcomings provided the impetus for evaluation of this proposal.

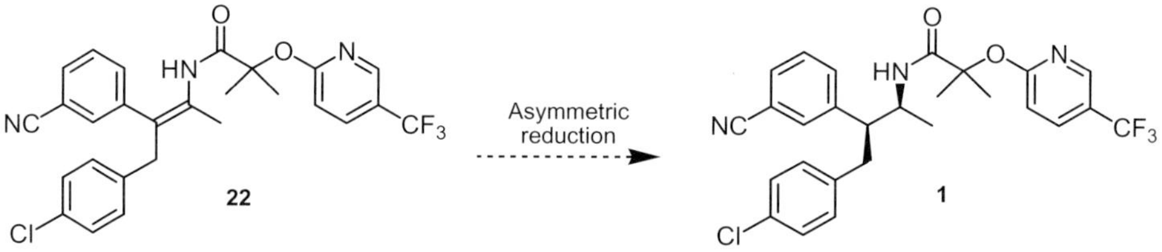

Scheme 9.15 Proposed asymmetric enamide reduction to generate taranabant.

The asymmetric reduction of enamides using hydrogenation conditions is a well documented reaction with a number of groups reporting excellent results [14]. *Syn* delivery of hydrogen from the same face of the molecule ensures that the enamide geometry defines the relative stereochemistry obtained, and a range of chiral catalysts have been developed to control the absolute stereochemistry. However, a survey of the literature indicates a dearth of successful asymmetric hydrogenations on tetrasubstituted enamides [15] as would be required to support the taranabant synthesis. While these substrates provide significant challenges as hydrogenation substrates, it is likely that the lack of information is also due to limited reliable methods to selectively generate a tetrasubstituted enamide bearing four distinct groups [16].

In recent years, cross-coupling methodology has emerged as a viable tool for enamide synthesis, and, indeed, there are a number of published protocols which employ palladium- or copper-catalyzed stereospecific amidations of vinyl halides [17]. For example, Buchwald and coworkers had recently shown that a copper-catalyzed cross-coupling of vinyl bromides or iodides proceeded with retention of stereochemistry (Scheme 9.16), though the only example using a tetrasubstituted vinyl halide, **23**, lacked the need for any stereochemical control in the halide portion [18]. Based on this it seemed feasible that the desired enamide **22** could potentially be assembled via a comparable coupling between amide **24** and a stereodefined vinyl halide such as **25**.

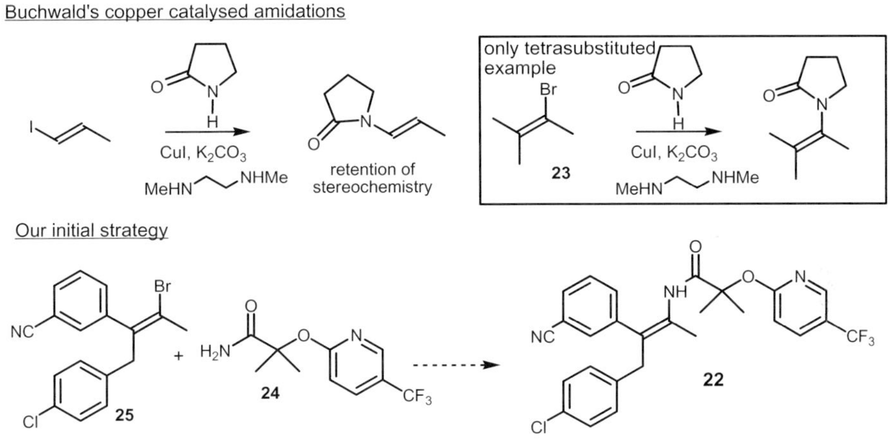

Scheme 9.16 Copper-catalyzed amidation of vinyl halides.

Unfortunately, it quickly became apparent that a shortfall in this proposal was an inability to prepare the desired vinyl halide **25** in a straightforward and selective manner [19]. In contrast, we reasoned that the selective formation of an enol sulfonate, such as the enol triflate **26a**, could be controlled by judicious tuning of enolization conditions starting from the corresponding ketone, and that such an enol sulfonate would possibly be a substrate for a palladium-mediated coupling (Scheme 9.17). In this way a common intermediate from the previously defined synthesis, that is, the racemic ketone *rac*-**13** or its cyano equivalent *rac*-**5** could be used to generate the required enamide.

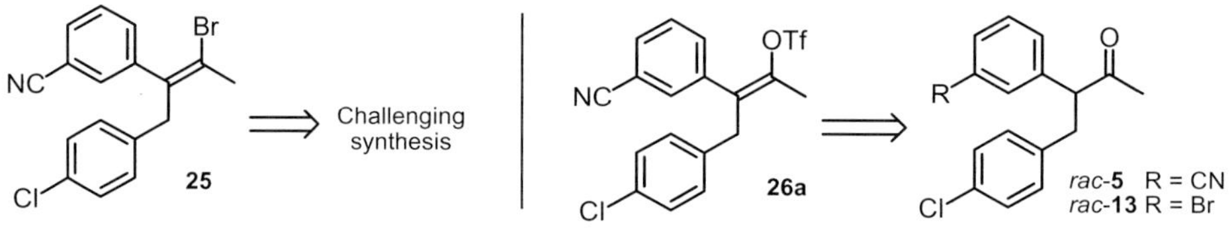

Scheme 9.17 Proposal to prepare and couple an enol sulfonate.

9.2.2.1 Enol Triflate Synthesis

Based on the known reactivity of enol triflates in a range of palladium-mediated couplings [20], and the ease of their formation, our work in this area started by considering the viability of a palladium-catalyzed amidation of an enol triflate, a reaction unknown at initiation of this work. While both the bromo- and cyano-derivatives of the methyl ketone (**5** and **13**) were viable substrates, to avoid any competitive reactions of the aryl bromide in the palladium processes we chose to focus our efforts on the cyanoketone *rac*-**5**. This compound was prepared in 95% yield from ketone *rac*-**13** using a comparable cyanation reaction to that utilized in Section 9.1, with the previous issue of racemization no longer relevant in the racemic series (Scheme 9.18). Moreover, the cyanoketone *rac*-**5** was found to be a crystalline solid, providing a key purification point to reject any earlier synthetic impurities, which had been lacking in our previous approach.

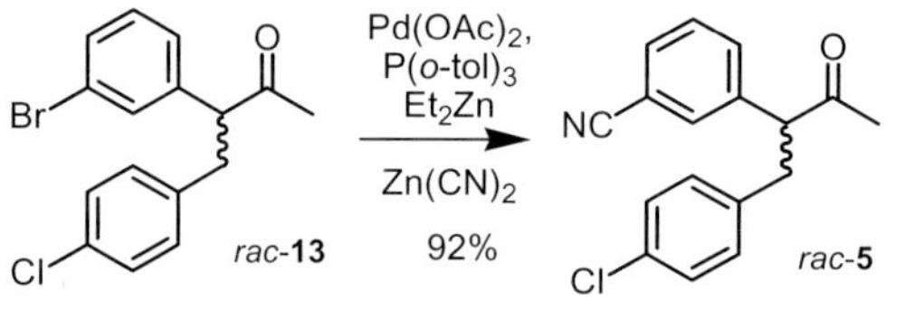

Scheme 9.18 Cyanation of ketone.

Enolization of *rac*-**5** with sodium hydride in THF followed by quench with *N*-phenyl-bis-(trifluoromethylsulfonimide) ($PhNTf_2$) afforded an approximately 1 : 1 mixture of enol triflates **26a** *(E)* and **26b** *(Z)* in good yield. Addition of DMPU to the reaction resulted in a dramatic increase in selectivity, to significantly favor the desired *E*-isomer **26a**. As the ratio of DMPU to solvent was increased, the ratio of **26a**:**26b** improved, up to the optimized ratio of 90 : 10 (Table 9.4). Under the hypothesis that DMPU was altering the aggregate of the enolate in solution, we sought to mimic this effect with solvents that were structurally similar to DMPU, but less expensive. Amide solvents such as NMP, DMAc or DMF afforded comparable ratios favoring **26a** while solvents such as acetonitrile or ethyl acetate did not exhibit a similar effect.

Employment of sodium hydride in a large scale process would present safety issues, so the choice of base was addressed and *tert*-butoxide bases were identified

Table 9.4 Solvent effect on enol triflate isomer ratio.

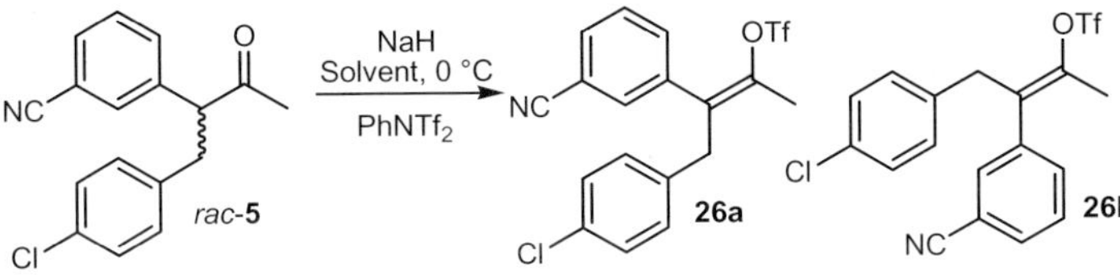

Entry	Solvent	Conversion	26a : 26b
1	THF	95	40 : 90
2	THF/DMPU (80/20)	95	83 : 17
3	THF/DMPU (20/80)	94	90 : 10
4	MTBE/DMPU (80/20)	95	82 : 18
5	DME/DMPU (80/20)	95	74 : 26
6	MeCN	90	53 : 47
7	EtOAc	87	44 : 56
8	DMF	91	89 : 11
9	NMP	94	90 : 10
10	DMAc	98	90 : 10

Table 9.5 Counterion and substituent effect on enolization selectivity.

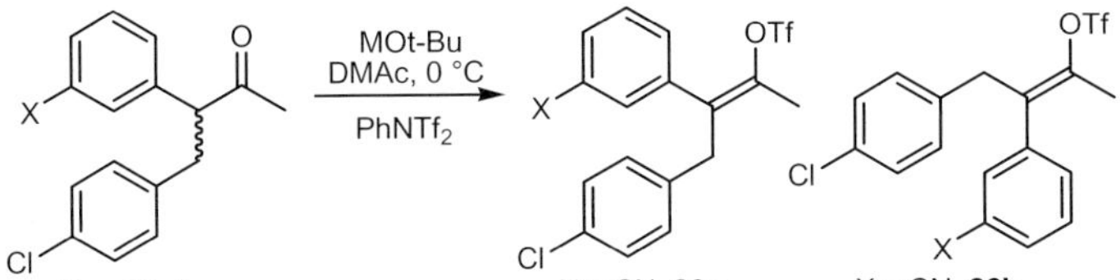

X = CN, **5**
X = CO_2Me
X = Br, **13**
X = H

X = CN, **26a**
X = CO_2Me, **a**
X = Br, **a**
X = H, **a**

X = CN, **26b**
X = CO_2Me, **b**
X = Br, **b**
X = H, **b**

M	X = CN (26a : 26b)	X = CO_2Me (a : b)	X = Br (a : b)	X = H (a : b)
Li	84 : 16	37 : 63	61 : 39	30 : 70
Na	90 : 10	65 : 35	80 : 20	57 : 42
K	95 : 5	75 : 25	85 : 15	80 : 20

as suitable alternatives for the enolization/quench procedure. It is interesting to note that as the counterion of the alkoxide base was changed from lithium through potassium, an increase in the *E*:*Z* isomer ratio was observed (Table 9.5). This trend was consistent for several *meta*-substituted analogs, sometimes resulting in a complete turnover in selectivity (X = CO_2Me, H). The exact reason for this trend is not clear, but may be related to differing bond length and/or changing basicity of the enolate with the different metals. Notably lower selectivities for the desired *E*-isomer were seen starting from the bromo-ketone **13**, providing a further driver for use of the cyano ketone **5** in preference to **13**. Although the highest *E*:*Z* selectivity for the enolization was obtained with KO*t*-Bu, the low solubility of the potassium enolate led to volume inefficiency, and use of NaO*t*-Bu in DMAc was found to be optimum, leading to a 90 : 10 ratio in favor of the desired **26a**. The isomers could be separated by column chromatography, affording an 85% yield of **26a**.

9.2.2.2 **Synthesis of a Model Enamide**

With a pure sample of enol triflate in hand we were ready to explore the proposed amidation reaction. In the first instance a model study was carried out using acetamide as the amide to probe conditions for both the coupling and subsequent hydrogenation, and also recognizing this could allow a synthesis of the previously prepared amine **2**. Using conditions developed by Buchwald for the amidation of aryl halides [21] [$Pd(OAc)_2$, Xantphos and Cs_2CO_3 in 1,4-dioxane at 80 °C] coupling of enol triflate **26a** with acetamide was attempted (Scheme 9.19). After 8 h complete conversion of the enol triflate to a 60 : 40 mixture of two new products was observed and, after chromatographic separation, these were identified as the two isomeric enamides **27a** and **27b**. This indicated that the coupling was indeed a viable process, but also suggested that isomerization of the enol triflate or product had taken place during the reaction. Indeed, resubjection of either of the purified enamides to the reaction conditions afforded a 60 : 40 mixture of *E*:*Z* isomers. Lowering the temperature to 40–50 °C led to slower isomerization, and at room temperature the enamides were relatively stable to the reaction conditions. Hence, for successful implementation, amidation at lower temperatures would be required. Reaction screening indicated that running the reaction at 30 °C allowed moderate conversion with minimal isomerization and a change of palladium source to $Pd_2(dba)_3$ gave a 95 : 5 ratio of enamide isomers with good conversion after 8 h. Even at this temperature product equilibration still occurred with a prolonged reaction time, and the result quoted above represented the best balance between conversion and isomerization. The desired enamide **27a** was found to be a crystalline solid, allowing rejection of the minor isomer, and could be isolated in 80% yield.

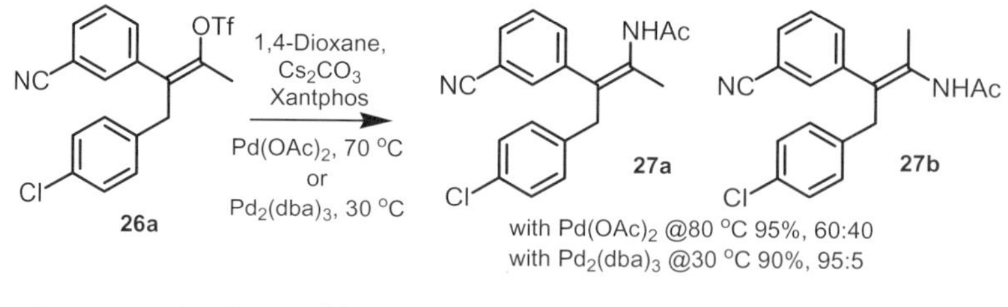

Scheme 9.19 Amidation of the enol triflate with acetamide.

Box 9.2 An evaluation of the utility of the enol triflate amidation

At the time of this work the palladium-mediated amidation of enol triflates had not been described, and, hence, the scope of the reaction was evaluated with a range of other enol triflates and amides (Scheme 9.20) [22, 23]. A number of key observations were made, some of which were to prove important during further development.

1) Highest yields are obtained where no isomerization is possible, and reactions can be pushed to completion without risk of competing isomerization, equations 1 versus equation 2.
2) A hindered enamide (such as that derived from coupling with t-butylamide) is not susceptible to isomerization, also giving a higher yield, equation 1.
3) Secondary amides do not undergo the amidation except when cyclic, equation 3.
4) The enol triflate moiety undergoes amidation in preference to any aryl bromide in either starting material, equations 4 and 5.
5) Carbamates and sulfonamides can also be used as coupling partners (not shown).

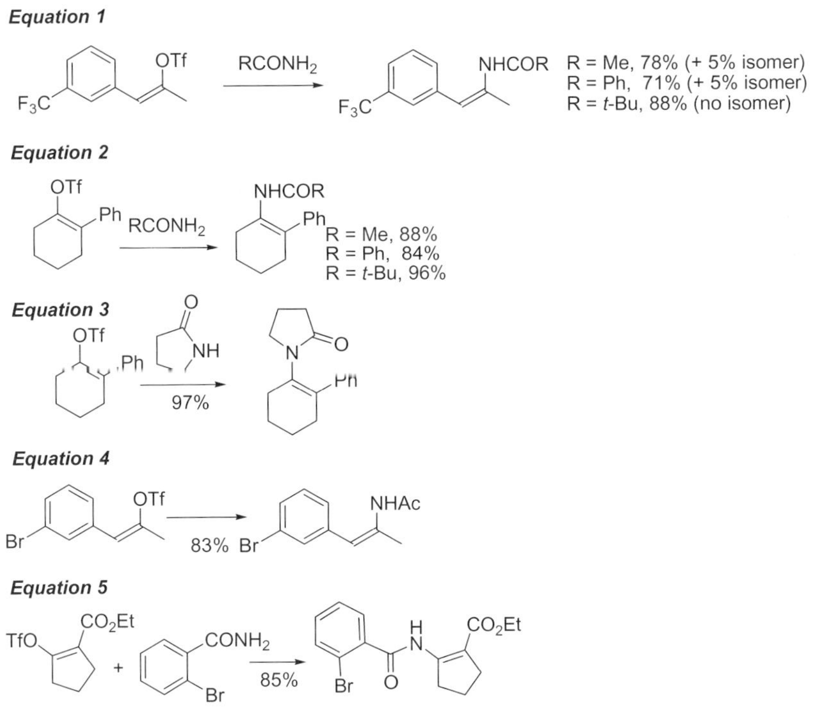

Scheme 9.20 Enol triflate amidations.

9.2.2.3 Preliminary Hydrogenation Studies

With enamide **27a** in hand, screening of potential hydrogenation conditions was initiated. The asymmetric hydrogenation was quickly confirmed to be a viable process using cationic rhodium catalysts with a variety of diphosphines such as **28** on a small scale (<100 mg reactions, >20 mol% catalysts). As anticipated the stereochemistry of the enamide was retained in the product, leading to the *syn*-isomer and modest conversions and enantiomeric excesses could be obtained (Scheme 9.21). The resulting amide **29** was hydrolyzed to penultimate amine **2** in 65% yield using HCl in dioxane at 100 °C for 48 h. The modest yield is attributed to competing nitrile hydrolysis to both amide and acid groups under the forcing conditions, however, both the ee and de of the amine mirrored that from the amide, indicating that no epimerization had occurred. The product amine **2** is the key coupling partner for the previously used amide bond formation to generate taranabant and, as such, this synthetic sequence represented a formal alternative route to **1**, and also achieved proof of concept for both the stereoselective synthesis of a tetrasubstituted enamide and its subsequent asymmetric hydrogenation.

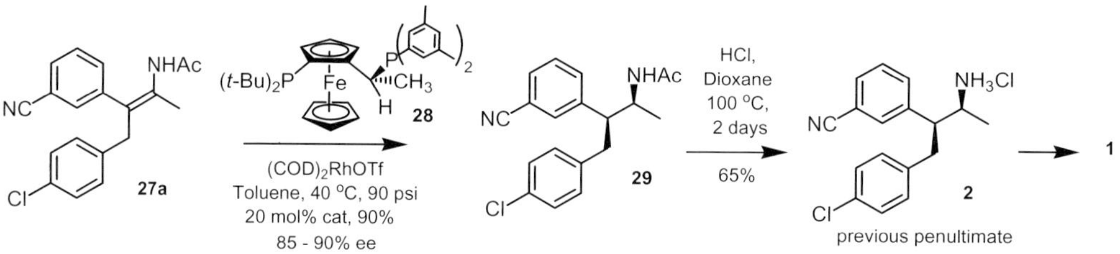

Scheme 9.21 Asymmetric hydrogenation of enamide **27a**.

Unfortunately, attempts to scale the key hydrogenation reaction beyond gram scale or reduce the catalyst loading led to lower conversions and/or enantiomeric excess. During this screening phase other enamides were also prepared and subjected to hydrogenation screens, leading to the observation that some enamides underwent a more facile reduction than others. A correlation was quickly realized: poor results were obtained on all enamides bearing the aryl nitrile group, regardless of the nature of the nitrogen "protecting group" (R in Scheme 9.22). In contrast the des-CN enamide **30** gave promising results under conditions where reduction of the closely related **27a** was ineffective.

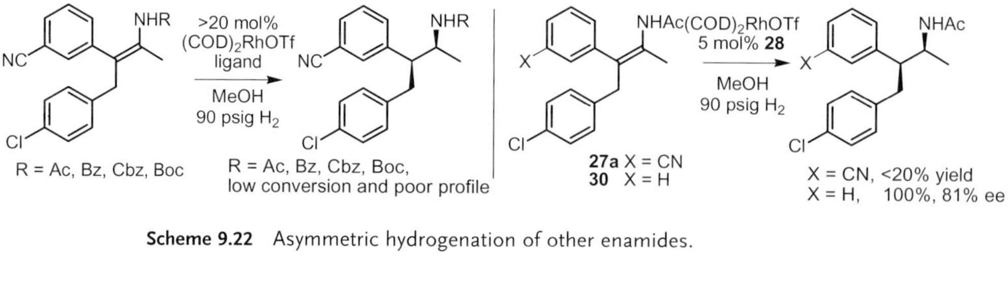

Scheme 9.22 Asymmetric hydrogenation of other enamides.

This led to the proposal that the aryl nitrile group could be compromising the hydrogenation, and, indeed, NMR studies suggested that the preferred mode of coordination of the catalyst to substrate is via the nitrile, rather than the enamide group. For example, treatment of a solution of a phosphine ligand and $(COD)_2RhBF_4$ with either enamide **27a** or a simple aryl nitrle (*m*-tolunitrile **31**) generated a new species with identical ^{31}P NMR spectra, implying a common coordinating group for both **27a** and **31** (Scheme 9.23). Additionally, inhibition of cationic rhodium catalysts by nitrile functionality in asymmetric hydrogenations has been observed previously during the reduction of an α,β-unsaturated nitrile, providing further proof as to the challenge of this process [24]. Despite the hydrogenation difficulties a new route to the taranabant penultimate **2** had been demonstrated, which could potentially be extended to a direct route to the final compound **1** via a hydrogenation of the more complex enamide **22**. The ability to avoid the shortcomings of the previous sequence remained attractive and, hence, the issues with the new approach were considered with a view to finding viable alternatives (Table 9.6).

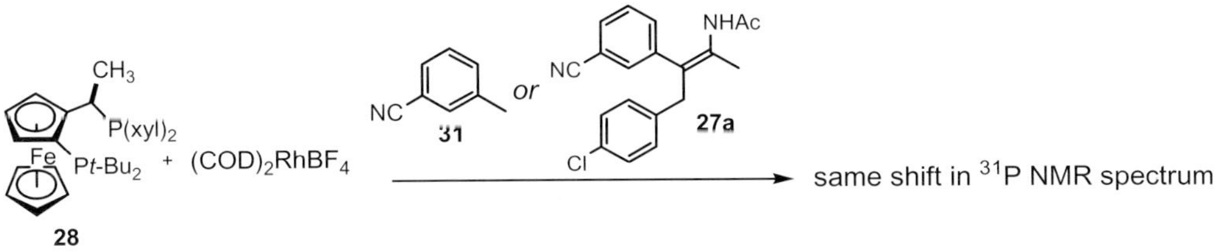

Scheme 9.23 Coordination of aryl nitriles to rhodium catalysts.

Table 9.6 Assessment of development targets for the enamide hydrogenation route.

Issue	Proposal
Use of $PhNTf_2$ for enol triflate formation. In addition to expense of this reagent, large scale availability was limited	Evaluate use of the related enol tosylate–which should be accessible from cheaper and more readily available reagents
Difficulty removing acetate protecting group from amide **29**	Amidation with functionalized amide, would obviating need to remove protecting group
Low yields and high catalyst loading in the hydrogenation reaction	Further screening of conditions, coupled with evaluation of substrates lacking the aryl nitrile group.

9.2.2.4 Formation of an Enol Tosylate

Enolization of cyanoketone *rac*-**5** under the previously optimized conditions, followed by reaction with *p*-toluenesulfonic anhydride afforded a 90:10 ratio of the two expected enol tosylates in 90% yield. An immediate advantage of the tosylates compared to the triflate was seen on isolation, when the desired compound **32** was found to be a crystalline solid. In this way an 85% isolated yield of **32** as a single isomer could be achieved without the need for chromatography (Scheme 9.24). Use of *p*-toluenesulfonyl chloride in place of the anhydride led to α-chlorination

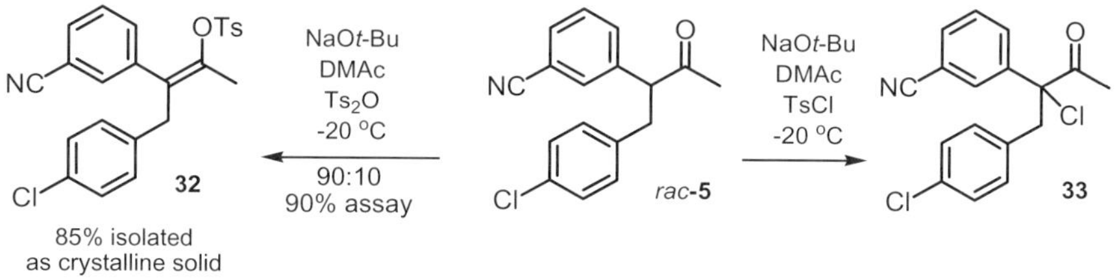

Scheme 9.24 Formation of the enol tosylate.

of the enolate to give **33** in low yield, with none of the desired enol tosylate being formed.

9.2.2.5 Amidation of the Enol Tosylate

With an efficient method to synthesize and isolate **32** in isomerically pure form, efforts were directed towards achieving an effective coupling with the primary amide side-chain **24**. This coupling was anticipated to pose a greater challenge than encountered previously, based on both the lower reactivity of the enol tosylate versus the triflate, and the greater steric bulk of the amide. Moreover, at the time this development started, the coupling of amides with vinyl tosylates was unprecedented [25]. To provide guidance around the reactivity of the amide portion an initial reaction with the previously employed enol triflate **26a** and amide **24** was carried out. This coupling proceeded smoothly using the $Pd_2(dba)_3$/Xantphos conditions to afford the desired enamide **22**, but did require a higher temperature to achieve complete conversion compared to when acetamide was used (Scheme 9.25). However, unlike the acetamide enamide **27a**, enamide **22** did not undergo double bond isomerization under the reaction conditions, even at this higher temperature, indicating that its greater steric bulk contributed to additional stability. A similar phenomenon had previously been seen when comparing amidations using *tert*-butylamide with acetamide (Scheme 9.21, equation 1).

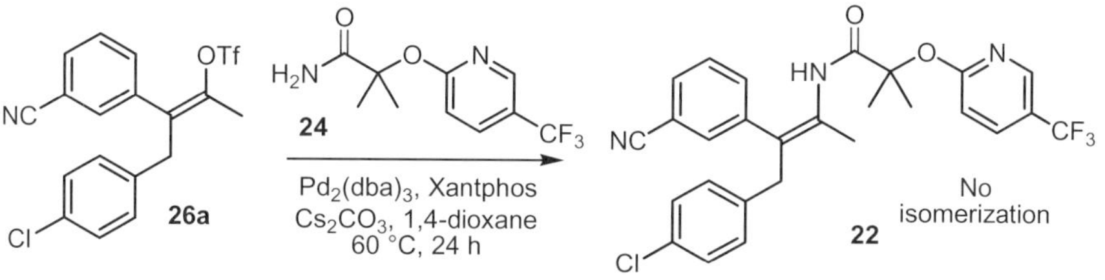

Scheme 9.25 Amidation of enol triflate **26a** with amide **24**.

Use of the above conditions in conjunction with the enol tosylate **32**, provided only low yields of **22**, prompting an extensive screening of structurally diverse phosphine ligands/solvents and palladium sources to attempt to define suitable conditions. Quite quickly a number of conditions were found to be effective, with chelating diphosphines being superior to monodentate phosphines (Table 9.7). In

Table 9.7 Pd-catalyzed amidation of enol tosylate 32.

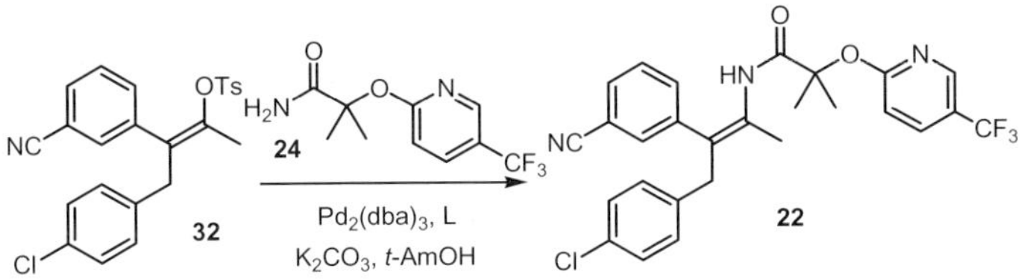

L	Assay yield
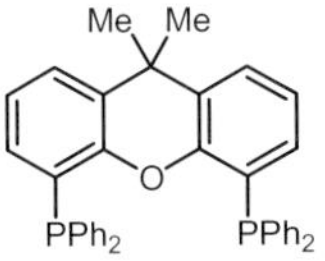	<0.1%
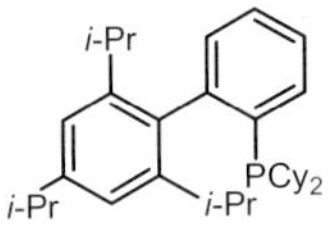	6%
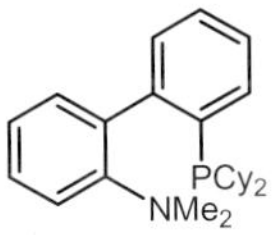	28%
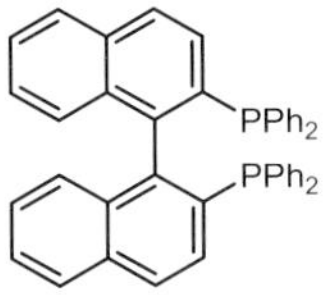	48%
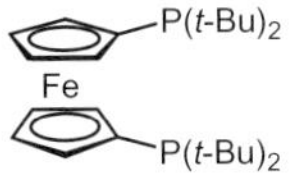	6%
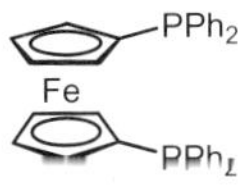	70%
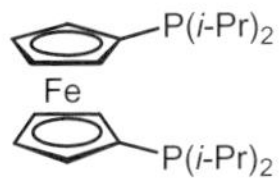 21	93%
	<1%
	37%
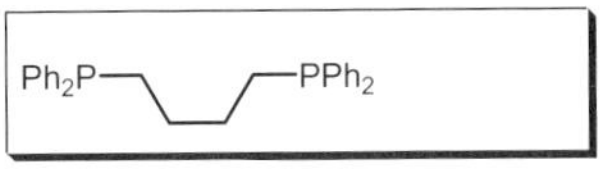	92%
	37%
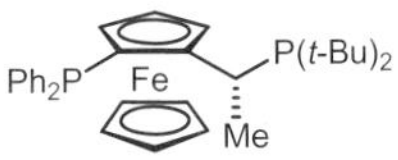	90%

All reactions were run using 2.5 mol% $Pd_2(dba)_3$, 10 mol% ligand, 2 equiv K_2CO_3, and 1.05 equiv of 24 in *tert*-amyl alcohol at 100 °C for 20 h.

particular, the affordable and readily available 1,4-bis(diphenylphosphino)butane (dppb) was optimal, affording a 92% assay yield of enamide **22**. Interestingly, a significant drop in the efficiency of the reaction was observed with analogs of dppb where the length of the tether was either increased or decreased, presumably indicating an optimum bite angle for this transformation.

The reaction conditions were optimized to afford clean coupling of enol tosylate **32** using only a slight excess of amide **24** (1.05 equiv) at 100 °C, 5 mol% $Pd_2(dba)_3$/dppb catalyst, and a toluene/*tert*-amyl alcohol solvent system. Even under the harsh reaction conditions required for complete conversion of the tosylate (100 °C, 20 h) no detectable *E*/*Z* isomerization was seen, providing further proof that the hindered nature of the enamide aids stability to isomerization. Treatment of the mixture with activated carbon (Darco KB-B) at the end of the reaction followed by isolation of the product by crystallization, afforded enamide **22** in 92% isolated yield.

Box 9.3 An evaluation of the utility of the enol tosylate amidation

The use of the enol tosylate compared to the enol triflate for the above amidation offers the advantage of a cheaper reagent to prepare the substrate and generation of a crystalline intermediate. Based on these positive attributes the scope of this reaction was explored (Scheme 9.26) [25, 26].

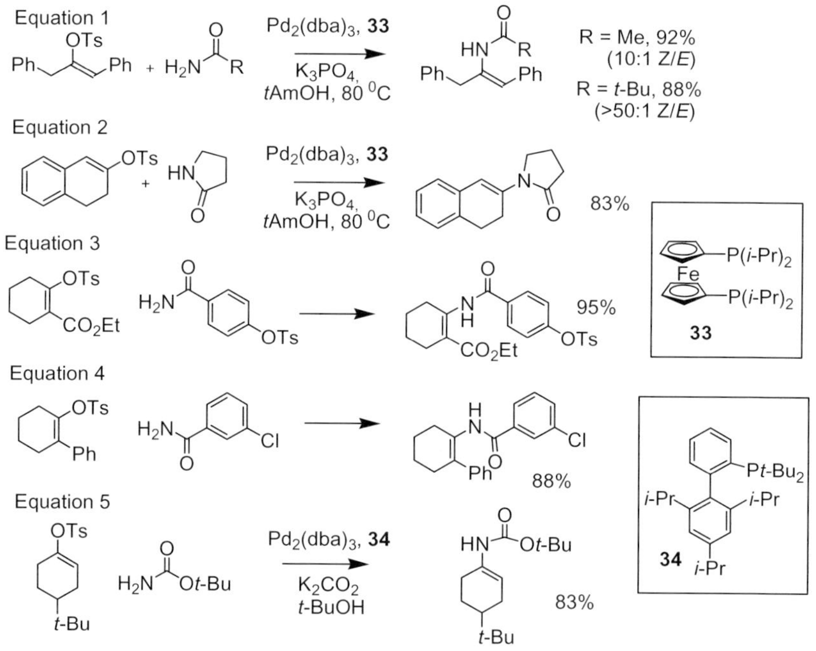

Scheme 9.26 Amidation of enol tosylates.

9.2.2.6 Asymmetric Hydrogenation of Enamide 22

With an efficient synthesis of the enamide in hand attention returned to the key hydrogenation. Despite our previous modest results for hydrogenation of substrates containing an aryl nitrile group, efforts were initially directed towards reduction of enamide **22**. Preliminary studies confirmed that the nitrile group of **22** reduced in preference to the enamide function under a range of hydrogenation conditions, leading to recovered starting material, aldehyde **35**, primary amine **36**, and reductive amination by-products, such as **37** (Scheme 9.27). All of these by-products acted as catalyst poisons in the hydrogenation and, in addition, both the starting enamide **22** and desired product **1** were shown to have a de-activating effect when added to other rhodium-catalyzed enamide hydrogenations.

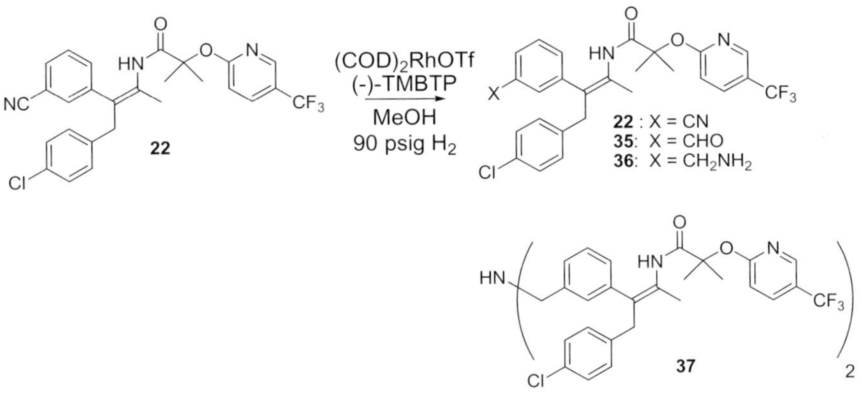

Scheme 9.27 Problems with hydrogenations of enamides bearing an aryl nitrile group.

The challenge was then clearly defined, to identify conditions that would preferentially reduce the enamide at both reasonable substrate concentrations and viable catalyst loadings and avoid the poisoning properties of the nitrile group. A comprehensive screen of solvents, ligands and additives was undertaken via high-throughput screening to attempt to address this goal. On small scale it was found that certain Lewis acidic additives helped to promote both the extent and chemoselectivity of the reduction, using the TMBTP ligand **38** giving modest yields and enantioselectivites for the desired product (Table 9.8). However, very dilute conditions were required to achieve these results and it did not prove possible to reduce the catalyst loading to an acceptable level or increase the concentration, leading to the proposal that a more reactive catalyst system was still required.

Further screening focusing on bis-phosphine ligands continued and, after significant experimentation, a unique set of conditions was identified whereby the reduction could be realized at lower catalyst loadings and higher concentrations. A catalyst derived from $(COD)_2RhBF_4$ and ligand **28** in 1,2-dichloroethane afforded nearly complete selectivity for enamide hydrogenation over nitrile reduction. Under optimized conditions enamide **22** was reduced directly to **1** in 85% ee and 90% yield using 2.5 mol% catalyst in 1,2-dichloroethane at 500 psi H_2 and 0.16 M

Table 9.8 An initial lead for the hydrogenation of enamide 22.

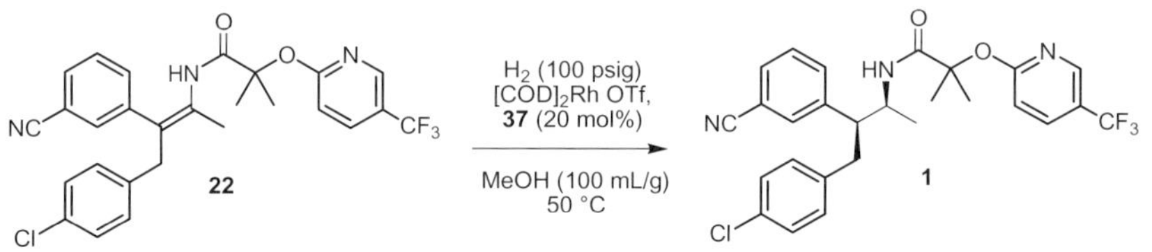

Additive	Conversion	CN reduction
None	5%	20%
BF_3-MeOH	72%	<1%
$In(OTf)_3$	76%	<1%
$Sc(OTf)_3$	86%	1%
TFA	57%	<1%

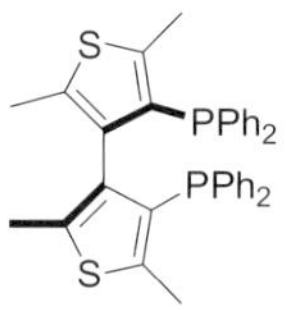

38 (-)-TMBTP

(Scheme 9.28). Despite the modest enantioselectivity, we knew from our previous work that upgrade of enantiomeric purity was possible via preferential removal of the less soluble racemic material by crystallization. In this manner, after a carbon treatment to achieve acceptable levels of residual rhodium, racemic **1** was removed by partial crystallization, followed by subsequent isolation of **1** as a crystalline solid in 72% yield from **22**, and in 98.5% ee.

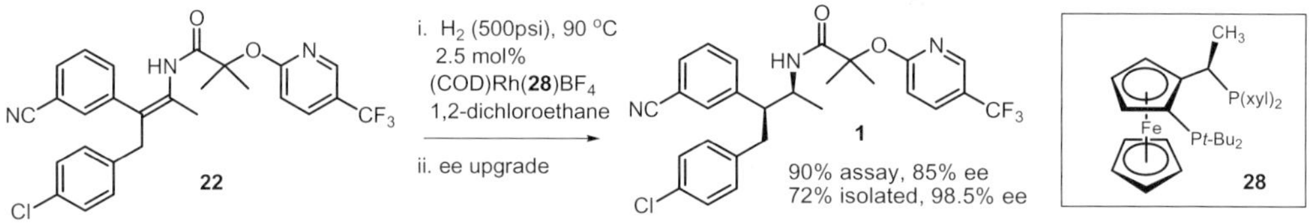

Scheme 9.28 Successful asymmetric hydrogenation of enamide **22**.

While the approach above constituted the shortest synthesis of taranabant to date, a number of issues were of concern to us in considering it for manufacturing scale implementation.

1) Cost implication of the relatively high rhodium catalyst loading.
2) Cost implications of the high hydrogenation pressure and limited number of facilities that such a reaction could operate in.
3) Yield loss due to modest chemical conversion and ee, and the need to remove both residual starting material and enantiomer from API.
4) Environmental and regulatory concerns about the use of a chlorinated solvent (1,2-dichloroethane) in the final step.
5) Use of a heavy metal-catalyzed procedure as the final step requiring extensive carbon/resin treatments to reduce the metal contamination in API.

Hence, in parallel with the above studies, hydrogenations of substrates lacking the nitrile group were explored with two possibilities being considered – use of the related bromo-enamide, and temporary protection of the nitrile group during the hydrogenation step.

9.2.2.7 Use of a Bromosubstituted-Enamide

It initially appeared that it should be feasible to carry out the asymmetric hydrogenation using bromo-enamide **39** and to introduce the nitrile group as the final chemical transformation (Scheme 9.29).

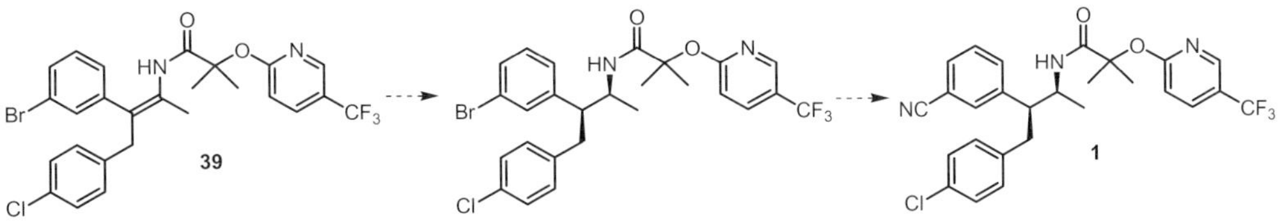

Scheme 9.29 Proposed Hydrogenation of bromo-enamide **39**.

Preparation of the bromo-tosylate **40** from ketone *rac*-**13** was possible using the same enolization and quench conditions as employed for ketone **5**, however, the procedure was less attractive for large scale use for two reasons. First, the enolization selectivity for the bromo-ketone **13** was lower than the related cyano-ketone **5** (see Table 9.5) and, secondly, the bromo-derivatives lacked the crystallinity of the cyano compounds, requiring column chromatography to separate the stereoisomers. To our surprise the bromo-tosylate **40** did not undergo the palladium-catalyzed amidation reaction under the optimized conditions or a range of other conditions (Scheme 9.30). While this was not due to competitive amidation of the aryl bromide moiety per se, the group did appear to have an inhibitory effect on the reaction, possibly due to preferential oxidative insertion into that group. The corresponding bromo-enol triflate **41** was prepared and also gave poor results on attempted coupling with the hindered amide, a surprising result given that in earlier studies a less hindered enol triflate had been shown to amidate in preference to an aryl bromide (see Scheme 9.20). In this case the loss in reactivity is felt to be due to the lower electron-withdrawing properties of the aryl bromide group when compared with the aryl nitrile. Based on the failure of either the triflate **41** or tosylate **40** to efficiently undergo amidation we, therefore, did not pursue hydrogenation studies of the bromo-enamide.

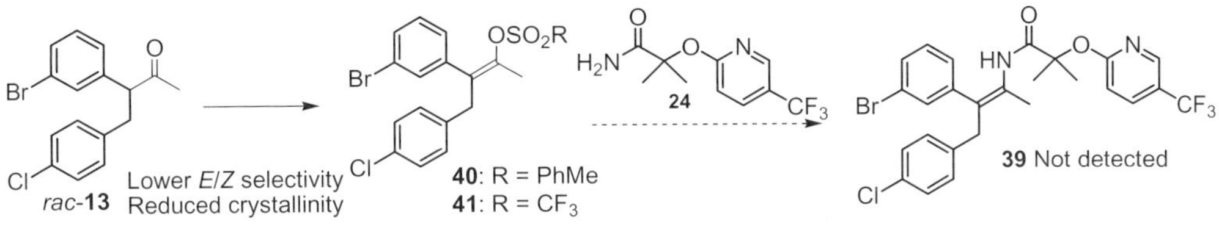

Scheme 9.30 Complications encountered during synthesis of bromo-enamide **39**.

9.2.2.8 Use of a "Nitrile Protected" Enamide

Based on the inability to efficiently prepare the bromo-enamide, our attention turned to a method to protect or "mask" the nitrile group during the hydrogenation. We felt that temporary conversion of the nitrile group in **22** to a primary amide might allow for realization of a more facile hydrogenation at lower catalyst loadings and in more benign solvents than previously achieved. While the addition of two steps to any process might not be ideal, it was felt that having additional crystalline intermediates prior to both the asymmetric hydrogenation and API steps would provide more purification opportunities to ensure high quality material entering both steps and also allow reduction of residual metals both pre- and post-hydrogenation. Given the ease of interconversion of nitrile and primary amide groups, the feasibility of this approach was investigated (Scheme 9.31). Hydration of the nitrile group in **22** using hydrogen peroxide in basic DMSO afforded the amide-enamide **42** as a highly crystalline solid in 95% yield and dehydration of the reduced amide **43** (initially prepared from hydration of **1**) was achieved by reaction with cyanuric chloride in 93% yield to give high purity API. Notably, the dehydration was totally selective for the primary amide with no reaction of the hindered secondary amide seen under these conditions. Moreover, during a study of the physical properties of the penultimate amide **43**, we discovered that a direct upgrade of enantiomeric purity was possible during the isolation, thereby reducing the cumbersome process that had previously been implemented at API stage involving crystallization of racemic material prior to isolation.

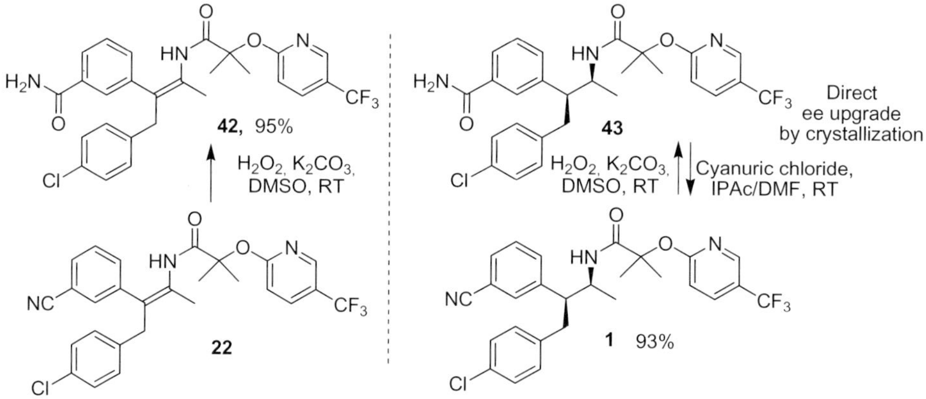

Scheme 9.31 Interconversion of primary amide and nitrile groups.

With all other pieces of the synthesis in place our attention now focused on the final piece in the jigsaw–the asymmetric hydrogenation of the amide enamide **42**. Screening of hydrogenation conditions rapidly led to identification of a number of conditions which allowed the desired hydrogenation to proceed at low catalyst loadings and in non-chlorinated solvents (Table 9.9).

A trend emerged in that the best conditions involved either the use of a Lewis acid additive ($BF_3{\bullet}OMe_2$) or a more polar solvent–such as trifluoroethanol (TFE), although the beneficial effect of TFE did depend on the ligand employed [27]. The

Table 9.9 Screen of hydrogenation conditions for enamide 42.

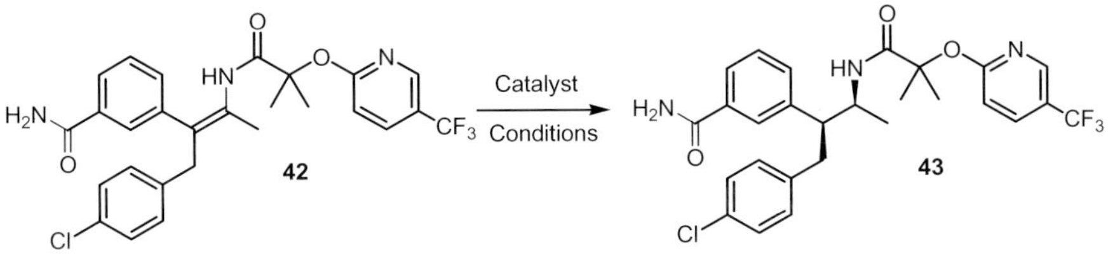

Ligand	mol% cat	Lewis acid	Solvent	psig H_2, T(°C)	ee	Conv.
38	1	None	MeOH	90, 50	89%	45%
38	0.5	40% BF_3-MeOH	MeOH	90, 50	88%	99%
38	0.5	8% BF_3-MeOH	IPA	150, 40	92%	100%
38	0.2	12% BF_3-MeOH	IPA	1000, 45	87%	100%
44	0.5	None	MeOH	150, 40	92%	58%
44	0.5	8% BF_3-MeOH	MeOH	150, 40	91%	78%
38	1	None	TFE	150, 40	44%	84%
44	0.2	None	TFE	75, 50	96%	100%
44	0.05	None	TFE	150, 60	95%	100%

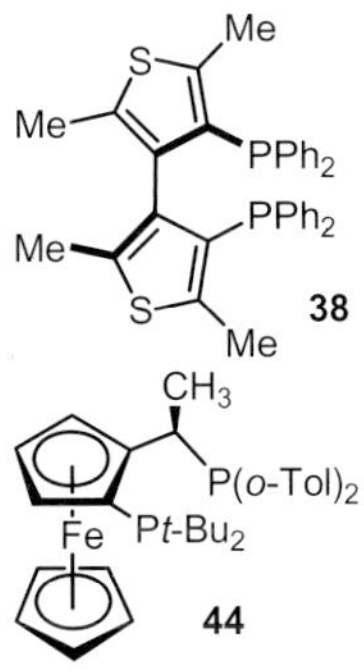

use of BF_3•OMe_2 gave low levels of methyl ester formation from the primary amide, an impurity that was found difficult to remove and, hence, optimization focused on ligand **44** in TFE.

Optimal conditions involved the use of 150 psi H_2 in TFE as solvent and only 0.05 mol% of a catalyst derived from $(NBD)_2RhBF_4$ and ligand **44**. This protocol gave essentially quantitative conversion to amide **43** as a single diastereoisomer in 96% ee. Moreover, as mentioned above, isolation of the product proceeded with an upgrade in enantiomeric purity, leading to a 90% isolated yield of **43** as a crystalline solid in >99.5%ee. Final dehydration of the primary amide in **43** proceeded in a straightforward manner to afford taranabant (**1**) in 79% yield over the three-step sequence from **22** (Scheme 9.32). Despite the additional chemical steps, this

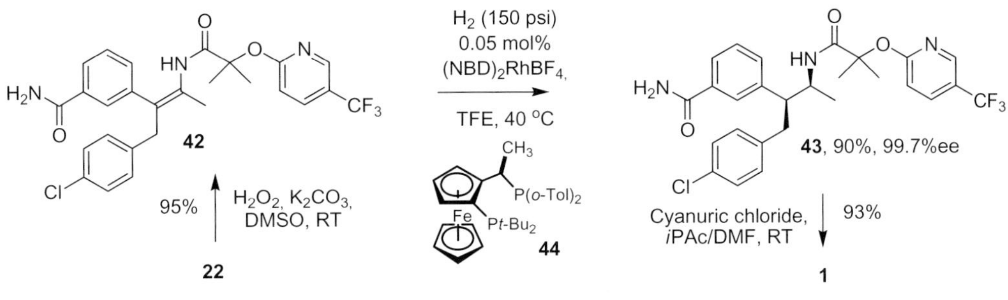

Scheme 9.32 Hydrogenation of enamide **42** and completion of synthesis.

alternate approach using the temporary "protection" of the nitrile group, proceeded in higher overall yield than the hydrogenation of enamide **22** and offers advantages of lower catalyst loading, more easily attainable hydrogen pressures and environmentally acceptable conditions coupled with higher purity final product. Additionally, the time- and solvent-consuming multi-operation ee upgrade of the API was replaced by a direct upgrade at the previous intermediate **43** [28].

With the successful demonstration of the above enamide hydrogenation a new asymmetric synthesis of taranabant had been achieved. The route utilized a selective enolization/tosylate formation to provide the enol tosylate **32** followed by a Pd-catalyzed coupling with primary amide **24** to afford the enamide **22**, containing all of the desired functionality for the final compound. To avoid the issues of nitrile inhibition in the hydrogenation enamide **42** was prepared and the chirality introduced by the use of a Rh-catalyzed asymmetric hydrogenation affording the penultimate **43** which was then converted to API (Scheme 9.33). The synthesis appeared to address the shortcomings of the previous approach including avoidance of sodium azide and the proprietary catalyst and also provided every intermediate from cyano-ketone **5** onwards as a crystalline solid, hence providing many opportunities for purification.

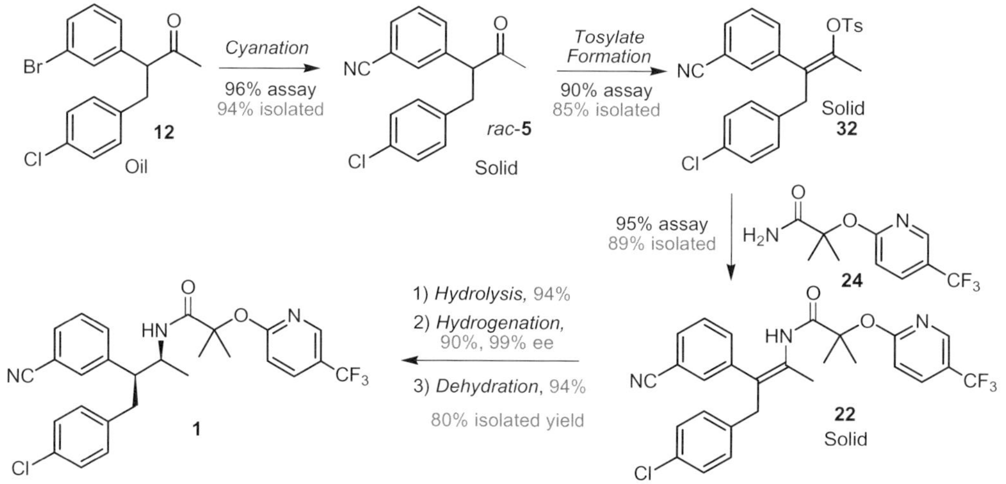

Scheme 9.33 Final enamide hydrogenation route to taranabant.

Box 9.4 Asymmetric hydrogenations of tetra-substituted enamides

The asymmetric hydrogenation of tetrasubstituted enamides remains a challenging goal in synthetic organic chemistry. In particular, reduction of an *acylic* enamide bearing *four distinct groups* (as in enamides **22** and **42**) has yet to be demonstrated by another group at the time of writing. In this section a brief outline of currently documented reductions of tetrasubstitued enamides is presented.

Cyclic tetrasubstituted enamides have been sucessfully reduced using both rhodium [15e] and ruthenium [15b,d] catalysis (Scheme 9.34).

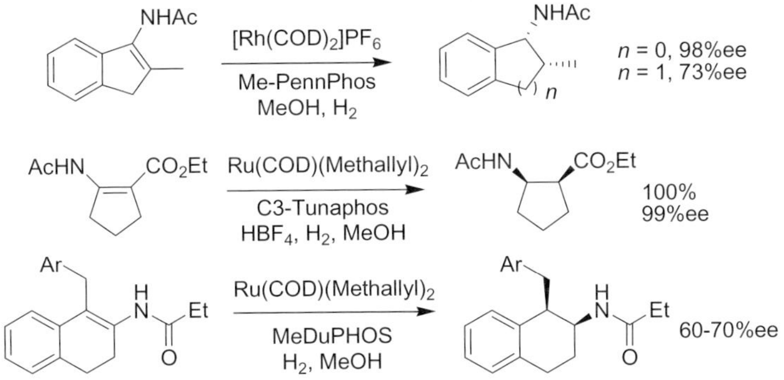

Scheme 9.34 Asymmetric hydrogenation of cyclic enamides.

The reduction of a structurally simple acyclic tetrasubstituted enamide has been achieved [15c], also using rhodium catalysis (Scheme 9.35).

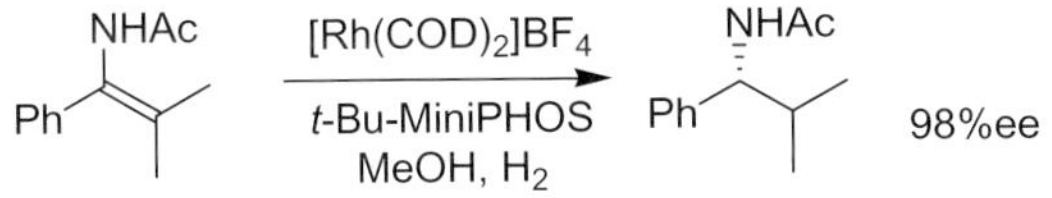

Scheme 9.35 Reduction of an acyclic tetrasubstituted enamide.

In other work at Merck [15a] the asymmetric hydrogenation of a tetrasubstituted acyclic ene-sulfonamide *E*-**45** bearing *four distinct groups* has been demonstrated, leading to good ee and dr in the reduction (Scheme 9.36). The related *Z*-isomer (*Z*-**45**) was more prone to double bond isomerization requiring a higher hydrogen pressure to ensure the reduction rate surpassed the isomerization rate.

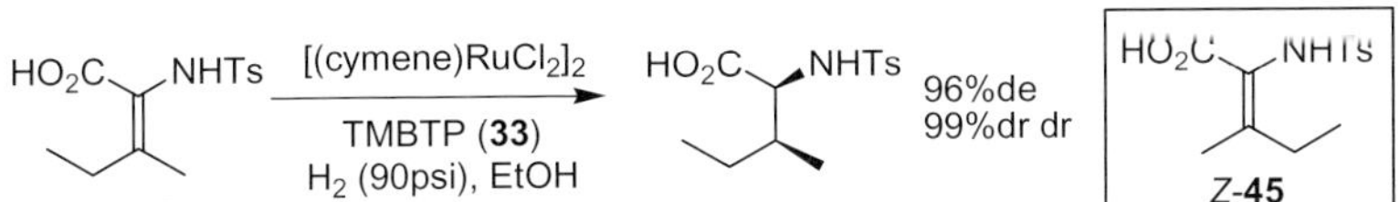

E-**45** *Z*-**45**

Scheme 9.36 Asymmetric hydrogenation of a tetrasubstituted ene-sulfonamide.

9.2.3
Evaluation and Route Selection

The team had now demonstrated two alternate asymmetric routes to taranabant; namely the DKR and the enamide hydrogenation. Notably each synthesis starts from the same intermediate, the racemic bromo-ketone *rac*-**13** and relies on an asymmetric hydrogenation to set both stereocenters (Scheme 9.37). Both routes had been demonstrated on >60 kg scale in internal and external pilot plant facilities and were shown to be robust and reliable at this scale.

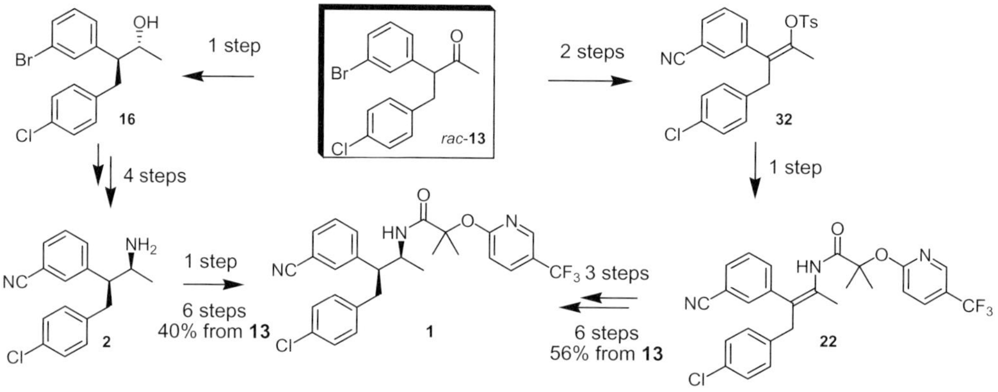

Scheme 9.37 Comparison of synthetic approaches to taranabant.

As the program moved forward to Phase III clinical studies an evaluation of the pros and cons of each approach was carried out to assist with route selection for long-term implementation (Table 9.10).

Based on the above considerations, the team felt that the enamide hydrogenation route addressed the major shortcomings of the DKR approach, particularly surrounding the lack of solid intermediates, the use of hazardous reagents and limited availability catalysts. Additionally, the improved yield and greater throughput due to the lack of separate upgrade steps afforded a cost benefit for the enamide route and this process was chosen as the long-term manufacturing route for taranabant.

Table 9.10 Comparison of synthetic approaches to taranabant.

Synthesis attribute	DKR route	Enamide hydrogenation route
Number of chemical steps	6	6
Yield from ketone *rac*-**13**	40%	56%
Additional linear steps	2, for ds and ee upgrade	None
Solid intermediates	All oils until amine **2**	All solids
Availability issues	Concern with xyl-BINAP DIPEN ligand	No known issues
Safety issues	Use of sodium azide	No known issues

9.3 Conclusion

The development of the synthesis of taranabant has been presented from the optimization of a Medicinal Chemistry route via implementation of a resolution, through development of a first generation asymmetric route, and, finally, the discovery of a new asymmetric route. By making the necessary synthetic changes as the project progressed, bulk drug needs were supplied with resources appropriate for the current development stage. During the course of our work fundamentally new chemistry was developed to allow synthesis of the required tetrasubstitued enamides via amidation of enol triflates and later enol tosylates, reactions not know in the literature at the time our work was initiated. Finally, the high yielding asymmetric hydrogenation of the two elaborated tetrasubstituted enamides remains among the most complex examples in the literature for such transformations.

Acknowledgments

I would like to thank all colleagues who worked on this project, whose names are listed in the references.

References

1 Olshansky, S.J., Passaro, D.J., Hershow, R.C., Layden, J., Carnes, B.A., Brody, J., Hayflick, L., Butler, R.N., Allison, D.B., and Ludwig, D.S. (2005) *N. Engl. J. Med.*, **352**, 1138.

2 Pertwee, R.G. (2000) *Expert Opin. Invest. Drugs*, **9**, 1553.

3 Lin, L.S., Lanza, T.J., Jewell, J.J.P., Liu, P., Shah, S.K., Qi, H., Tong, X., Wang, J., Xu, S.S., Fong, T.M., Shen, C.-P., Lao, J., Xiao, J.C., Shearman, L.P., Stribling, D.S., Rosko, K., Strack, A., Marsh, D.J., Feng, Y., Kumar, S., Samuel, K., Yin, W., der Ploeg, L.V., Mills, S.G., MacCoss, M., Goulet, M.T., and Hagmann, W.K. (2006) *J. Med. Chem.*, **49**, 7584.

4 Liu, P., Lanza, T.J., Jewell, J.P., Jones, C.P., Hagmann, W.K., and Lin, L.S. (2003) *Tetrahedron Lett.*, **44**, 8869.

5 Maligres, P.E., Waters, M.S., Fleitz, F., and Askin, D. (1999) *Tetrahedron Lett.*, **40**, 8193.

6 (a) Tschaen, D.M., Desmond, R., King, A.O., Fortin, M.C., Pipik, B., King, S., and Verhoeven, T.R. (1994) *Synth. Commun.*, **24**, 887; (b) Marcantonio, K., Frey, L.F., Liu, Y., Chen, Y., Strine, J., Phenix, B., Wallace, D.J., and Chen, C.-y. (2004) *Org. Lett*, **6**, 3723.

7 Wiss, J., Fleury, C., and Onken, U. (2006) *Org. Process Res. Dev.*, **10**, 349.

8 (a) Staudinger, H., and Meyer, J. (1919) *Helv. Chim. Acta*, **2**, 635; (b) Mungall, W.S., Greene, G.L., Heavner, G.A., and Letsinger, R.L. (1975) *J. Org. Chem.*, **40**, 1659; (c) Scriven, E.F.V., and Turnbull, K. (1988) *Chem. Rev.*, **88** (2), 297.

9 Noyori, R., Ikeda, T., Ohkuma, T., Wdhalm, M., Kitamura, M., Takaya, H., Akutagawa, S., Sayo, N., Saito, T., Taketomi, T., and Kumobasyashi, H. (1989) *J. Am. Chem. Soc.*, **111**, 9134.

10 For recent reviews of DKR reactions see: (a)Perllissier, H. (2003) *Tetrahedron*, **59**, 8291; (b) Huerta, F.H., Minidis, A.B.E., and Bäckvall, J.E. (2001) *Chem. Soc. Rev.*, **30**, 321.

11 Chung, J.Y.L., Mancheno, D., Dormer, P.G., Variankaval, N., Ball, R.G., and Tsou, N.N. (2008) *Org. Lett*, **10**, 3037.

12 Makina, K., Goto, T., Hiroki, Y., and Hamada, Y. (2006) *Tetrahedron Asym.*, **19**, 2816.

13 Chen, C.-y., Frey, L.F., Shultz, S., Wallace, D.J., Marcantonio, K., Payack, J.F., Vazquez, E., Springfield, S.A., Zhou, G., Liu, P., Kieczykowski, G.R., Chen, A.M., Phenix, B.D., Singh, U., Strine, J., Izzo, B., and Krska, S. (2007) *Org. Process Res. Dev.*, **11**, 616.

14 (a) Blaser, H.-U., Malan, C., Pugin, B., Spindler, F., Steiner, H., and Studer, M. (2003) *Adv. Synth. Catal.*, **345**, 103; (b) Gridnev, I.D., Yamanoi, Y., Higashi, N., Tsuruta, H., Yasutake, M., and Imamoto, T. (2001) *Adv. Synth. Catal.*, **343**, 118; (c) Burk, M.J., Gross, M.F., and Martinez, J.P. (1995) *J. Am. Chem. Soc.*, **117**, 9375. (d) Sawamura, M., Kuwano, R., and Ito, Y. (1995) *J. Am. Chem. Soc.*, **117**, 9602.

15 For some examples of hydrogenations of tetrasubstituted enamides see: (a)Shultz, C.S., Dreher, S.D., Ikemoto, N., Williams, J.M., Grabowski, E.J.J. Krska, S.W. Sun, Y., Dormer, P.G., and DiMichele, L. (2005) *Org. Lett.*, **7**, 3405; (b) Tang, W., Wu, S., and Zhang, X. (2003) *J. Am. Chem. Soc.*, **125**, 9570; (c) Gridnev, I.D., Yasutake, M., Higashi, N., and Imamoto, T. (2001) *J. Am. Chem. Soc.*, **123**, 5268; (d) Dupau, P., Bruneau, C., and Dixneuf, P.H. (2001) *Adv. Synth. Catal.*, **343**, 331; (e) Zhang, Z., Zhu, G., Jiang, Q., Xiao, D., and Zhang, X. (1999) *J. Org. Chem.*, **64**, 1774.

16 (a) Burke, M.J., Casy, G., and Johnson, N.B. (1998) *J. Org. Chem.*, **63**, 6086; (b) Neugnot, B., Cintrat, J.-C., and Rousseau, B. (2004) *Tetrahedron*, **60**, 3575; (c) Brice, J.L., Meerdink, J.E., and Stahl, S.S. (2004) *Org. Lett.*, **6**, 1845; (d) Harrison, P., and Meek, G. (2004) *Tetrahedron Lett.*, **45**, 9277; (e) Zhao, H., Vandenbossche, C.P., Koenig, S.G., Singh, S.P., and Bakale, R.P. (2008) *Org. Lett.*, **10**, 505.

17 (a) Ogawa, Y., Kiji, T., Hayami, K., and Suzuki, H. (1991) *Chem. Lett.*, 1443; (b) Shen, R., and Porco, J.A. (2000) *Org. Lett.*, **2**, 1333; (c) Coleman, R.S., and Liu, P.-H. (2004) *Org. Lett.*, **6**, 577; (d) Xianhau, P., Cai, Q., and Ma, D. (2004) *Org. Lett.*, **6**, 1809; (e) Cesati, R.R., III, Dwyer, G., Jones, R.C., Hayes, M.P., Yalamanchili, P., and Casebier, D.S. (2007) *Org. Lett.*, **9**, 561.

18 Jiang, L., Job, G.E., Klapars, A., and Buchwald, S.L. (2003) *Org. Lett.*, **5**, 3667.

19 For non-selective conversions of ketones to a mixture of vinyl halides see: (a) Spaggiari, A., Vaccari, D., Davoil, P., Torre, G., and Prati, F. (2007) *J. Org. Chem.*, **76**, 2216; (b) Furrow, M.E., and Myers, A.G. (2004) *J. Am. Chem. Soc.*, **126**, 5436.

20 (a) Stille, J.K. (1986) *Angew. Chem. Int. Ed. Engl*, **25**, 508; (b) Scott, W.J., and McMurray, J.E. (1988) *Acc. Chem. Res.*, **21**, 47.

21 Yin, J., and Buchwald, S.L. (2000) *Org. Lett.*, **2**, 1101.

22 Wallace, D.J., Klauber, D.J., Chen, C.-y., and Volante, R.P. (2003) *Org. Lett.*, **5**, 4749.

23 For other groups' work see: (a) Willis, M.C., and Brace, G.N. (2002) *Tetrahedron Lett.*, **43**, 9085; (b) Movassahi, M., and Oundrus, A.E. (2005) *J. Org. Chem.*, **70**, 8638; (c) Willis, M.C., Brace, G.N., and Holmes, I.P. (2005) *Angew. Chem.*, **117**, 407, *Angew. Chem. Int. Ed.*, 2005, **44**, 403.

24 Burk, M.J., de Koning, P.D., Grote, T.M., Hoekstra, M.S., Hoge, G., Jennings, R.A., Kissel, W.S., Le, T.V., Lennon, I.C., Mulhern, T.A., Ramsden, J.A., and Wade, R.A. (2003) *J. Org. Chem.*, **68**, 5731.

25 For subsequent work see: Willis, M.C., Brace, G.N., and Holmes, I.P. (2005) *Synthesis*, 3229.

26 Klapars, A., Campos, K.R., Chen, C.-y., and Volante, R.P. (2005) *Org. Lett.*, **7**, 1185.

27 For a review on the use of fluorinated alcohols in homogeneous catalysis see: Shuklov, I.A., Dubrovina, N.V., and Borner, A. (2007) *Synthesis*, 2925.

28 Wallace, D.J., Campos, K.R., Shultz, C.S., Klapars, A., Zewge, D., Crump, B.R., Phenix, B.D., McWilliams, J.C., Krska, S., Sun, Y., Chen, C.-y., and Spindler, F. (2009) *Org. Process Res. Dev*, **13**, 84.

Index

Note: Page entries in **bold** indicate tables, *italics* indicate figures and the suffix *n* indicates a footnote.

The Art of Process Chemistry. Edited by Nobuyoshi Yasuda

ISBN: 978-3-527-32470-5